Aquaculture Genetics

NIPA® GENX ELECTRONIC RESOURCES & SOLUTIONS P. LTD.
New Delhi-110 034

About the Editors

Mr. Durgesh Kumar Verma completed his B.F.Sc. & M.F.Sc. in Fisheries Resource Management from the Acharya Narendra Deva University of Agriculture and Technology, Kumarganj, Ayodhya, India. He has authored and published several research and review papers, popular articles, chapters, and books. He is an active and dedicated member of several professional societies engaged in fisheries R&D in the country. He has received several accolades, such as the Junior Researcher Award, Young Scientist Award, and Best Paper Presentation awards, at national and international meetings. Presently, he is working as a young researcher at the ICAR-Central Inland Fisheries Research Institute, Prayagraj, Uttar Pradesh, India.

Dr. Sachin O. Khairnar is an Associate Professor working at the College of Fisheries, Kishanganj, under Bihar Animal Sciences University, Patna, India. He earned his B.F.Sc and M.F.Sc from the College of Fisheries, Ratnagiri (BSKKV, Dapoli), and a PhD in Aquaculture from the Ocean University of China, Qingdao, under the prestigious Chinese Government Scholarship awarded by the Chinese Scholarship Council, Beijing. Before taking on his current role, he made significant contributions to leading institutions, including the College of Fisheries (GADVASU), Ludhiana, Central Institute of Freshwater Aquaculture, Bhubaneswar, and Fishery Survey of India, Mumbai. He has been working for more than 13 years on sustainable ornamental aquaculture and has developed low-cost rearing technologies for ornamental fish. He has authored over 50 peer-reviewed research articles of national and international repute and guided over 30 M.F.Sc. and Ph.D. scholars as a major and minor advisor. In recognition of his contributions, he has received several honours and accolades, including an Honorary Fellowship awarded by the Society of Life Sciences in Satna.

Mrs. Aishwarya Sharma is Assistant Professor with B.FSc, M.FSc & Ph.D. (ongoing) specializing in Fish Genetics & Breeding, affiliated with the Department of Fish Genetics & Breeding at College of Fisheries, Kishanganj, Bihar. She has more than five years of experience in the field of Fish Genetics & Biotechnology R&D and education. She has published several research and review papers, popular articles, book chapters, and manuals. She is a recipient of an SRF (Ph.D.) scholarship from the Indian Council of Agriculture Research, New Delhi, India. She has received several accolades, such as the Best Thesis Award, Best Paper and Poster Presentation awards, and other awards at national and international conferences. At present, she is working in the area of selective breeding to study the underlying genetic variations of economic traits in the important ornamental fish species of Bihar.

Aquaculture Genetics
Breeding, Conservation and Sustainable Production

Durgesh Kumar Verma
ICAR-Central Inland Fisheries Research Institute
Prayagraj, Uttar Pradesh

Sachin O. Khairnar
Associate Professor
College of Fisheries, Kishanganj
Bihar Animal Sciences University, Bihar

Aishwarya Sharma
Assistant Professor
Department of Fish Genetics & Breeding
College of Fisheries, Kishanganj
Bihar Animal Sciences University, Bihar

NIPA® GENX ELECTRONIC RESOURCES & SOLUTIONS P. LTD.
New Delhi-110 034

NIPA® GENX ELECTRONIC RESOURCES & SOLUTIONS P. LTD.

101,103, Vikas Surya Plaza, CU Block
L.S.C. Market, Pitam Pura, New Delhi-110 034
Ph : +91-11-43860225, Mob.: +91 9717133558, 9540816132
E-mail: newindiapublishingagency@gmail.com
Website: www.nipaersources.com

Print ISBN: 978-93-58874-45-7
ebook ISBN: 978-93-58878-24-0

Composed and Designed by NIPA®.

Preface

Genetic studies on fish began late, as there was significant scope to improve aquaculture production by managing husbandry practices. However, recent studies have demonstrated the potential for achieving substantial gains in aquaculture and fisheries management by applying genetic tools. Keeping this in mind, the book is designed to give rapid and easy access to all the recent advancements to ensure they exploit their potential to a greater extent. This book offers a comprehensive overview of fish genetics and biotechnology aimed at students, researchers, and policymakers.

This book touches on different topics, such as the history of genetics in aquaculture, sex determination mechanisms, and ethical considerations, particularly regarding monosex practices. It highlights the importance of biosecurity and biotechnological applications for aquatic health management, alongside discussions on the role of biotechnology in climate change adaptation. It also covers the key advancements, including chromosome manipulation, gene editing, and genome sequencing, for their potential to enhance genetic improvement and productivity. The use of molecular breeding and markers is also featured, emphasizing their importance in sustainable aquaculture.

This compilation aims to foster an understanding of how genetics and biotechnology can transform aquaculture while ensuring environmental stewardship. We appreciate the contributions of our experts and encourage readers—from researchers to industry stakeholders to engage with this material as we explore the future of sustainable fisheries and aquaculture.

Editors

Contents

1

Biotechnological Applications of Transgenic Fish
Current and Future Status

Sahana Ghosh

National High School (C.B.S.E.), Kolkata, West Bengal (700 019), India

Abstract

Transgenic organisms are defined as being introduced artificially with a transgene and incorporated stably into their genomes. There are various aquatic organisms in which the trans-genesis process has been applied like fish, crustaceans, molluscs, etc. Presently scientists can transfer a particular gene for a desirable and beneficial trait from one organism to another with more precision and in less time than traditional breeding. To date, over thirty-five species have been genetically engineered in laboratories. The transgenic fish are efficient feed converters. Salmon is the first transgenic fish that was used as food. Various techniques are involved in aqua-transgenesis. They include – microinjection, electroporation, particle gun bombardment, liposome-mediated gene transfer and so on (setac. onlinelibrary.wiley.com). Through all these techniques a fish is genetically transformed so that it has desirable and beneficial trait. Numerous significant biotechnological uses for fish transgenesis are anticipated to be beneficial across a wide range of industries. The majority of genetically engineered fish serve purposes in fundamental studies of genetics and development. Various research teams have been creating genetically modified zebrafish to identify aquatic pollution. The lab that created GloFish initially aimed for them to alter their colour in response to pollutants, acting as environmental indicators. Transgenic fish have an improved growth rate than their non-transgenic counterparts. Moreover, they are disease-resistant and can be used as models for performing various scientific experiments in the field of environmental toxicology and biotechnology. All these beneficial traits are making transgenic fish more valuable in the biotechnological field. To fully realize the potential of transgenic fish technology in aquaculture, certain ethical and ecological issues must be scientifically resolved.

Keywords: *Aqua-transgenesis, Genetically modified organisms, Microinjection, Biotechnological potential*

Introduction

Various new approaches to improve aquatic organisms genetically have been developed into a distinct field of study in recent years. One of them is transgenesis. Aqua-transgenesis is the process in which new genes are transferred into aquatic organisms like fish, crustaceans and molluscs to improve their genetic traits. The term genetically modified fish (GM fish) describes organisms that fall within the taxonomic clade comprising the classes Agnatha (jawless fish), Chondrichthyes (cartilaginous fish) and Osteichthyes (bony fish). Their genetic material (DNA) has been modified through genetic engineering methods. The primary aim of such alterations is to incorporate a novel trait into the fish that is absent in species in their natural state. An organism characterized by the presence of a foreign or modified gene in its genome is termed a transgenic organism or Genetically Modified Organism (GMO). Recently, genetic engineering has evolved into one of the most transformative technologies. Presently, scientists can transfer a particular gene for a desirable and beneficial trait from one organism to another with more precision and in less time than traditional breeding (Saxena and Jha, 2013). Genetically modified fish are engineered to function as indicators of environmental pollutants and to aid in aquaculture food production. In 2015, the US Food and Drug Administration (FDA) approved the Aqua-Advantage salmon for commercial production, sale and consumption, thus becoming the first genetically modified animal authorized for human consumption. The first recorded instances of transgenic production in aquatic species are in rainbow trout (1984) and goldfish (1985) (Betsy and Siva, 2023). To date, over 35 species have been subjected to genetic engineering within laboratory environments. In January 1996, history was made when the first genetically engineered salmon, called “AquAdvantage salmon,” was successfully raised in a commercial hatchery in Loach Fyne Scotland.

Techniques Involved in Aqua-transgenesis

Microinjection

- Microinjection is a technique used to introduce substances into fish eggs for gene transfer and transgenesis (Csenki *et al.,* 2019).
- A solution containing foreign DNA is injected through a hollow needle which is inserted into the fish egg through the micropyle. The egg can be fertilized and inactive during the procedure, and then activated after injection (Davies, 2022).
- This method has been utilized for studying the effects of contaminants on embryonic development in Zebrafish (Schubert *et al.,* 2014).

- Microinjection has a few advantages which include high efficiency, specificity, and controlled dosage of all components. However, the microinjection process is laborious, technically challenging, and limited to a single cell per injection (Yan *et al.,* 2022).

Electroporation

- In this technique, high-voltage electric shocks serve as a method for introducing DNA into cellular structures (Potter, 2003)
- This technique employs a sequence of brief electric pulses to facilitate the penetration of the cell membrane, thereby enabling the creation of temporary pores on the surface of the targeted cells. Through these temporary pores, the transgene is introduced into the cytoplasm where it is then delivered to the nucleus by the cellular machinery.
- The advantage of electroporation is that it is fast, less costly, can be applied for several cell types and simultaneously a large number of cells can be treated. However, there are certain disadvantages related to this technique including cell damage and non-specific transport of material into and out of the cell during the time of electro-permeability (Vry *et al.,* 2010).

Particle gun bombardment

- This method is also known as biolistics.
- This is a method of genetic transformation that uses high-velocity micro-projectiles to introduce DNA and other substances into cells and tissues (Kikkert *et al.,* 2005).
- The advantage of particle gun bombardment is that it is highly efficient, can help in knock out of desired genes and is cell wall independent. However, there are certain disadvantages related to this technique including its infrastructure, high cost of the equipment and required reagents (Sreenikethanam *et al.,* 2022).

Liposome-mediated gene transfer

- This technique uses liposomes to deliver DNA into cells. Large DNA constructs have been introduced into fish embryos using this technique in the field of aqua-transgenesis.
- This technique is employed to incorporate substantial DNA constructs into the zygotes of African Catfish (Bowen *et al.,* 2020).
- This method provides protection from external enzymes. However, there are certain disadvantages related to this technique including requirement

of cell wall deficient cells, risk of digestion of genetic material after release and need of optimized preparation of liposomes (Sreenikethanam *et al.,* 2022).

Sperm-mediated gene transfer

- The transgenic approach described here utilizes the inherent ability of sperm cells to attach to and absorb foreign DNA, allowing for its delivery into an oocyte during fertilization, which leads to the development of genetically modified animals,
- Some fishes that are produced by this method are- Zebrafish, rainbow trout, catfish, tilapia, and salmon (Zouridakis, 2019).
- One of the primary benefits of this method is its affordability and simplicity of use. However, it is important to note that its limitations include suboptimal efficiency and a lower rate of foreign DNA integration into the sperm's genetic material (Park *et al.,* 2018).

Some Examples of Fish Species

Salmon: AquAdvantage salmon, a genetically modified variant, exhibits a significantly enhanced growth rate and size compared to its wild-type Atlantic salmon progenitor, potentially achieving double the weight in a shorter period to reach maturity (Saxena and Jha, 2013).

Tilapia: Tilapia exhibits significant sensitivity to low temperatures, making it best suited for environments where temperatures exceed 60° F. It ranks as the fifth most crucial species in aquaculture due to its substantial size, swift growth rate, and favourable taste (Saxena and Jha, 2013).

Zebrafish: They are very popular as aquarium fish and play a vital role as model organisms in research. Males exhibit gold stripes interspersed within the blue stripes, whereas females display silver stripes within the blue stripes (Saxena and Jha, 2013).

Table 1. Transgenic fishes under development for use in aquaculture, Source (FAO, 2000)

Species	Gene(s) Introduced	Desired effect	Country
Atlantic Salmon	AFP-salmon GH	Cold tolerance, Increased growth and feed efficiency	Canada
Coho Salmon	AFP-chinook salmon GH	Increased growth	Canada
Chinook Salmon	AFP-salmon GH	Increased growth and feed efficiency	New Zealand
Rainbow Trout	AFP-salmon GH	Increased growth and feed efficiency	Canada
Cutthroat Trout	AFP-chinook salmon GH	Increased growth	Canada
Tilapia	AFP-salmon GH	Increased growth and feed efficiency	Canada, UK
Tilapia	CMV-tilapia GH	Increased growth	Cuba
Tilapia	Tilapia insulin gene	Production of human insulin	Canada
Salmon	Rainbow trout lysozyme, Flounder pleurocidin	Disease resistance	United States
Striped Bass	Insect cecropin	Disease resistance	United States
Mud Loach	Loach and mouse MT- Mud loach GH	Increased growth and feed efficiency	China, Korea
Channel Catfish	RSVLTR-GH	Increased growth	United States
Common Carp	Salmon and human GH	Increased growth, disease resistance, tolerance of low dissolved oxygen	China, United States
Indian Major Carps	Human GH	Increased growth	India
Goldfish	AFP-GH	Increased growth	China
Abalone	Various promoters coho salmon GH	Increased growth	United States
Oysters	Various promoters coho salmon GH	Increased growth	United States

Biotechnological potential of transgenic fishes

Numerous significant biotechnological applications of transgenic fish hold potential value across various sectors (Tonelli *et al.,* 2017). These genetically modified fish have been engineered to exhibit enhanced growth rates, improved cold tolerance, and greater resistance to diseases. Furthermore, specific transgenic fish have been created to serve as bio-indicators and bio-reactors (Tonelli *et al.,* 2017). Their applications extend to the development of pharmaceuticals for human ailments and research in environmental toxicology, among other uses. They are also known to improve the efficiency of food conversion (Tonelli *et al.,* 2017). Transgenic fish technology represents a significant biotechnological advancement that has the potential to improve both the global output and quality of aquatic food products.

Applications of Transgenic Fish Technology in Research

Transgenic fish have been utilized in research across five major domains-

1. Enhancing the attributes of fish available in the market
2. Their utilization as bioreactors for the synthesis of proteins that hold significant biomedical value
3. Their utilization as indicators for identifying aquatic pollutants
4. Creating animal models that are not derived from mammals
5. Functional genomic investigations

The majority of genetically modified fish are utilized primarily in fundamental research about genetics and developmental biology. The two species most frequently altered are zebrafish and medaka, as they possess transparent chorions (shells), exhibit rapid development, and allow for easy observation and micro-injection of transgenic DNA into their one-celled embryos (Saran *et al.,* 2019). The creation of transgenic rainbow trout has been initiated to enhance the investigation of muscle development. The incorporation of the GFP gene results in the expression of green fluorescence in fast-twitch muscle fibres during the early stages of development. This characteristic persists throughout the organism's lifespan. There is a suggestion that these transgenic fish may function as indicators of aquatic contaminants or other developmental factors. They can be engineered with reporter genes that activate in response to chemical exposure. When these fish are exposed to contaminated water, the resulting reporter gene activity can be assessed, providing a measure that is directly proportional to the concentration of the chemicals in the environment (Winn *et al.,* 2001). Genetic engineering has been applied to grass carp, introducing a transgene that produces human lactoferrin. This modification has led to a doubling of their survival rate compared to control fish when subjected to *Aeromonas* bacteria and the Grass carp haemorrhage virus (Zhong *et al.,* 2002). Zebrafish possess the remarkable ability to regenerate their organ tissues. Additionally, they play a significant role in drug discovery processes. Researchers are exploring genetically modified zebrafish to gain insights into diseases and failures of human organ tissues. For instance, these zebrafish are employed to enhance the understanding of heart tissue repair and regeneration, thereby aiding in the research and development of potential treatments for cardiovascular diseases (Angom *et al.,* 2024).

Applications of Transgenic Fish Technology in Recreation and Game Fishing

Certain transgenic fish have been developed for keeping as pets.

- GloFish represents a proprietary advancement that enables genetically modified fish, including tetras, barbs, and zebrafish, to exhibit proteins derived from jellyfish and sea coral. This modification results in the fish displaying vibrant fluorescent colours such as red, green, or orange when observed under ultraviolet light (Saxena and Jha, 2013).
- The fish, initially developed and patented for scientific research purposes at the National University of Singapore, were later acquired by a Texas-based company, Yorktown Technologies.
- Various other transgenic strains of ornamental fish include Medaka, recognized for their enduring transparency, as well as pink-bodied transgenic angelfish and lion head fish that exhibit the red fluorescent protein derived from Acropora coral (Wakamatsu *et al.,* 2001).
- The successful micro-injection and expression of the ocean and pout type III antifreeze protein transgene in goldfish have been achieved. The transgenic individuals exhibited enhanced cold tolerance compared to the control group (Wang *et al.,* 1995).

Applications of Transgenic Fish Technology in the Detection of Aquatic Pollution

Several scientific groups have been advancing the genetic modification of zebrafish to serve as indicators of water pollution. The laboratory that developed the GloFish initially aimed for these fish to display colour changes in response to pollutants, and function as environmental sentinels (Dubinska-Magiera *et al.,* 2016). A range of genetic methodologies has been utilized to incorporate target DNA into zebrafish for environmental monitoring purposes. These methodologies encompass micro-injection, electroporation, particle gun bombardment, and liposome-mediated gene transfer. Among these techniques, micro-injection is particularly notable for being the most commonly employed method for generating transgenic zebrafish, as it yields the highest survival rates (Saxena and Jha, 2013).

Application in the Food and Feeding Industry

A significant field of study concerning genetically modified fish centers on enhancing food production by altering the expression of growth hormones (GH). The growth enhancements vary among species; some fish exhibit a weight increase of up to twofold, while others attain weights nearly 100 times greater than their wild-type equivalents at the same age. This research has led to notable growth advancements in several species, including salmon, trout, and tilapia. In particular, salmon have shown an 11-fold increase in growth, whereas mud loach has reached a remarkable 30-fold increase relative to their wild-type counterparts (Leggatt *et al.,* 2017). The evolution of transgenic fish has attained a stage where several species are prepared for commercial release in various countries. Genetically modified tilapia is currently being promoted in Cuba, while genetically modified carp is available in the People's Republic of China. Additionally, genetically modified salmon is marketed in both the United States and Canada. In 2014, it was reported that applications for the approval of transgenic fish as food products were submitted in Canada, China, Cuba, and the United States (Leggatt *et al.,* 2017). The overproduction of growth hormone (GH) by the pituitary gland significantly accelerates growth rates, mainly by increasing the fish's consumption of food, while also enhancing feed conversion efficiency by 10 to 15%. Another approach to augmenting meat production in genetically modified (GM) fish is referred to as "double muscling". This occurrence results in a phenotype similar to that of Belgian Blue cattle in rainbow trout. It is achieved through the use of transgenes that express follistatin, which suppresses myostatin and promotes the development of two muscle layers (Shoyombo *et al.,* 2022).

AquAdvantage Salmon

- The Aquadvantage salmon, developed by AquaBounty, obtained authorization from the United States Food and Drug Administration (FDA) in November 2015 for its commercial production, distribution, and consumption. This approval marks the inaugural instance of a genetically modified animal being approved for human consumption (Grossman *et al.,* 2016).
- This fish is primarily an Atlantic salmon that has undergone genetic modification through the introduction of a particular gene complex. This construct comprises a gene that regulates growth hormone derived from a Chinook salmon, along with a promoter sequence sourced from an ocean pout. Consequently, the genetically modified salmon can produce growth hormone continuously throughout the year, unlike wild-type Atlantic salmon, which experience a seasonal pause in production. Wild-type salmon typically take between 24 to 30 months to achieve a market weight of 4 to 6 kilograms, while genetically modified salmon can attain this weight in just 18 months (Grossman *et al.,* 2016).
- AquaBounty claims that their genetically modified salmon can be cultivated nearer to consumer markets with enhanced efficiency,

necessitating 25% less feed to reach market weight in comparison to Atlantic salmon, which are presently raised in remote coastal regions. This practice is believed to be more beneficial for the environment, as it minimizes waste and reduces transportation costs.

- To reduce the likelihood of genetically modified fish interbreeding with wild salmon, all food-producing fish are female, and triploid, and 99% are engineered to be reproductively sterile. These fish are raised in a facility located in Panama, which employs physical barriers and geographical containment measures, including river and ocean temperatures that are excessively high for AquaAdvantage to have a substantial impact on the environmental conditions in the United States (Grossman *et al.,* 2016).

Ethical Issues Concerned with Transgenic Fish Technology

Transgenic fish engineered for increased growth hormone display considerable differences in growth rates among the mosaic founder fish. This disparity is a result of the uneven distribution and varying proportions of transgenic cells within their bodies. Fish that attain significant growth rates, along with their offspring, may display a morphological anomaly similar to acromegaly observed in humans. This condition is characterized by a disproportionately large head to the body and a prominent operculum. As the fish age, this condition can deteriorate, affecting their ability to feed and often resulting in death (Ormandy *et al.,* 2011). Genetically modified Coho salmon exhibit morphological alterations and modified allometric relationships that result in diminished swimming capabilities. Additionally, these fish display atypical behaviours, including heightened activity levels related to feeding and swimming. Other transgenic fish species, such as Coho salmon, also demonstrate reduced swimming proficiency, which is likely attributable to their body shape and muscle composition (Ormandy *et al.,* 2011). Triploid fish that have undergone genetic modification and are more likely to suffer from temperature-related stress, have an increased incidence of deformities, such as those in the eyes and lower jaws, and are less aggressive when compared to diploid fish.

Concerns regarding the welfare of genetically modified fish also include heightened stress levels in environments with insufficient oxygen, due to an increased oxygen demand. Research indicates that the impact of hypoxia on Coho salmon is especially significant in transgenic strains. It has been proposed that this heightened vulnerability to reduced oxygen levels stems from the presence of an extra chromosome set, which necessitates a larger nucleus. This, in turn, leads to an overall increase in cell size and a reduced surface area-to-volume ratio (Ormandy *et al.,* 2011).

Ecological Issues Concerned with Transgenic Fish Technology

Transgenic fish are generally derived from wild populations. They possess a significant ability to interbreed with both their species and wild relatives, which raises considerable concerns about their potential to thrive in natural ecosystems if they escape from biotic or abiotic containment. Numerous concerns have been expressed about the potential impacts of genetically modified organisms (GMOs) entering the natural environment. In the case of polyploidy organisms, these worries encompass issues such as the degree of sterility, interference with reproductive processes, and competition for resources without aiding future generations. For transgenic fish, the primary concerns revolve around the characteristics of the genotype, the nature and role of the genes involved, the possibility of pleiotropic effects, interactions with the overall genome, the stability of the genetic construct, and the ability of the DNA construct to transpose within or across different genomes (Tiedje *et al.,* 1989). A research study employing relevant life history data from the Japanese medaka predicts that a transgene introduced into a natural population via a small number of transgenic fish will expand due to enhanced mating advantages. The reduced viability of the offspring is expected to result in the eventual local extinctions of both populations (Tiedje *et al.,* 1989). Genetically modified Coho salmon display a heightened propensity for risk-taking behaviours and a more efficient strategy for utilizing scarce food resources compared to their wild-type equivalents. These transgenic specimens show improved feeding efficiency and accelerated growth rates, leading to a considerably larger body size in contrast to non-transgenic salmon. When placed in the same habitat and competing for varying food levels, transgenic salmon consistently outspace non-transgenic salmon in growth. In scenarios where food is scarce, dominant individuals, which are predominantly transgenic, exhibit aggressive and cannibalistic behaviours towards their peers, thus monopolizing the limited food supply. In settings characterized by limited food resources, populations of transgenic salmon often face significant declines or even compete towards extinction, whereas non-transgenic salmon populations generally demonstrate a greater capacity for survival. This observation has prompted the hypothesis that these genetically modified fish might possess a superior likelihood of enduring adverse conditions compared to their wild-type counterparts. Additionally, there have been documented instances of successful artificial hybridization between two loach species that do not hybridize in their natural habitats (Doyle *et al.,* 2004).

Conclusion

Aqua-transgenesis is the process in which new genes are transferred into aquatic organisms to improve their genetic traits. The technology of transgenic fish holds significant promise for transforming the aquaculture sector by incorporating advantageous genetic characteristics into finish or shellfish, thereby enabling the production of superior transgenic fish strains for aquaculture purposes. There are various techniques involved in aqua-transgenesis that help scientists transfer a particular gene for a desirable and beneficial trait from one organism to another with greater precision and in less time than traditional breeding. These characteristics may include increased growth promotion, enhanced efficiency in food conversion, resistance to a range of known diseases, tolerance to low oxygen conditions, and the ability to withstand sub-zero temperatures. These desirable traits make transgenic fish more valuable in the biotechnological field. Transgenic fish have been used in the detection of aquatic pollution, in recreation and game fishing as well as in research. To achieve the full potential of transgenesis and other biotechnological applications in aquaculture, several critical scientific advancements must be made. These advancements encompass the identification of genes linked to desirable traits in aquaculture, the creation of precise gene transfer technologies, the discovery of suitable promoters to control the expression of transgenes at optimal levels during particular developmental phases, and the assessment of the safety and environmental implications associated with transgenic fish. Therefore, it is essential to address all these issues through scientific inquiry to unlock the complete potential of transgenic fish technology in aquaculture.

References

Angom, R.S., & Nakka, N.M.R. (2024). Zebrafish as a model for cardiovascular and metabolic disease: The future of precision medicine. Biomedicines, 12(3), 693.

Betsy, C.J., & Siva, C. (2023). Transgenic Fish Production. Springer Science and Business media LLC.

Bowen, J., Schloop, A. E., Reeves, G. T., Menegatti, S., & Rao, B. M. (2020). Discovery of membrane-permeating cyclic peptides via mRNA display. Bioconjugate chemistry, 31(10), 2325-2338.

Csenki, Z., Garai, E., Risa, A., Cserháti, M., Bakos, K., Márton, D., ... & Urbányi, B. (2019). Biological evaluation of microbial toxin degradation by microinjected zebrafish (Danio rerio) embryos. Chemosphere, 227, 151-161.

Daneluz, L. O., Acosta, I. B., Nunes, L. S., Blodorn, E. B., Domingues, W. B., Martins, A. W., ... & Campos, V. F. (2020). Efficiency and cell viability implications using tip type electroporation in zebrafish sperm cells. Molecular biology reports, 47, 5879-5887.

Davies, P.L. (2022). Reflections on antifreeze proteins and their evolution. Biochemistry and Cell Biology.

De Vry, J., Martínez-Martínez, P., Losen, M., Temel, Y., Steckler, T., Steinbusch, H. W., ... & Prickaerts, J. (2010). In vivo electroporation of the central nervous system: a non-viral approach for targeted gene delivery. Progress in neurobiology, 92(3), 227-244.

Doyle, D., & Kelso, T. (2004). Gengtically engineered salmon, ecological risk, and environmental policy. Bulletin of Marine Science, 74(3), 509-528.

Dubińska-Magiera, M., Daczewska, M., Lewicka, A., Migocka-Patrzałek, M., Niedbalska-Tarnowska, J., & Jagla, K. (2016). Zebrafish: a model for the study of toxicants affecting muscle development and function. International journal of molecular sciences, 17(11), 1941.

Grossman, M.R. (2016). Genetically engineered animals in the United States: the Aqu AdvantageSalmon. Eur. Food & Feed L. Rev, 11, 190.

Kikkert, J.R., Vidal, J.R., Reisch, B.I., (2005). Stable transformation of plant cells by particle bombardment/biolistics. Methods Mol Biol, 286, 61-78.

Lavitrano, M., Busnelli, M., Cerrito, M.G., Giovannoni, R., Manzini, S., & Vargiolu, A. (2006). Sperm-mediated gene transfer. Reproduction, fertility, and development, 18(1-2), 19-23.

Leggatt, R.A., Sundstrom, L.F., Woodward, K., & Devlin, R.H. (2017). Growth-enhanced transgenic coho salmon (Oncorhynchus kisutch) strains have varied success in simulated streams: Implications for risk assessment. PloS one, 12(1), e0169991.

Ormandy, E.H., Dale, J., & Griffin, G. (2011). Genetic engineering of animals: ethical issues, including welfare concerns. The Canadian Veterinary Journal, 52(5), 544.

Park, M. H., Kim, M. S., Yun, J. I., Choi, J. H., Lee, E., & Lee, S. T. (2018). Integrin heterodimers expressed on the surface of porcine spermatogonial stem cells. DNA and cell biology, 37(3), 253-263.

Potter, H., (2003). Transfection by Electroporation. Current Protocols in Molecular Biology, 62, 9.3.1-9.3.6.

Saran, M., & Sunita. (2019). Genetically modified fish: An overview. International Journal of Veterinary Sciences and Animal Husbandry, 4(3), 17-18.

Saxena, A., & Jha, P.A. (2013). Transgenic fish: techniques, potential and prospects. Trends Fish Res, 2, 7-16.

Schubert, S., Keddig, N., Hanel, R., & Kammann, U. (2014). Microinjection into zebrafish embryos (Danio rerio)- a useful tool in aquatic toxicity testing? Environmental Sciences Europe, 32.

Shoyombo, A. J., Abdulmojeed, Y., Alabi, O. O., Popoola, M. A., Okon, E. M., & Arije, D. O. (2022). Applications of myostatin in poultry and aquaculture-a review. The Open Agriculture Journal, 16(1).

Sreenikethanam, A., Raj, S., Banu, R.J., & Gugulothu, P. (2022). Genetic Engineering of Microalgae for Secondary Metabolite Production: Recent Developments, Challenges, and Future Prospects. Frontiers in Bioengineering and Biotechnology, 10, 836056.

Tiedje, J. M., Colwell, R. K., Grossman, Y. L., Hodson, R. E., Lenski, R. E., Mack, R. N., & Regal, P. J. (1989). The planned introduction of genetically engineered organisms: ecological considerations and recommendations. Ecology, 70(2), 298-315.

Tonelli, F. M., Lacerda, S. M., Tonelli, F. C., Costa, G. M., de França, L. R., & Resende, R. R. (2017). Progress and biotechnological prospects in fish transgenesis. Biotechnology Advances, 35(6), 832-844.

Tonelli, F.M.P., Lacerda, S.M.S.N., Tonelli, F.C.P., Costa, G.M.J., de Franca, L.R., & Resende, R.R. (2017). Progress and biotechnological prospects in fish transgenesis. Biotechnology advances, 35(6), 832-844.

Wakamatsu, Y., Pristyazhnyuk, S., Kinoshita, M., Tanaka, M., & Ozato, K. (2001). The see-through medaka: a fish model that is transparent throughout life. National Academy of Sciences of the United States of America, 98(18), 10046-10050.

Wang, R., Zhang, P., Gong, Z., & Hew, C.L. (1995). Expression of the antifreeze protein gene in transgenic goldfish (Carassius auratus) and its implication in cold adaptation. Molecular marine biology and biotechnology, 4(1), 20-26.

Wang, Y., Hamid, N., Jia, P. P., & Pei, D. S. (2021). A comprehensive review on genetically modified fish: key techniques, applications and future prospects. Reviews in Aquaculture, 13(3), 1635-1660.

Winn, R.N. (2001). Transgenic fish as models in environmental toxicology. ILAR Journal, 42(4), 322-329.

Yan, J., Kang, D.D., Turnbull, G., & Dong, Y. (2022). Delivery of CRISPR-Cas9 system for screening and editing RNA binding proteins in cancer. Advanced Drug Delivery Reviews, 180, 114042.

Zhong, J., Wang, Y., & Zhu, Z. (2002). Introduction of the human lactoferrin gene into grass carp (Ctenopharyngodon idellus) to increase resistance against GCH virus. Aquaculture, 214(1-4), 93-101.

Zhu, Z., Li, G., He, L., & Chen, S. (1985). Novel gene transfer into the fertilized eggs of gold fish. A angew. Icthyology, 1, 31-34.

Zouridakis, G. (2019). Biomedical Technology and Devices Handbook. CRC Press.

2

Genetic and Hormonal Approaches to Monosex Aquaculture Innovations and Implications

Alwinpeter M., Afrith M., Ashish Patel, Abinesh C., Ishita Bambhaniya and Pinak Bamaniya*

College of Fisheries Science, Kamdhenu University, Veraval, Gujarat (362 265) India

Abstract

Monosex culture in aquaculture refers to the practice of cultivating either all-male or all-female populations of fish, which has gained traction due to its numerous advantages over mixed-sex farming. This method is particularly prevalent in species such as tilapia, where male fish are favoured for their superior growth rates and feed conversion efficiency. Monosex populations can be developed through various techniques, including selective breeding, genetic manipulation (such as chromosome manipulation and CRISPR technologies), hormonal treatments for sex reversal, and hybridization methods. The benefits of monosex culture include increased yields, improved product quality, and reduced management challenges associated with reproduction. By focusing on one sex, aquaculturists can optimize stocking densities and resource utilization, leading to enhanced economic returns. However, the implementation of monosex culture is not without challenges. Ethical concerns regarding genetic and hormonal interventions, regulatory hurdles, environmental risks associated with escaped organisms, and high initial investment costs can impede its widespread adoption.

Keywords: *Hormonal treatments, Genetic manipulation, CRISPR technology, Sex reversal and Selective breeding*

Introduction

Aquaculture, the farming of aquatic life like fish and shellfish, has become essential for global food supply. It helps meet the rising demand for seafood, supports countless livelihoods, and promotes sustainable development. As wild fish populations dwindle due to overfishing and environmental damage, aquaculture provides a sustainable way to ensure food security and economic growth. Monosex culture, the practice of raising only one sex of a species, is particularly useful for fish like tilapia and prawns, where one sex grows faster and more efficiently. By focusing on single-sex populations, aquaculture can boost production, simplify management, and increase overall yield. Monosex culture is important because it can improve profits while protecting the environment

Historical Development of Monosex Practices

The development of the monosex culture began in the mid-20th century when scientists started to investigate ways to control the sex ratios of fish populations. Early methods focused on selective breeding to improve traits like growth rate and size. Over time, advancements in genetic manipulation and hormonal treatments have revolutionized the field, allowing for more precise control over sex determination. Hormonal treatments like 17 alpha methyltestosterone for sex reversal in tilapia during the 1970s, significantly improved production efficiency. The development of genetic technologies, such as chromosome manipulation and CRISPR, has further propelled monosex culture into mainstream aquaculture practices. Today, monosex culture is recognized as a cornerstone of sustainable aquaculture, with ongoing research aimed at refining techniques and addressing ethical considerations associated with genetic interventions.

Scientific Basis for Monosex Culture

Monosex culture in aquaculture is a scientifically driven practice of cultivating a single sex of a species to enhance productivity and efficiency. This approach leverages sexual dimorphism, where one sex typically grows faster or larger, ensuring better growth performance and feed conversion rates. It also prevents uncontrolled reproduction, which can lead to overcrowding and resource competition. Techniques such as hormonal manipulation, genetic approaches like YY supermale technology, and selective breeding are commonly employed to produce monosex populations (Miao *et al.*, 2023). These methods minimize energy loss associated with reproduction and ensure uniform harvests, optimizing economic returns. While monosex culture offers ecological benefits, such as reducing risks of non-native population establishment from escapees, it also raises ethical concerns over hormonal treatments and genetic diversity. Nevertheless, it remains a cornerstone of sustainable and efficient aquaculture practices.

Methods for Monosex Population Development

Monosex population development plays a pivotal role in industries such as aquaculture, agriculture, and pest management, providing a means to improve productivity, manage reproduction, and optimize resource utilization. This process focuses on producing populations that consist predominantly of one sex, which is often advantageous due to specific biological traits such as growth rates, reproduction efficiency, or reduced competition. For example, in aquaculture, male tilapia are preferred as they grow faster and larger than females, making monosex male populations more desirable for commercial farming.

Hormonal techniques are widely used in aquaculture to induce sex reversal. By administering hormones like androgens or estrogens during early development, fish can be manipulated to become a specific sex. For instance, young fish may be treated with methyltestosterone to produce phenotypic males, even in genetically female individuals. Additionally, environmental factors such as temperature, light, or pH can be manipulated to influence the natural processes of sex determination. For example, temperature changes during critical developmental periods have been shown to skew sex ratios in reptiles such as turtles and certsin fish species. Hybridization techniques, on the other hand, involve crossbreeding between species or strains to naturally produce offspring of a desired sex. Certain hybrids inherently favor one sex due to the genetic interactions between the parental species, which can result in populations dominated by either males or females. Each of these methods has its own set of applications, benefits, and ethical considerations, depending on the target species and industry goals. By leveraging these techniques, industries can achieve more sustainable practices, maximize yields, and tailor production to meet specific economic or ecological demands.

The method for producing YY-type ultra-male Tilapia seeds is the YY/GMT technology, which stands for Genetically Male Tilapia Technology. This approach involves using a specific sex determination mechanism in Tilapia mossambica, where males have the XY genotype and females have the XX genotype. The process begins by mating an XY male with an XX female, resulting in the production of YY-type ultra-male Tilapia (YY ♂). By conducting test crosses, further YY ♂ individuals are obtained, leading to a 100% male offspring when crossed with other YY ♂.

To produce YY female fish, a sex-reversal technique is used, involving the administration of female hormones to induce sex change. This method is time-consuming and resource-intensive, limiting its widespread use in tilapia breeding.

In many aquaculture species, all-female or sterile populations are preferred due to their superior growth rates and uniformity. The researchers previously created a hybrid carp by crossing two different carp species. They then developed a method to produce all-female and all-male diploid fish. By crossing these fish, they were able to generate fertile all-female diploid fish and sterile all-female triploid hybrids. This technique offers a scalable solution for producing all-female triploid populations, which is highly beneficial for the aquaculture industry.

Monosex Culture Practices

Gynogenesis

Gynogenesis is a technique that produces offspring with only maternal DNA, allowing for the creation of all-female populations. This involves fertilizing eggs with inactive sperm and then shocking the eggs to prevent the second cell division, resulting in a diploid offspring. Meiotic gynogenesis, in particular, uses UV-irradiated sperm to trigger egg activation and then applies a shock to retain the second polar body. This process is similar to triploid induction, where a shock is used to prevent the expulsion of the second polar body, leading to an extra set of chromosomes.

Mitotic gynogenesis involves inhibiting the first cell division to produce double diploid offspring. This is typically achieved through physical shocks like cold or pressure. While meiotic gynogenesis might result in fewer homozygous diploids, it often leads to higher survival and fertility rates. Both meiotic and mitotic gynogenesis processes are illustrated in Figures 1 and 2. Efforts to induce gynogenetic diploids in European loach have been made using various shock treatments on irradiated sperm-fertilized eggs. Additionally, spontaneous gynogenesis has been observed in some fish, amphibians, and reptiles.

Gynogenesis and hormone sex reversal are techniques used in aquaculture to create single-sex populations, which can improve economic benefits and increase production. In gynogenesis, male genetic material is destroyed using radiation, and the irradiated sperm is used to fertilize eggs. Various shock treatments, such as heat, cold, or chemical shocks, are applied to prevent the second cell division, resulting in a diploid offspring with only maternal DNA. This technique can be used to manipulate fish chromosomes and study sex determination. By eliminating paternal genetic material and diploidizing the haploid egg, gynogenesis can create single-sex populations.

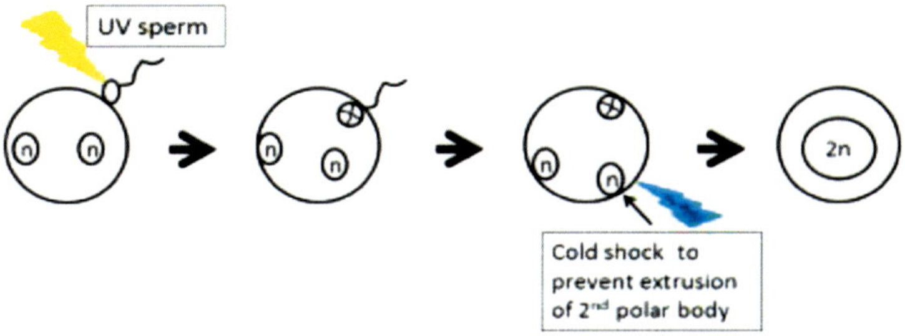

Fig. 1: Mentioned above illustrates the process of meiotic gynogenesis. UV-irradiated sperm is used to activate the egg, and a cold shock is applied to prevent the second polar body from being expelled. This results in the formation of a diploid gynogens.

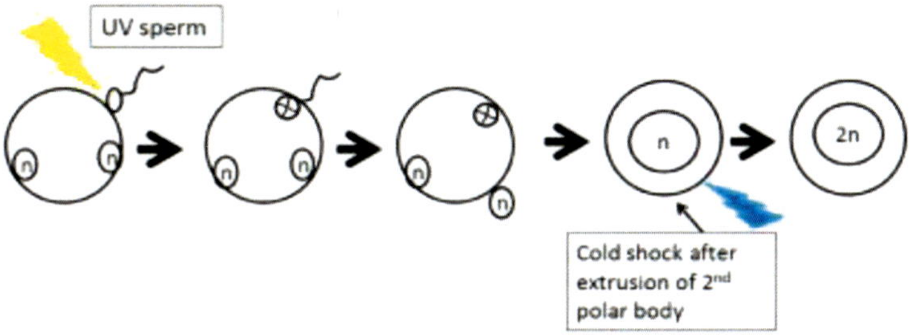

Fig. 2: Mentioned above depicts the following mitotic gynogenesis process. Here the UV-irradiated sperm is used to activate the egg, and a cold shock is being applied after the second polar body is expelled. This leads to the formation of diploid gynogen.

Androgenesis

Androgenesis is a more challenging technique than gynogenesis and has limited practical applications in aquaculture. This technique produces offspring with only paternal DNA, a process known as all-paternal inheritance. It involves fertilizing an egg with its maternal DNA removed using UV radiation with normal sperm. This creates a zygote with solely paternal DNA

Once fertilization is complete, a heat shock is applied to the fertilized egg to prevent chromosome separation. This ensures that the egg retains two sets of chromosomes, both from the male parent. This process results in a diploid androgenetic organism with only male-derived genetic material.

Androgenesis has significant applications in aquaculture, particularly in creating cryopreserved gene banks for fish species. This technique is also used to produce YY 'supermales' in XY-based sex determination systems, such as tilapia. YY males can rapidly generate all-male populations, which grow faster and have more uniform traits. This offers an alternative to hormonal sex reversal for producing monosex populations. For ZW sex determination systems, androgenesis can be combined with sex-reversal techniques to create monosex strains. This approach complements androgenesis and provides additional methods for controlling sex ratios in aquaculture.

Applications of Monosex Culture in Aquaculture Species

***Nile Tilapia* (*Oreochromis niloticus*):** Monosex male production in Nile tilapia is among the most extensively studied and applied examples of monosex culture. Male tilapia grows 15–20% faster than females due to differences in energy allocation—males invest more in growth, whereas females expend energy on reproduction. Techniques like hormonal treatment with 17α-methyltestosterone are used during early larval stages to ensure male populations. Genetic methods, such as the production of YY males, involve selectively breeding "super males" (YY genotype) with XX females, producing offspring that are nearly 100% male. This practice not only boosts production efficiency but also prevents overpopulation in farming systems, which is a common issue with mixed-sex tilapia populations.

***Catfish* (*Clarias gariepinus and Ictalurus punctatus*):** In catfish farming, males are often preferred for their larger size and faster growth. In African catfish (*Clarias gariepinus*), monosex male populations are produced using hormonal sex reversal or hybridization techniques. For instance, hybrid catfish (crosses between male blue catfish and female channel catfish) are highly sought after for their fast growth and resilience. These hybrids are predominantly male and are better suited for intensive farming conditions. Additionally, in channel catfish (*Ictalurus punctatus*), hormonal treatments or selective breeding programs help produce male-biased populations, further enhancing growth rates and yield.

***Carp Species (Cyprinus carpio and Ctenopharyngodon idella*):** Carps, widely farmed in Asia and Europe, benefit greatly from monosex culture. In common carp, monosex production through selective breeding ensures uniform size and reduces market inconsistencies. For grass carp, monosex culture ensures predictable vegetation consumption and improved growth. These methods are especially useful in integrated farming systems where carp coexist with other species, preventing overpopulation and resource competition

Giant Freshwater Prawns (Macrobrachium rosenbergii): In giant freshwater prawn farming, males grow significantly larger than females and are more economically valuable. Selective breeding, hormonal treatments, and environmental manipulations are used to produce all-male populations. For instance, RNA interference (RNAi) technology can be applied to silence the insulin-like androgenic gland hormone, effectively converting genetic females into functional males. This innovative approach has been successful in boosting production yields while maintaining genetic diversity for future breeding. In

India, the production of all-male and all-female scampi production; amongst the all-male showed higher growth productivity and cost effectiveness than all-female.

Pacific White Shrimp (Litopenaeus vannamei): Although shrimp farming traditionally involves mixed-sex populations, the monosex production of male shrimp is gaining attention due to their faster growth rates and better feed conversion ratios. Hormonal and environmental manipulations, along with selective breeding, are being explored to produce male-biased populations. Monosex shrimp culture not only improves production efficiency but also reduces variability in size at harvest, which is essential for meeting market demands.

Mollusks (e.g., Oysters and Scallops): Monosex culture in mollusks like oysters (*Crassostrea gigas*) and scallops (*Pecten maximus*) is less common but has specific applications. Female oysters, for example, are often preferred in some markets due to their larger size and better meat quality. Techniques such as triploidy induction (e.g., creating sterile populations) are used to produce larger and faster-growing individuals, indirectly benefiting monosex production by ensuring consistent quality.

Rainbow trout: All-female populations of rainbow trout, both diploid and triploid, are highly promising for commercial trout farming. They exhibit rapid growth rates and eliminate the presence of males, which mature earlier and have poorer flesh quality. All-female triploidy (AFT) production is particularly beneficial for trout farmers aiming to produce larger fish for specific markets.

Brook trout: Brook trout, a North American species, were initially farmed in both Canada and the United States. Monosex female brook trout have been successfully produced using the gynogenetic sex reversal techniques, which involve treatments with methyltestosterone through immersion or oral administration.

Future Aspects in Monosex Aquaculture

Aquaculture, while beneficial, can have significant negative environmental impacts. 1 The use of monosex fish, like GMT, raises concerns about potential interactions with wild populations. While GMT males are genetically normal, they could potentially breed with native fish in suitable areas. To mitigate this risk, the DFID Fish Genetics Research Programme is developing sterile fish using antisense DNA technology. These sterile fish would have minimal long-term environmental impact and help prevent genetic mixing between wild and farmed populations, a problem observed in Atlantic salmon populations.

Recent advances in genetic technology have opened up new possibilities for controlling sex by directly targeting genes involved in sex determination and reproductive maturation. Two promising techniques are antisense RNA gene knockdown and the insertion of transgenes for transgenic sterilization. While these methods are not yet widely used in commercial applications due to technical and regulatory challenges, research has begun on species like tilapia, rainbow trout, and common carp.

Researchers have focused on key reproductive hormone genes like gonadotropin-releasing hormone (GnRH), which plays a crucial role in stimulating the release of other hormones that influence gonadal development. In rainbow trout, experiments using a single-stranded antisense RNA silencing technique targeting the Atlantic salmon GnRH3 gene successfully downregulated GnRH mRNA but did not significantly affect other hormone levels or completely prevent maturation. Similarly, in tilapia, females carrying a GnRH3 construct showed reduced fertility compared to control females.

While these initial studies did not achieve complete sterility, they demonstrate the potential of genetic techniques to control sexual processes in aquaculture species in the future.

Advantages of Monosex Culture in Aquaculture

The practice of cultivating fish populations, either exclusively male or female, is referred to as monosex culture. This technique has gained significant traction within the aquaculture industry due to its numerous advantages. By focusing on single-sex populations, aquaculturists can optimize production efficiency, reduce management complexities associated with reproduction, and enhance overall yield. This specialized approach offers a multitude of benefits, including improved growth rates, enhanced feed conversion efficiency, and reduced competition for resources. Furthermore, monosex culture can help mitigate potential environmental impacts by preventing the unintended release of reproductive individuals into natural water bodies, thereby minimizing the risk of genetic interbreeding and ecological disruption

Increased Growth Rates and Yield

One of the primary benefits of monosex culture is the increased growth rates observed in the selected sex. For instance, male tilapia are known to grow faster than females because their metabolic energy is directed towards growth rather than reproduction. This results in higher yields per unit area, making monosex populations particularly attractive for aquaculture operations (Fregene *et al.*, 2024). Studies have shown that mono-sex tilapia can reach market sizes more quickly and uniformly compared to mixed-sex populations, which often exhibit slower growth due to reproductive energy expenditure (Abd El-Hack *et al.,* 2022).

Improved Product Quality and Marketability

Monosex culture not only enhances growth but also improves product quality. Fish produced in a monosex system tend to be more uniform in size and weight, which is preferred by consumers and enhances marketability. Consistency in product quality can lead to better pricing and increased demand in the marketplace (Dunham *et al.,* 2001). Moreover, with a focus on one sex, producers can target specific market preferences more effectively.

Reduced Reproduction-Related Management Issues

Managing mixed-sex populations can lead to complications such as over population and competition for resources. Monosex culture mitigates these issues by preventing unwanted breeding, which can result in stunted growth and increased mortality rates among fish due to overcrowding. By controlling reproduction, aquaculturists can maintain optimal stocking densities and improve overall health conditions within the culture system (Ventura, 2018).

Better Resource Utilization

Monosex culture allows for better resource utilization, including feed, space, and energy. Since all individuals are of the same sex, feeding strategies can be optimized based on their specific growth patterns without the need to account for reproductive needs. This leads to more efficient feed conversion ratios (FCR) and reduces waste, ultimately lowering operational costs (Fregene *et al.,* 2024). Additionally, higher stocking densities are feasible with monosex populations, maximizing space usage while maintaining fish health.

Challenges and Limitations of Monosex Culture in Aquaculture

While monosex culture in tilapia production offers numerous advantages, such as improved growth rates, enhanced feed conversion efficiency, and reduced competition for resources, it also presents several challenges and limitations that must be carefully considered for sustainable implementation. These challenges include ethical concerns related to the manipulation of natural reproductive processes, regulatory hurdles associated with the use of hormones and genetic engineering techniques, potential environmental risks, such as the escape of genetically modified fish into natural water bodies, and economic constraints, including the high costs of implementing monosex culture technologies.

Ethical Concerns Surrounding Genetic and Hormonal Interventions

The use of genetic and hormonal techniques to produce monosex populations raises significant ethical concerns. Hormonal sex reversal methods, such as the use of methyltestosterone, have been criticized for potential impacts on fish welfare and human health perceptions regarding hormone residues in fish products (Das *et al.*, 2022). Additionally, the manipulation of genetic traits could lead to unforeseen ecological consequences if these organisms escape into natural environments.

Regulatory Issues and Market Acceptance

Monosex culture often faces regulatory hurdles that can impede its adoption. Different countries have varying regulations regarding the use of hormones and genetically modified organisms (GMOs) in aquaculture. This inconsistency can create barriers for producers seeking to enter new markets or expand their operations. Furthermore, consumer acceptance of products derived from genetically modified or hormonally treated fish remains a contentious issue, potentially affecting market demand (Fregene *et al.*, 2024).

Environmental Risks

There are notable environmental risks associated with monosex culture. The escape of monosex or genetically modified organisms into wild ecosystems can disrupt local biodiversity and alter ecological balances. For example, if genetically modified tilapia were to escape into natural waters, they could outcompete native species or disrupt local food webs. This risk necessitates stringent management practices to prevent escapes and monitor environmental impacts.

High Initial Investment and Technical Expertise Requirements

Implementing monosex culture systems often requires a high initial investment in technology and infrastructure, as well as ongoing operational costs. Additionally, the need for specialized knowledge in hormonal applications and breeding techniques can pose a barrier for many aquaculture practitioners. Farmers may require training and technical support to effectively manage these systems, which can be a significant hurdle in regions with limited access to resources or expertise (Das *et al.*, 2022).

References

Abd El-Hack, M. E., El-Saadony, M. T., Nader, M. M., Salem, H. M., El-Tahan, A. M., Soliman, S. M., & Khafaga, A. F. (2022). Effect of environmental factors on growth performance of Nile tilapia (Oreochromis niloticus). International Journal of Biometeorology, 66(11), 2183-2194.

Amoussou, T. O., Karim, I. Y. A., Dayo, G. K., Kareem, N., Toko, I. I., Chikou, A., & Toguyéni, A. (2019). An insight into advances in fisheries biology, genetics and genomics of African tilapia species of interest in aquaculture. Aquaculture Reports, 14, 100188.

Arai, K. (2001). Genetic improvement of aquaculture finfish species by chromosome manipulation techniques in Japan. Aquaculture, 197(1-4), 205-228.

Azhaguchamy, M. (2013). Induction of gynogenesis in Asian Catfish, Clarias batrachus, by using UV irradiated heterologous sperm of Heteropneustes fossilis. World Aquaculture Society, 2013-meeting abstract.

Beardmore, J. A., Mair, G. C., & Lewis, R. I. (2001). Monosex male production in finfish as exemplified by tilapia: applications, problems, and prospects. Reproductive biotechnology in finfish aquaculture, 283-301.

Budd, A. M., Banh, Q. Q., Domingos, J. A., & Jerry, D. R. (2015). Sex control in fish: approaches, challenges and opportunities for aquaculture. Journal of Marine Science and Engineering, 3(2), 329-355.

Das, B., Chakraborty, S., Sarker, B., Hossain, M.A., Alam, M.M.M, Mian, S., & Iqbal, M.M. (2022). Status and problems of monosex tilapia (Oreochromis niloticus) seed production using androgen hormone in Bangladesh. Sustainable Aquatic Research,1(2), 133-146.

Devlin, R. H., & Nagahama, Y. (2002). Sex determination and sex differentiation in fish: An overview of genetic, physiological and environmental influences. Aquaculture, 208, 191–364pp.

Devlin, R. H., & Nagahama, Y. (2002). Sex determination and sex differentiation in fish: an overview of genetic, physiological, and environmental influences. Aquaculture, 208(3-4), 191-364. https://doi.org/10.1016/S0044-8486(02)00057-1.

Dunham, R. A., Majumdar, K., Hallerman, E., Bartley, D., Mair, G., Hulata, G., ... & Hörstgen-Schwark, G. (2000, February). Review of the status of aquaculture genetics. In Aquaculture in the third millennium. Technical Proceedings of the Conference on Aquaculture in the Third Millennium, Bangkok, Thailand (Vol. 20, p. 25).

Emefe, O., & Sorhue, U. G. (2014). Gynogenesis: An effective way of controlling fish population and increasing profit in aquaculture.

Ezaz, M. T., Myers, J. M., Powell, S. F., McAndrew, B. J., & Penman, D. J. (2004). Sex ratios in the progeny of androgenetic and gynogenetic YY male Nile tilapia, Oreochromis niloticus L. Aquaculture, 232(1-4), 205-214. https://doi.org/10.1016/j.Aquaculture.2003.08. 001.

Fregene, B. T., Karisa, H. C., Bolorunduro, P., & Olaniyi, A. (2024). Extension manual on monosex tilapia production and management. Gates Open Res, 8(67), 67.

Gupta, M. V., Bartley, D. M., & Acosta, B. O. (Eds.). (2004). Use of genetically improved and alien species for aquaculture and conservation of aquatic biodiversity in Africa (Vol. 68). WorldFish.

Haffray, P., Petit, V., Guiguen, Y., Quillet, E., Rault, P., & Fostier, A. (2009). Successful production of monosex female brook trout Salvelinus fontinalis using gynogenetic sex-reversed males by a combination of methyltestosterone immersion and oral treatments. Aquaculture, 290(1-2), 47-52.

Hidayah, M., Ikhwanuddin, M., Shahreza, M. S., Abol-munafi, A. B., & Fariduddin, O. (2012). Triploidy Induction Using cold shock treatment in Banana shrimp, Penaeus merguiensis (de Man, 1888). Borneo Science, 30, 64–69pp.

Hussain, M. G. (2004). Farming of Tilapia: Breeding plans, mass seed production and aquaculture techniques (p. 149pp). Book, 984-32-1839-6.

Komen, H., & Thorgaard, G. H. (2007). Androgenesis, gynogenesis and the production of clones in fishes: A review. Aquaculture, 269, 150–173pp.

Li, T., Jia, Y., Gao, J., Zhao, J., Jiang, W., Gu, Z., & Yang, R. (2018). Optimization of shock conditions to induce gynogenesis in topmouth culter (Culter alburnus). Aquaculture, 490, 344–349pp.

Manan, H., Hidayati, A. N., Lyana, N. A., Amin-Safwan, A., Ma, H., Kasan, N. A., & Ikhwanuddin, M. (2022). A review of gynogenesis manipulation in aquatic animals. Aquaculture and Fisheries, 7(1), 1-6.

Miao, L., Tang, X. N., Li, M. Y., Wang, T., Wang, S., Zhang, X. L., & Chen, J. (2014). Artificial gynogenesis in Pseudosciaena crocea (Perciformes, Sciaenidae) with heterologous sperm and its verification using microsatellite markers. Aquaculture Research, 45, 1253–1259pp.

Miao, M., Li, S., Yuan, J., Liu, P., Fang, X., Zhang, C., ... & Li, F. (2023). CRISPR/Cas9-mediated gene mutation of EcIAG leads to sex reversal in the male ridgetail white prawn Exopalaemon carinicauda. Frontiers in Endocrinology, 14, 1266641. doi: 10.3389/fendo.2023.1266641. PMID: 38075036; PMCID: PMC10703455.

Myers, J. M., Penman, D. J., Basavaraju, Y., Powell, S. F., Baoprasertkul, P., Rana, K. J., ... & McAndrew, B. J. (1995). Induction of diploid androgenetic and mitotic gynogenetic Nile tilapia (Oreochromis niloticus L.). Theoretical and Applied Genetics, 90, 205-210.

Nan'er, C., Feng, L., & Gall, G. A. E. (1991). Artificial induction of gynogenesis in Chinese shrimp, Penaeus orientalis kishinouye. In Genetic in in aquaculture IV (p. 314pp). Elsevier publication. Institute of Oceanography,. Academia Sinica, Qingdao, China. Published in Book title.

Pandian, T. J., & Koteeswaran, R. (1998). Ploidy induction and sex control in fish. Hydrobiologia, 384, 167–243pp.

Sarder, M. R. I., Penman, D. J., Myers, J. M., & McAndrew, B. J. (1999). Production and propagation of fully inbred clonal lines in the Nile tilapia (Oreochromis niloticus L.). Journal of experimental zoology, 284(6), 675-685. https://doi.org/10.1002/(SICI)1097-010X(19991101) 284:6<675::AID-JEZ9>3.0.CO;2-D.

Sheehan, R. J., Shasteen, S. P., Suresh, A. V., Kapuscinski, A. R., & Seeb, J. E. (1999). Better growth in all-female diploid and triploid rainbow trout. Transactions of the American Fisheries Society, 128(3), 491-498.

Suzuki, R., Oshiro, T., & Nakanishi, T. (1985). Survival, growth and Fertility of Gynogenetic diploids induced in the cyprinid loach, Misgurnus anguillicaudatus. Aquaculture, 48, 45–55pp.

Ventura, T. (2018). Monosex in aquaculture. Marine organisms as model systems in biology and medicine, 91-101.

Yamamoto, E. (1999). Studies on sex-manipulation and production of cloned populations in hirame, Paralichthys olivaceus (Temminck et Schlegel). Aquaculture, 173, 235–246pp.

Zonneveld, N., & Van Zon, H. (2019). The biology and culture of grass carp (Ctenopharyngodon idella), with special reference to their utilisation for weed control. In Recent advances in aquaculture (pp. 119-191). Routledge.

3

Integrative Approaches to Understanding Sex Determination and Differentiation in Fish

Panchakarla Sedyaaw* and V.R. Bhatkar

College of Fisheries, Ratnagiri, Dr. Balasaheb Sawant Konkan Krishi Vidyapeeth Dapoli, Maharashtra (415 712), India

Abstract

Sex determination and differentiation in fish are complex processes driven by genetic, genomic, and endocrine factors. Fish have a remarkable variety of sex-determination systems compared to mammals. These systems range from purely genetic ones like XY/XX and ZZ/ZW to environmental ones influenced by social and environmental factors like temperature. Fish sex determination is genetically controlled by master sex-determining genes, such as foxl2 in females and dmrt1 in males. However, through genomic studies, scientists have discovered QTLs and numerous other regulatory elements that control sex differentiation. During the differentiation process, endocrine factors are important because sex steroids, such as estrogens and androgens, dictate gonad differentiation. The expression of the aromatase enzyme, which catalyses the conversion of androgens to estrogens, will decide whether testes or ovaries develop; its lack of expression often determines the development of testes. In addition, hormone cues usually synergise with other genetic pathways to control sex plasticity in some species. Understanding how genetic, genomic, and endocrine pathways play together in sex determination/differentiation will have far-reaching implications for aquaculture and conservation activities because it will help regulate sex ratios and improve the productivity and sustainability of fish populations.

Keywords: *Sex determination, Genomics, Endocrinology, Sex differentiation, QTL.*

Introduction

Sexual development in animals is considered one of the most complex research areas. This complexity arises from the underlying genetic structures that determine gender. Additionally, sexual development is viewed as one of the most intricate traits that have evolved, making it essential to studying evolution and population genetics (Schartl *et al.*, 2018). According to evolution, fish have a link between invertebrates and higher-stacked vertebrates. According to Smith & Wootton, (2016), Kang *et al.* (2017), nearly all known mechanisms of sex determination observed in vertebrates have been identified in fish. As a result, finding the fish sex determination genes and their role in fish biology has excellent significance for the purely biological science of sex determination in animals. The energy from food is consumed into growth before fish become sexually mature. Since there is a range of ages for sexual maturity among different fish species, this explains the bilateral sexual growth differences in the species. The common carp (*Cyprinus carpio*) and the half-smooth tongue sole (*Cynoglossus semilaevis*) are two species that show a pattern in which females mature later than males, leading to larger females.

Sex-determining genes and their action mechanisms are crucial for guiding sex control breeding strategies in animals like the Nile tilapia (*Oreochromis niloticus*) and yellow catfish (*Pelteobagrus fulvidraco*), where males are categorized as more advanced than females Shao and Chen 2012; Chen *et al.*, 2013; Long *et al.*., 2015; Mei and Gui. 2020). However, over ten years ago, the most evolved animals were the primary research subjects on the genes that determine sex and the mechanisms by which they work. According to Sinclair *et al.*, SRY, also known as Sinclair, is the first human sex-determining gene found on the Y chromosome. in mice) and subsequently turned out to be a gene that determines maleness.

From that moment onward, the SRY gene has been established as the principal determinant of male sex in various animal species. It affects sex differentiation by modulating multiple downstream transcription factors, including Sox9, Fgf9, and Dmrt1. Thus far, the action mechanism established regarding SRY controls the transcription of Sox9 by direct contact with a motif in the promoter region of the TESCO gene (Sekido and Lovell-Badge 2008).

With SOX9, SRY also triggers the expression of Fg, f9, which helps turn gonadal-supporting cell precursors into Sertoli cells and promotes testicular development (Wilhelm *et al.*, 2007). Furthermore, SRY causes ovarian gene expression to decrease by blocking the Wnt signalling pathway (Li *et al.* 2014). It is still unknown, however, precisely how the SRY protein activates the Sox9 gene. All SRY proteins are characterized by having a DNA binding domain. It is important to note that the structure of the SRY protein differs between mice and humans, with mice possessing distinct non-TD N-terminal and C-terminal regions and an additional glutamine-rich peptide (Sekido,

2010). A human SRY gene may initiate male development in XX mice, suggesting that human SRY could interact with a specific protein chaperone. A sex chromosome system exists in birds, similar to that in mammals; however, the specific sex-determining genes in birds remain unidentified. The W chromosome in chickens is approximately 6 MB long and primarily composed of heterochromatin. Sex-determining genes are absent because all 25 ApW C genes and 116 ApW non-coding RNAs identified on the W chromosome showed alleles on the Z chromosome with comparable expression levels (Warren *et al.,* 2017). Prioritization was given to other genes for determining sex in birds, even though DMRT1 is located on the Z chromosome. For example, Dmrt1 is overexpressed in familial male yolk sac embryo gender chromosome (X) diabetes, but embryogenesis starts from distinct germ layers (Lambeth *et al.,* 2014). Dmrt1 depletion by RNA interference forces males to lay eggs of the female chicken strain (Smith *et al.*, 2009a). In contrast, Pinzgauer chicken lines with only a fragment Z1 stem chromatic era edited with the aid of C-RisPR/Cas9 reverse transcription fill Z2-dwarfs Z0 at female sexual organs appeal (Ioannidis *et al.*, 2021). However, genetically modified hens did not exhibit female traits. Sex differentiation in birds is fundamentally cell-autonomous and will not be XY-dependent.

Activating avian-specific genes that govern female development is a critical initial step in forming gonads. This process is essential for the proper differentiation of the gonadal structures and influencing the growth of somatic cells distinct from the gonadal tissues. The interplay between these genetic determinants is crucial for ensuring appropriate sexual differentiation and the overall developmental trajectory of the organism. Neither candidate DMRT1's properties enable expression confined to the genitourinary system. HINTW is the only known gene in which the Z chromosome allele differs from a single W chromosome allele. HINTW proteins are believed to be homologous without the catalytic domain, while HINTZ proteins bind nucleic acids. According to Bellotti *et al.* (2017), HINTW has long been considered a negative dominant regulator of male H-1 development. On the other hand, because HINTW does not distinguish between males and females, the observation of a residual HINTW protein emphasizes the gender-neutral character of the result (Smith *et al.,* 2009b). Furthermore, it's critical to remember that super expression is still required for hints.

The sex determinant DM-W has been independently developed from DMRT1 in eight closely related Xenopus species, which is a fantastic phenomenon. DM-W, a gene that encodes a protein lacking the transcriptional activation domain, prevents DMRT1 from binding, in contrast to DMRT1. Consequently, the inhibition of male development regulates female development (Yoshimoto *et al.*, 2008). In terms of sex differentiation, amphibians, like reptiles, appear to be both temperature-sensitive and hormone-controlled. Even in the amphibians examined in the field, sex changes may take place.

Sex Determination in Fish

Fish species account for the genetic makeup of over fifty per cent of all vertebrates. According to Arkhipchuk (1995), there are various mechanisms through which sex differentiation occurs in fish; however, only a limited number of fish species have been recognized as having heteromorphic sex chromosomes. The majority of sex-determining systems include more than the commonly known XX/XY and ZZ/ZW systems such as XX/XO, ZZ/ZO, XX/XYY2, X1X2X1X2/X1X2Y, X1X2X1X2/X1X2X1 among others which are special types of sex-determining chromosomes. The way sex is determined in an organism is different not only across different species but genera too. For instance, members of the same genus, such as *Oryzias dancena* and *Oryzias hubbsi* have distinct sex chromosome systems (O. dancena has XX/XY form chromosome, whereas O. hubbsi has ZZ/ZW chromosome.) (Takehana *et al.,* 2007. Most salmonid species use a "sdy" marker to identify sex, though it can be found on different chromosomes (Yano *et al.,* 2013; Bertho and associates 2020). According to Gammerdinger and Kocher (2018), one lineage of the family East African stiltfish has up to 12 sex-determining loci that come from 12 distinct chromosomes.

A further pair of chromosomes, known as the B chromosome, is present in both the male and female genomes and may contain genes that determine sex (Ding *et al.,* 2021). This makes it extremely difficult to isolate the genes that determine sex in fish (Bao *et al.,* 2019). Fish exhibit a significantly larger variety of sexes compared to other vertebrate groups. This diversity in sex-determining systems across different fish species offers extensive reference materials for studying the mechanisms behind sex determination in other species and their respective determining genes (Guiguen *et al.*, 2019).

Genetic Sex Determination

Gonadal sexual determination (GSD) involves a complex interplay of biological processes that regulate sex determination and differentiation. For example, Bull proposes that vertebrate genetic systems operate through polygenic mechanisms, where dominant genes interact with autosomal control genes and sex chromosome-specific genes. These methods have been reported in teleost fish, demonstrating that genes involved in sex determination can be found throughout the genome, on specific chromosomes, or clustered inside a single genomic locus (Devlin & Nagahama, 2002).

Monogenic systems: Sex chromosomes

Not only do teleosts have sex-determining factors, but closely related species also do. Transposition of an existing sex-determining gene, the synthesis of new sex-determining genes on autosomes, and combinations of sex chromosomes and autosomes are some of the strategies used to quickly replace teleost's sex chromosomes (Doorn and Kirkpatrick 2007; Ross *et al.,* 2009). According to Devin and Nagahama (2002), in monogenic systems, sex determination is controlled by a single gene on one chromosome, with genes on other chromosomes having little effect. Cytogenetic hybridization and fluorescence in situ hybridization techniques can be utilized to differentiate between species of teleost fishes that possess heteromorphic sex chromosomes, which are relatively uncommon among these species. For example, this method has been applied to *Aulopus japonicus* (Ota *et al.*, 2003), Riobuglus catfishes, *Liobagrus marginatus*, and *Liobagrus styani* (Chen *et al.*, 2008), as well as the four-spine stickleback, *Apeltes quadracus* (Peichel *et al.*, 2004).

Polygenic Systems

Sex is determined in polygenic systems by the combined genetic contributions of all pertinent factors rather than by a single locus (Bull, 1985). The Nile tilapia, *Oreochromis niloticus* is a species that demonstrates genetic sex determination mechanisms, such as XY or ZW systems. This species may significantly impact autosomal components (Scott *et al.*, 1989). The study of passive autosomal transformation, referred to as a "smash," in common carp (Cyprinus carpio) with the XY sex-determination framework has highlighted the role of autosomal effects in sexual determination. Individuals who are homozygous for this alteration exhibit masculinization, suggesting that a single autosomal trait can lead to sexual inversion in this species (Komen *et al.*, 1992).

The DMY gene, also known as Dmrt1bY, located on the Y chromosome, is recognized as the primary sex-determining gene for this species. When XY individuals possess mutations that disable the DMY gene, their female traits become more pronounced. Conversely, the presence of the DMY gene on the Y chromosome causes XX individuals to develop testes and become functional males (Matsuda *et al.*, 2007). Additionally, the autosomal duplication of the Dmrt1 gene has been shown to enhance the functionality of the DMY gene (Kondo *et al.*, 2006). This gene likely belongs to the highly conserved DM domain family, which is found across all vertebrates.

Gonadal Development

Medaka has been broadly investigated for morphological characteristics of gonadal advancement. The Adriatic sturgeon *Acipenser naccarii* (Grandi and Chicca, 2008), rainbow trout, zebrafish, the viviparous eel Zoarces, the inland *Gambusia affinis*, the cichlid *Cichlasoma dimerus*, and *Gasterostus aculeatus* (Lewis *et al.,* 2008). In viviparous species, the two gonadal primordia fuse during development to form a single gonad, in contrast to the majority of teleosts, which have fused adult gonads. Two leaves are attached to the base of the ovaries, which are a single organ in these animals.

Ovarian Differentiation

The majority of teleost fishes undergo ovarian development, which begins with the growth of germ and basal cells. This is followed by the division of the first oocyte and the eventual formation of the ovarian cavity. Ovarian separation in most teleosts starts with the development of meiotic figures within a limited number of ovarian germ cells (Patino and Takashima, 1995). Moreover, the number of undifferentiated cells destined to become ovaries becomes more pronounced compared to the reproductive glands, which are meant to develop into testes. Consequently, the extent of ovarian separation tends to increase organically. For instance, female Medaka embryos develop more rapidly than male germ cells (Saito *et al.*, 2007).

In the three-spined stickleback, *Gasterosteus aculeatus* there is a significant increase in primordial germ cells (PGCs) among the early embryos of females, which have not yet differentiated into reproductive glands. This growth in PGCs can facilitate the transformation of undifferentiated reproductive glands into ovarian tissue (Lewis *et al.*, 2008).

During ovarian separation, physical cells and embryos work together to form the follicles. This process involves a layer of granulosa cells enveloping the outer layer of oocytes and the theca cells. Certain species have cystic structures, suggesting an internal depth that forms during ovarian separation as part of their developmental preparation. In some species, such as *Oryzias latipes* and *Carassius auratus*, the development of ovarian depth is associated with the proliferation of pre-embryonic cells. In contrast, others, like *Oryzias niloticus*, produce meiotic oocytes with relatively little development in ovarian depth. The differentiation of the ovary from the gonad can be noted in specific varieties (Nakamura *et al.*, 1998).

Testicular Differentiation

The testes form later than the ovaries, and interstitial separation occurs sometime after the germ cells are placed in the testes. The germ cells remain dormant for a long period. Spermatogenesis starts essentially afterwards after bringing forth in other species, such as *O. mossambicus* and *O. niloticus* give birth to killifish offspring 50 days later (Matsuda, 2005). One of the characteristics that demonstrate the beginning of testicular separation is the arrangement of efferent conduits, which are recognized by the nearness of modest holes inside stromatic tissues. Stromatic cell amassing marks the starting of testicular advancement in Poecilia reticulata; these cells in the long run advance into sperm conduits and interstitial tissue. In Zoarces viviparous, testicular germ cells are inactive and don't start meiosis all through embryonic improvement. However, other postnatal anatomical changes include the development of testicular lobules, the presence of blood vessels in the interstitial portion of the testis, and the initial appearance of the vas deferens—all of which are indicators of testicular cleavage (Rasmussen *et al.,* 2006).

Molecular Bases of Sexual Determination and Differentiation

Many qualities have been captured in the sexual development and self-confidence of vertebrate gonads (Pieau *et al.*, 2001). Much of the information on the atomic tools underlying these exercises comes from studies of warm-blooded organisms, highlighting the different properties these morphologies contain, some of which have special abilities. The most extensively researched are the following: SF-1; Wt1, Wilms tumor suppressor-1; Sox-9, HMG gene related to Sry-9; Dmrt1, dual sex gene and mab-3-1 related gene; Amh or Mis, anti-Mullerian hormone; Gata-4, GATA-4 power protein; Dax1, dosage-dependent reversion, X-linked intrinsic adrenal hypoplasia; Wnt4, wingless MMTV integration site family, member 4; Fox, and Sry (sex position on the Y chromosome), which determines the sex of warm-blooded animals.

Genes Involved in Sexual Determination and Differentiation in Teleosts

Various qualitative families associated with sexual safety in other vertebrates exhibit differences in perspective, indicating that the elements of sexual safety may be less stringent (Hofsten and Olsson, 2005). Several sex-determining qualities in teleosts act so also to those in warm-blooded animals, even though their part isn't completely caught on. Be that as it may, agreeing to distributed thinks about, no teleost partner of the mammalian Sry quality has been found to however, with the special case of the DMY quality in medaka.

Sf-1

This is one of the main qualities seen in the beginning of the Adrenal of mammals, and there is a place for the NRA51 super family receipt. SF-1 and the quality associated with it seem to be transmitted by steroid-produced cells and are related to sexual activities. The qualities associated with SF-1 differ from various types of cerebrovascular accidents, as indicated by the research of Higa *et al.* (2000) and studies on Danio Fish. The zebrafish genes Ff1b and Ff1d perform functions akin to those of SF-1. Ff1d is expressed in both male and female interrenal cells, but it is thought to be more common in the male gonad, particularly in Leydig and Sertoli cells (Von Hofsten *et al.,* 2005). SF-1, a characteristic associated with sexual differentiation and determination in mammals, has been suggested to have a similar function in zebrafish, possibly affecting the regulation of Amh (Hofsten *et al.,* 2005).

Sox9

A translation factor that typically possesses the HMG box and DNA-binding motif is significantly associated with sexual differentiation in vertebrates. Sox9 may play a crucial role in testicular differentiation and cartilage formation, as supported by various experimental studies conducted on animals. The Sox9 gene was identified in a few teleost species, including rainbow trout, tilapia (Ijiri *et al.*, 2008), zebrafish, guppy, common carp, and swamp eel. In such species, Sox9 has been connected to certain occasions of sexual separation and is likely included in the preparation.

Dmrt1

Among the numerous strategies of sexual assurance and separation that exist nowadays, quality Dmrt1, which encodes for a translation figure, is moderated in vertebrates and spineless creatures. The quality observed has been demonstrated to manifest within the gonads of various teleost species and has been associated with both factors (Matsuda, 2005). Its expression starts after that of Sox9 but comes before morphological sex differentiation in Oncorhynchus mykiss. Dmrt1 is only expressed in male gonads and is known to play a major role in testicular sex differentiation (Vizziano *et al.*, 2007). Essentially, within the case of tilapia, Dmrt1 is basically detailed to be communicated with a specific design in XY gonads, showing that this quality plays an important part in testicular separation, though it isn't communicated in XX gonads. Sexual inversion experiments involving hormone treatment substantiated the substantial function of Dmrt1 in the process of testicular separation within this species, which is notable. Dmrt1 is not expressed in XY individuals subjected to estrogenic treatment; however, it is expressed in XX individuals of the protogynous wrasse, *Halichoeres tenuispinis*. Recent studies have indicated that the gene sequence of Dmrt1 is also specific to males (Jeong *et al.*, 2009).

Cyp19 (aromatase)

Certain teleost species play a crucial role in ovarian development through the function of Cyp19a1. In rainbow trout, the expression of P450arom is at its peak in XX individuals before sexual differentiation and subsequently re-emerges during the process of ovarian differentiation. So also, in tilapia, P450arom is female-specific, with expression starting on or around day 5 after bring forth; be that as it may, this happens sometimes recently when the gonad is anatomically separated. A zebrafish-based demonstration has been created to clarify sexual assurance and separation, with P450arom at the centre. Cyp19a1 is only expressed in the gonads and is connected with sexual gonadal differentiation, whereas Cyp19a2 is widely distributed in the brain and appears to be indirectly engaged in sexual differentiation via the hypothalamus-hypophysis-gonad axis (Sawyer *et al.*, 2006). Cyp19a1 is critical for ovarian development in various teleost species. In rainbow trout, P450arom expression is greatest in XX animals before sexual differentiation and again during ovarian differentiation.

In tilapia, the enzyme P450arom is specific to females, with its expression beginning around day 5 after birth. This expression often occurs just after the gonads become anatomically differentiated. Similarly, a zebrafish-based study has been developed to investigate sexual determination and differentiation, with P450arom playing a central role. P450arom maintains the high levels of estrogen necessary for ovarian development in female embryos. In male embryos, the expression of this enzyme is suppressed, likely due to the activity of genes involved in testicular differentiation.

Foxl2

The Foxl2 quality of the family of translation components named fork head is female-specific in creatures and seen within the early stages of gonadal advancement. This design of dimorphic expression is held all through vertebrate taxa, and its part is still intaglio amid ovarian improvement, especially granulosa cell separation. Foxl2 may do this by controlling Cyp19 translation (Liu *et al.*, 2007). Foxl2 expression is sexually dimorphic in teleosts, with the ovaries being more critical than guys in species such as tilapia and rainbow trout. This bolsters the concept that Foxl2 plays a key part in ovarian differentiation in teleosts because it does in other vertebrate taxa. In tilapia and medaka, Foxl2 was first identified in the somatic cells surrounding the germ cells of the gonads that had commenced the differentiation process toward being ovaries.

The only known differences pertain to HINTW, specifically between the Z-chromosome allele and the W-chromosome allele. The HINTZ protein functions as a nucleic-acid binding protein, whereas the HINTW protein is a homolog of HINTZ and is presumed to lack a catalytic domain. Due to this, HINTW has been viewed as a dominant, consistently negative regulator of male H-1 development (Bellott *et al.* 2017). By contrast, the sexless nature of the result is highlighted by the presence of HINTW protein, since HINTW did not change from male to female (Smith *et al.*, 2009b).

Dax1

The gene Dax1 (dosage-sensitive sex reversal, adrenal hypoplasia critical region on chromosome X, gene 1) encodes an unusual member of the nuclear receptor family for steroid hormones that have been linked to the X chromosome in humans and mice. In mammals, Dax1 acts as a negative regulator of steroidogenesis by inhibiting Sf 1 transcriptional activity (Stop and Jameson, 2005). It has been separated and characterised in other randomly mated species such as tilapia, rainbow trout, medaka, and European sea bass (*Dicentrarchus labrax*; Nakamoto *et al.*, 2007). Most of these species have communicated a practical agent of this quality, and articulation has been dissected during the gonadal turn of events, yet its significance during sexual resolution is unclear.

Approaches to Study Genomics of Fish Sex Determination and Differentiation

There are a few ways to examine the hereditary qualities of angle sex assurance and separation. This theme has been secured in impressive detail somewhere else. A fast overhaul is given within the taking after section. Sex-associated indicators can aid in identifying sex-determining locations or genes, but significantly separated genomic regions are unique due to the physical separation of sex chromosomes.

As a result, screening these districts has not continuously been as compelling as combining medium- to high-density hereditary maps with a legitimately chosen family. Moreover, since angle species show such solid synteny, comparative genomics can offer assistance in pinpointing the area of the sex-determining locale or quality. As far as the examination of female vs. male sex chromosomes under species, the availability of monosex populaces frequently influences the technique utilized to investigate the hereditary qualities of angle sex assurance and separation. Monosex fish populations have been established in species with a XX/XY sex-determination system (e.g., rainbow trout or Nile tilapia *Oreochromis niloticus*) by mating normal females to sex-reversed genotypic females transformed into phenotypic males by androgen therapy (Fig. 1) (Piferrer *et al.*, 2012).

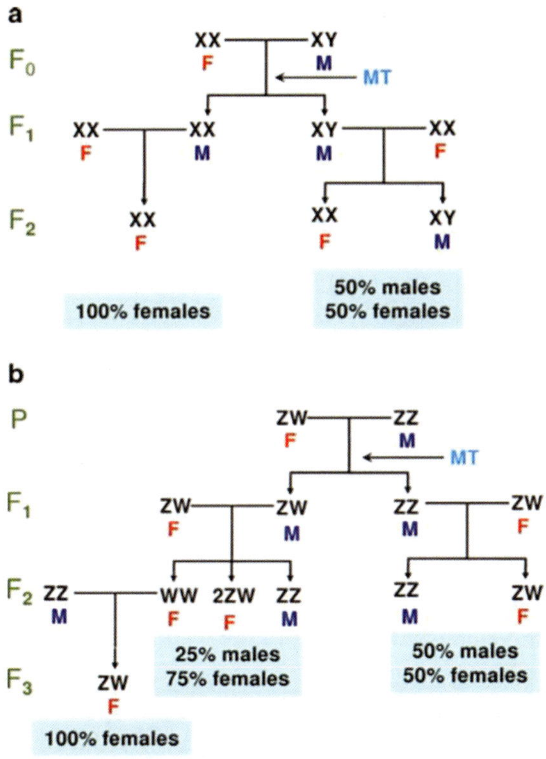

Fig. 1: Methods to produce all-female stocks based on endocrine therapy

Genetic Variation Within Species: sex Determination as a Complex Trait

Sex determination (SD) in ichthyological studies has traditionally focused on identifying the primary sex-determining characteristics predicted by earlier models. However, modern research consistently indicates that other minor hereditary and environmental factors also play a role in SD, even in species with well-characterized sex-determining traits, such as the dmY gene in Medaka (Matsuda *et al.*, 2002). Therefore, more efforts should be made to explore this diversity to gain a more comprehensive understanding of sex determination. Additionally, data available remotely suggests that:

i) Even though natural components may impact the sex proportion in most considered species, hereditary components more often than not constitute the major SD calculate;

ii) Major genes, those clarifying the next extent of characteristic phenotypic fluctuation, frame the premise of SD in numerous species, which is likely because of the various levelled nature of sex advancement pathway and

iii) Interaction of other qualities partaking in SD with the major components and with the environment.

Endocrine Regulation of Fish Sex Differentiation

Yamamoto's tests within the early 1950s illustrated that steroids are viable particles capable of actuating phenotypic sex reversal in angle (Yamamoto, 1969). Yamamoto's classification of androgens and oestrogens as "andro-inducers" (male inducers) and "gyno-inducers" (female inducers) has aroused discussion about their potential role as common inducers of gonadal separation. The first suspicion stemmed from steroids' ability to cause significant masculinisation or feminisation in a wide range of animal species after treatment. Additional research, such as sex-steroid measures, steroid digesting systems, and, most recently, steroid-related quality expression and transcriptome considerations, demonstrated that male and female angle-separating gonads have varied capacities for sex hormone union.

Ovarian separation is by and large characterized by the exceptionally early onset of estrogen generation, whereas testicular separation is by the nonattendance of estrogen blend and the blend of particular male androgens to be specific, 11-oxygenated androgens. These findings support Yamamoto's hypothesis that the balance of oestrogens and androgens (11-oxygenated androgens) determines the outcome of gonadal sex separation, rather than their presence or absence. Be that as it may, in Chinook salmon, by blocking estrogen union with specific enzymatic inhibitors avoiding the movement of the aromatase (Cyp19a1a) protein, it has been conceivable to appear for the primary time the total phenotypic masculinization of hereditarily female populaces (Piferrer *et al.*, 1994). Since then, the masculinising effects of aromatase inhibitors (AI) have been demonstrated in a variety of other species of angle, implying that the simple absence of oestrogens is all that is required for sex separation to occur as male, which would not require any kind of androgen stimulatory activity but rather the absence of oestrogen to continue. According to Guiguen *et al.* (2010), the female sex separation requires the presence of oestrogens, while the male sex separation requires their absence.

Conclusion

The integrative approach to sex determination and differentiation in fish demonstrates a remarkable diversity of mechanisms and pathways. Unlike mammals, which typically exhibit a rigid, binary system of sex determination, fish utilize a wide range of systems, from strictly genetic to environmentally influenced methods. For instance, key sex-determining genes like dmrt1 and foxl2 have been identified. These discoveries, along with data from whole-genome sequencing, have significantly enhanced our understanding of the genomic framework underlying sex determination. Many studies suggest that sex determination in most fish may be polygenic or complex, with some species exhibiting notably distinct sex chromosomes, while others show little differentiation between them. This complexity extends to the endocrine system as well; the direction of gonadal differentiation is influenced critically by sex steroids such as estrogens and androgens. The enzyme aromatase is responsible for synthesizing estrogen, which promotes ovarian development, while its inhibition leads to male differentiation. Temperature has also been shown to affect both genetic and endocrine pathways, where interactions between the two can result in sex reversal or other forms of phenotypic plasticity. Understanding these intricate processes is essential for applications in aquaculture and fish conservation. Manipulating sex ratios through genetic or endocrine approaches could enhance the efficiency of farming species where one sex grows faster or holds greater commercial value. Additionally, insights into sex determination mechanisms will assist conservation programs in managing fish populations in response to environmental changes that may impact specific species. Overall, ongoing research in these areas holds significant promise for advancing our knowledge of fish biology and improving aquaculture management practices.

References

Arkhipchuk, V. V. (1995). Role of chromosomal and genome mutations in the evolution of bony fishes. Hydrobiological Journal, 31(1), 55-65.

Bao, L., Tian, C., Liu, S., Zhang, Y., Elaswad, A., Yuan, Z., & Liu, Z. (2019). The Y chromosome sequence of the channel catfish suggests novel sex determination mechanisms in teleost fish. BMC biology, 17, 1-16.

Bellott, D. W., Skaletsky, H., Cho, T. J., Brown, L., Locke, D., Chen, N., & Page, D. C. (2017). Avian W and mammalian Y chromosomes convergently retained dosage-sensitive regulators. Nature genetics, 49(3), 387-394.

Bertho, S., Herpin, A., Schartl, M., & Guiguen, Y. (2021). Lessons from an unusual vertebrate sex-determining gene. Philosophical Transactions of the Royal Society B, 376(1832), 20200092.

Bull, J. J. (1985). Sex determining mechanisms: an evolutionary perspective. Experientia, 41, 1285-1296.

Chen, J., Hu, W., & Zhu, Z. (2013). Progress in studies of fish reproductive development regulation. Chinese Science Bulletin, 58, 7-16.

Chue, J., & Smith, C. A. (2011). Sex determination and sexual differentiation in the avian model. The FEBS Journal, 278(7), 1027-1034.

Devlin, R. H., & Nagahama, Y. (2002). Sex determination and sex differentiation in fish: an overview of genetic, physiological, and environmental influences. Aquaculture, 208(3-4), 191-364.

Ding, M., Li, X. Y., Zhu, Z. X., Chen, J. H., Lu, M., Shi, Q., & Gui, J. F. (2021). Genomic anatomy of male-specific microchromosomes in a gynogenetic fish. PLoS genetics, 17(9), e1009760.

Gammerdinger, W. J., & Kocher, T. D. (2018). Unusual diversity of sex chromosomes in African cichlid fishes. Genes, 9(10), 480.

Grandi, G., & Chicca, M. (2008). Histological and ultrastructural investigation of early gonad development and sex differentiation in Adriatic sturgeon (Acipenser naccarii, Acipenseriformes, Chondrostei). Journal of Morphology, 269(10), 1238-1262.

Guiguen, Y., Fostier, A., & Herpin, A. (2018). Sex determination and differentiation in fish: genetic, genomic, and endocrine aspects. Sex control in aquaculture, 35-63.

Guiguen, Y., Fostier, A., Piferrer, F., Chang, C.F. (2010). Ovarian aromatase and estrogens: a pivotal role for gonadal sex differentiation and sex change in fish. General and Comparative Endocrinology 165 (3), 352–366.

Higa, M., Kanda, H., Kitahashi, T., Ito, M., Shiba, T., & Ando, H. (2000). Quantitative analysis of fushi tarazu factor 1 homolog messenger ribonucleic acids in the pituitary of salmon at different prespawning stages. Biology of reproduction, 63(6), 1756-1763.

Ijiri, S., Kaneko, H., Kobayashi, T., Wang, D. S., Sakai, F., Paul-Prasanth, B., & Nagahama, Y. (2008). Sexual dimorphic expression of genes in gonads during early differentiation of a teleost fish, the Nile tilapia Oreochromis niloticus. Biology of reproduction, 78(2), 333-341.

Ioannidis, J., Taylor, G., Zhao, D., Liu, L., Idoko-Akoh, A., Gong, D., & Clinton, M. (2021). Primary sex determination in birds depends on DMRT1 dosage, but gonadal sex does not determine adult secondary sex characteristics. Proceedings of the National Academy of Sciences, 118(10), e2020909118.

Jeong, H. B., Park, J. G., Park, Y. J., Takemura, A., Hur, S. P., Lee, Y. D., & Kim, S. J. (2009). Isolation and characterization of DMRT1 and its putative regulatory region in the protogynous wrasse, Halichoeres tenuispinis. Gene, 438(1-2), 8-16.

Kang, Y., Guan, G. J., & Hong, Y. H. (2017). Insights of sex determination and differentiation from medaka as a teleost model. Yi Chuan= Hereditas, 39(6), 441-454.

Komen, J., De Boer, P., & Richter, C. J. J. (1992). Male sex reversal in gynogenetic XX females of common carp (Cyprinus carpio L.) by a recessive mutation in a sex-determining gene. Journal of Heredity, 83(6), 431-434.

Kondo, M., Hornung, U., Nanda, I., Imai, S., Sasaki, T., Shimizu, A., & Schartl, M. (2006). Genomic organization of the sex-determining and adjacent regions of the sex chromosomes of medaka. Genome Research, 16(7), 815-826.

Lambeth, L. S., Raymond, C. S., Roeszler, K. N., Kuroiwa, A., Nakata, T., Zarkower, D., & Smith, C. A. (2014). Over-expression of DMRT1 induces the male pathway in embryonic chicken gonads. Developmental Biology, 389(2), 160-172.

Lewis, Z. R., McClellan, M. C., Postlethwait, J. H., Cresko, W. A., & Kaplan, R. H. (2008). Female-specific increase in primordial germ cells marks sex differentiation in threespine stickleback (Gasterosteus aculeatus). Journal of morphology, 269(8), 909-921.

Li, Y., Zheng, M., & Lau, Y. F. C. (2014). The sex-determining factors SRY and SOX9 regulate similar target genes and promote testis cord formation during testicular differentiation. Cell reports, 8(3), 723-733.

Liu, Z., Wu, F., Jiao, B., Zhang, X., Hu, C., Huang, B., & Wang, D. (2007). Molecular cloning of doublesex and mab-3-related transcription factor 1, forkhead transcription factor gene 2, and two types of cytochrome P450 aromatase in Southern catfish and their possible roles in sex differentiation. Journal of endocrinology, 194(1), 223-241.

Long, J. A., Zheng, S. Q., Wang, X. S., Zhang, S. A., & Wang, D. S. (2020). Role of TGF-β signaling pathway in sex determination and differentiation in fish.

Martins, R. S., Deloffre, L. A., Mylonas, C. C., Power, D. M., & Canário, A. V. (2007). Developmental expression of DAX1 in the European sea bass, Dicentrarchus labrax: lack of evidence for sexual dimorphism during sex differentiation. Reproductive Biology and Endocrinology, 5, 1-13.

Matsuda, M. (2005). Sex determination in the teleost medaka, Oryzias latipes. Annu. Rev. Genet., 39(1), 293-307.

Matsuda, M., Nagahama, Y., Shinomiya, A., Sato, T., Matsuda, C., Kobayashi, T., & Sakaizumi, M. (2002). DMY is a Y-specific DM-domain gene required for male development in the medaka fish. Nature, 417(6888), 559-563.

Matsuda, M., Shinomiya, A., Kinoshita, M., Suzuki, A., Kobayashi, T., Paul-Prasanth, B., & Nagahama, Y. (2007). DMY gene induces male development in genetically female (XX) medaka fish. Proceedings of the National Academy of Sciences, 104(10), 3865-3870.

Mei, J., & Gui, J. F. (2015). Genetic basis and biotechnological manipulation of sexual dimorphism and sex determination in fish. Science China Life Sciences, 58, 124-136.

Nakamoto, M., Wang, D. S., Suzuki, A., Matsuda, M., Nagahama, Y., & Shibata, N. (2007). Dax1 suppresses P450arom expression in medaka ovarian follicles. Molecular Reproduction and Development, 74(10), 1239-1246.

Nakamura, M., Kobayashi, T., Chang XiaoTian, C. X., & Nagahama, Y. (1998). Gonadal sex differentiation in teleost fish.

Ota, K., Tateno, Y., & Gojobori, T. (2003). Highly differentiated and conserved sex chromosome in fish species (Aulopus japonicus: Teleostei, Aulopidae). Gene, 317, 187-193.

Park, S. Y., & Jameson, J. L. (2005). Minireview: transcriptional regulation of gonadal development and differentiation. Endocrinology, 146(3), 1035-1042.

Peichel, C. L., Ross, J. A., Matson, C. K., Dickson, M., Grimwood, J., Schmutz, J., & Kingsley, D. M. (2004). The master sex-determination locus in threespine sticklebacks is on a nascent Y chromosome. Current biology, 14(16), 1416-1424.

Pieau, C., Dorizzi, M., & Richard-Mercier, N. (1999). Temperature-dependent sex determination and gonadal differentiation in reptiles. Cellular and Molecular Life Sciences CMLS, 55, 887-900.

Piferrer, F., Ribas, L., & Díaz, N. (2012). Genomic approaches to study genetic and environmental influences on fish sex determination and differentiation. Marine biotechnology, 14, 591-604.

Piferrer, F., Zanuy, S., Carrillo, M., Solar, II., Devlin, RH., Donaldson, E. (1994). Brief treatment with an aromatase inhibitor during sex differentiation causes chromosomally female salmon to develop as normal, functional males. Journal of Experimental Zoology 270 (3), 255–262.

Rasmussen, T. H., Jespersen, Å., & Korsgaard, B. (2006). Gonadal morphogenesis and sex differentiation in intraovarian embryos of the viviparous fish Zoarces viviparus (Teleostei, Perciformes, Zoarcidae): A histological and ultrastructural study. Journal of Morphology, 267(9), 1032-1047.

Ross, J. A., Urton, J. R., Boland, J., Shapiro, M. D., & Peichel, C. L. (2009). Turnover of sex chromosomes in the stickleback fishes (Gasterosteidae). PLoS genetics, 5(2), e1000391.

Saito, D., Morinaga, C., Aoki, Y., Nakamura, S., Mitani, H., Furutani-Seiki, M., Tanaka, M. (2007). Proliferation of germ cells during gonadal sex differentiation in medaka: Insights from germ cell-depleted mutant zenzai. Developmental biology, 310(2), 280-290.

Sawyer, S. J., Gerstner, K. A., & Callard, G. V. (2006). Real-time PCR analysis of cytochrome P450 aromatase expression in zebrafish: gene specific tissue distribution, sex differences, developmental programming, and estrogen regulation. General and comparative endocrinology, 147(2), 108-117.

Schartl, M., Schories, S., Wakamatsu, Y., Nagao, Y., Hashimoto, H., Bertin, C., & Herpin, A. (2018). Sox5 is involved in germ-cell regulation and sex determination in medaka following co-option of nested transposable elements. BMC biology, 16, 1-17.

Scott, A. G., Penman, D. J., Beardmore, J. A., & Skibinski, D. O. F. (1989). The 'YY'supermale in Oreochromis niloticus (L.) and its potential in aquaculture. Aquaculture, 78(3-4), 237-251.

Sekido, R. (2010). SRY: A transcriptional activator of mammalian testis determination. The international journal of biochemistry & cell biology, 42(3), 417-420.

Sekido, R., & Lovell-Badge, R. (2008). Sex determination involves synergistic action of SRY and SF1 on a specific Sox9 enhancer. Nature, 453(7197), 930-934.

Shao, C., Chen, S., 2012. Sex-determining genes and its association with mechanism of sex chromosome evolution in vertebrate. J. Agri. Biotechnol. 20, 1463–1474 (in Chinese).

Sinclair, A. H., Berta, P., Palmer, M. S., Hawkins, J. R., Griffiths, B. L., Smith, M. J., & Goodfellow, P. N. (1990). A gene from the human sex-determining region encodes a protein with homology to a conserved DNA-binding motif. Nature, 346(6281), 240-244.

Smith, C., & Wootton, R. J. (2016). The remarkable reproductive diversity of teleost fishes. Fish and Fisheries, 17(4), 1208-1215.

Smith, C. A., Roeszler, K. N., Ohnesorg, T., Cummins, D. M., Farlie, P. G., Doran, T. J., & Sinclair, A. H. (2009). The avian Z-linked gene DMRT1 is required for male sex determination in the chicken. Nature, 461(7261), 267-271.

Smith, C. A., Roeszler, K. N., & Sinclair, A. H. (2009). Genetic evidence against a role for W-linked histidine triad nucleotide binding protein (HINTW) in avian sex determination. International Journal of Developmental Biology, 53(1), 59.

Takehana, Y., Naruse, K., Hamaguchi, S., & Sakaizumi, M. (2007). Evolution of ZZ/ZW and XX/XY sex-determination systems in the closely related medaka species, Oryzias hubbsi and O. dancena. Chromosoma, 116, 463-470.

Van Doorn, G. S., & Kirkpatrick, M. (2007). Turnover of sex chromosomes induced by sexual conflict. Nature, 449(7164), 909-912.

Vandeputte, M., Dupont-Nivet, M., Chavanne, H., & Chatain, B. (2007). A polygenic hypothesis for sex determination in the European sea bass Dicentrarchus labrax. Genetics, 176(2), 1049-1057.

Vizziano, D., Randuineau, G., Baron, D., Cauty, C., & Guiguen, Y. (2007). Characterization of early molecular sex differentiation in rainbow trout, Oncorhynchus mykiss. Developmental dynamics: an official publication of the American Association of Anatomists, 236(8), 2198-2206.

von Hofsten, J., & Olsson, P. E. (2005). Zebrafish sex determination and differentiation: involvement of FTZ-F1 genes. Reproductive Biology and Endocrinology, 3, 1-11.

Warren, W. C., Hillier, L. W., Tomlinson, C., Minx, P., Kremitzki, M., Graves, T., & Cheng, H. H. (2017). A new chicken genome assembly provides insight into avian genome structure. G3: Genes, Genomes, Genetics, 7(1), 109-117.

Wilhelm, D., Palmer, S., & Koopman, P. (2007). Sex determination and gonadal development in mammals. Physiological reviews, 87(1), 1-28.

Yamamoto, T.O. (1969). Sex Differentiation. In: Hoar, W.S. and Randall, D.J. (eds). Fish Physiology. Academic Press, New York, NY, pp. 117–175.

Yano, A., Nicol, B., Jouanno, E., Quillet, E., Fostier, A., Guyomard, R., & Guiguen, Y. (2013). The sexually dimorphic on the Y-chromosome gene (sdY) is a conserved male-specific Y-chromosome sequence in many salmonids. Evolutionary applications, 6(3), 486-496.

Yoshimoto, S., Okada, E., Umemoto, H., Tamura, K., Uno, Y., Nishida-Umehara, C., & Ito, M. (2008). A W-linked DM-domain gene, DM-W, participates in primary ovary development in Xenopus laevis. Proceedings of the National Academy of Sciences, 105(7), 2469-2474.

4

Molecular Approaches to Biosecurity Innovations in Aquaculture Disease Management

***K.G. Pithiya*[1], *S.R. Lende*[2], *H.M. Zankat*[1] and *A.B. Bamaniya*[1]**

[1]*Department of Aquaculture, College of Fisheries Science, Kamdhenu University Himmatnagar, Gujarat (362265), India*

[2]*Assistant Professor, Centre of Excellence in Aquaculture, Kamdhenu University Ukai, Gujarat (382010), India*

Abstract

Infectious diseases in aquaculture present substantial challenges to the growth and sustainability of the industry particularly sustainability, especially as farmed seafood is increasingly recognized and acknowledged as a viable feasible solution to global food security and poverty alleviation. The implementation of effective biosecurity programs is critical to the prevention, management, and eradication of diseases, protecting aquatic species and preserving ecosystems. Biosecurity entails a holistic approach that incorporates physical, operational, and biological measures. Furthermore, biotechnology significantly enhances biosecurity through advancements in genetic modification, molecular diagnostics, probiotics, immunostimulants, vaccines, and nutrigenomics. These innovations not only enhance the health and resilience of aquatic species but also contribute to the promotion of sustainable aquaculture practices and the efficacy of disease management strategies.

Introduction

Infectious diseases in aquaculture are becoming increasingly important to address to support the growth and sustainability of the industry. This is especially true as many view farmed seafood production as a key solution for feeding the world's growing population and reducing poverty in numerous countries (FAO, 2016). It is essential to establish biosecurity programs that effectively prevent, manage, and eliminate diseases within aquaculture operations. Biosecurity plays a critical role in safeguarding agriculture and aquaculture operations from the devastating effects of disease outbreaks. These programs are critical to safeguarding the health of aquatic species, protecting ecosystems, and ensuring the sustainability of the industry. At first, implementing biosecurity measures in aquaculture may seem like a simple concept. The well-known saying "prevention is better than cure" is especially relevant to aquaculture diseases, given the significant economic consequences of disease outbreaks. Effective biosecurity practices can help prevent these costly issues from arising. Biosecurity is a comprehensive and coordinated strategy that includes policies and regulations designed to assess and manage risks across various sectors, such as food safety, animal health, and plant health, while also addressing related environmental risks. Biosecurity measures focus on preventive practices such as quarantine, hygiene protocols, and environmental management.

A goal of biosecurity is to prevent new diseases from entering native populations, achieved through quarantine, screening, and surveillance. Quarantine isolates individuals or groups exposed to a disease but not yet showing symptoms. Screening involves testing potentially exposed individuals to detect infections, while surveillance monitors ongoing outbreaks to identify new cases and limit spread (Oidtmann *et al.,* 2011).

Another crucial aspect is preventing disease transmission within populations. This can be done through vaccination, hygiene practices, and personal protective equipment. Vaccination is a powerful tool in disease prevention, while hygiene practices like handwashing, cleaning, and waste disposal help reduce transmission. Personal protective equipment such as masks and gloves also play a significant role in controlling the spread of diseases (Snieszko, 1970; Honda & Iwata, 2016).

Biotechnology offers advanced solutions that support the sustainable growth of aquaculture, fisheries, and the food industry. Genetic modification and biotechnology have significant potential to enhance both the quality and quantity of fish raised in aquaculture. Biotechnological tools, including molecular diagnostic techniques, vaccines, and immunostimulants, are becoming increasingly popular for enhancing disease resistance in fish and shellfish species. These technologies are crucial in reducing the risks of disease outbreaks and ensuring the health and safety of aquatic species.

Categories of Biosecurity in Aquaculture

Aquaculture biosecurity aims to prevent the introduction and transmission of pathogens within and across aquaculture facilities, and it can be divided into three main categories: physical, operational, and biological. These biosecurity measures are essential for ensuring the health and welfare of aquatic species and for stopping the spread of diseases within aquaculture systems (Hasimuna *et al.,* 2020; Aly & Fathi, 2024).

Physical Aquaculture Biosecurity

To stop infections from entering aquaculture systems, physical biosecurity measures are essential. lowering the possibility of disease outbreaks and preserving aquatic animals' health and welfare (Dong, & Zhou, 2023). To stop possible disease vectors from entering the aquaculture system, physical barriers including fences, nets, and screens are used as part of physical biosecurity measures. Predators like birds, animals, and reptiles can transport and spread diseases, so fencing and netting are frequently employed to keep them out. (Panigrahi, 2012). The type and strength of the barrier, the size of the openings, and the location of the aquaculture facility are some of the variables that affect how well physical barriers work to keep viruses out of aquaculture systems. (Bondad-Reantaso *et al.*, 2012). Similarly, the location of the aquaculture facility, particularly its proximity to wild fish populations and the type of holding units used, can influence the risk of disease transmission by affecting interactions between wild and cultured fish Similar to this, by influencing interactions between wild and cultured fish, the aquaculture facility's location specifically, how close it is to wild fish populations and the kind of holding units it uses can alter the likelihood of disease transmission. (Bondad-Reantaso *et al.*, 2012).

In addition to preventing the introduction of pathogens into aquaculture systems, physical barriers can also serve to prevent farmed fish from escaping, thereby reducing the risk of spreading diseases to wild fish populations (Dong & Zhou, 2023). Escapees can carry diseases, parasites, and other pathogens that may infect wild fish, potentially leading to severe ecological and economic consequences (Naylor *et al.*, 2005). Physical biosecurity measures also include the use of closed systems, such as recirculating aquaculture systems (RAS) and flow-through systems, which effectively block the entry of external pathogens, invasive species, and other disease carriers into aquaculture environments (Rurangwa & Verdegem, 2015). RAS is a particularly efficient and sustainable system, as it recycles water through treatment processes, allowing for the cultivation of high-density fish populations in a controlled, closed environment. This setup significantly reduces the risk of disease transmission from external sources, making it a valuable biosecurity strategy (Schreier *et al.*, 2010).

Operational Aquaculture Biosecurity

Operational biosecurity measures are essential for ensuring the health and welfare of aquatic species and preventing disease transmission within aquaculture systems (Oidtmann *et al.*, 2011). These measures include effective feed management, routine monitoring of water quality, and the enforcement of rigorous hygiene and sanitation protocols (Hasimuna *et al.*,

2020). Feed management is a key component of biosecurity, as using high-quality, pathogen-free feed helps sustain the health and immunity of aquatic animals (Bera *et al.*, 2018). Proper feed storage, routine checks for spoilage or contamination, and species-specific formulations are essential practices for effective feed management (White *et al.*, 2018).

Maintaining water quality is also vital for the well-being of farmed fish and shellfish. Regularly monitoring key water parameters like temperature, pH, dissolved oxygen, and ammonia levels can help detect issues before they escalate into major disease outbreaks. Poor water quality can stress aquatic animals and increase their vulnerability to infections. Technologies like recirculating aquaculture systems (RAS) and water treatment processes can maintain water quality and reduce disease risks (Ahmed & Turchini, 2021). Strict hygiene and sanitation protocols are fundamental to preventing pathogen spread between and within aquaculture facilities. Regular cleaning and disinfecting of equipment, tanks, and vehicles, as well as training personnel in proper hygiene practices such as handwashing and the use of personal protective equipment (PPE), help minimize cross-contamination. PPE like gloves and boots prevent the transfer of pathogens between workers and aquatic animals (Muthu *et al.*, 2020).

Moreover, emerging technologies such as sensors, drones, and artificial intelligence (AI) are increasingly used to enhance the effectiveness of operational biosecurity. Sensors can continuously monitor water quality in real-time (Mustapha *et al.*, 2021), while drones provide aerial surveillance of aquaculture farms to detect risks (Ubina *et al.*, 2021). AI analyses data from these technologies to create predictive models, identifying potential disease outbreaks before they occur. AI has great potential to enhance biosecurity measures.

Biological Aquaculture Biosecurity

Biological biosecurity measures are essential strategies for controlling and preventing disease spread in aquaculture. One of the most effective measures is the use of disease-resistant species. Specific pathogen-free (SPF) and specific pathogen-resistant (SPR) species possess genetic characteristics that reduce their vulnerability to diseases.

Specific Pathogen-Free (SPF) refers to aquatic species that are bred or selected to be free from certain pathogens. These species are carefully maintained in controlled environments, such as quarantine facilities, to ensure they are not exposed to diseases. The use of SPF stocks is an important biosecurity measure in aquaculture, as it helps prevent the introduction and spread of harmful pathogens within farming systems, leading to healthier fish populations and reduced disease risks. SPF fish are typically used in breeding programs or as a basis for stocking farms, contributing to the overall sustainability and productivity of aquaculture operations. Specific Pathogen-Resistant (SPR) refers to aquatic species that have been selectively bred or genetically modified to possess resistance to specific pathogens. These species have genetic traits that make them less susceptible to infections from certain diseases, enhancing their survival and health in aquaculture environments. The use of SPR fish helps reduce the need for antibiotics or other chemical treatments, improving disease management and biosecurity in fisheries. By incorporating SPR species into aquaculture, the risk of disease outbreaks is lowered, leading to more sustainable and efficient farming practices.

Another key biological biosecurity measure is the use of probiotics. Probiotics are live microorganisms that help boost the immunity of aquatic animals, reducing their susceptibility to infections (Austin *et al.*, 2022). Incorporating probiotics into aquaculture has been demonstrated to enhance the growth rate and survival of fish, shrimp, and other aquatic species. Probiotics also used for the disease treatment. The use of immunostimulants is another crucial biological biosecurity measure in aquaculture. Immunostimulants are substances that boost the immune response of aquatic animals, enhancing their resistance to diseases (Austin *et al.*, 2022). Research has shown that immunostimulants can improve the survival and growth rates of aquatic species in various aquaculture systems.

Overall, biological biosecurity measures play a vital role in ensuring the sustainability of aquaculture. The adoption of disease-resistant species, vaccinations, probiotics, bacteriophages, immunostimulants, and other microbial-based strategies can strengthen the immune systems of aquatic animals, minimize the risk of disease outbreaks, and enhance the overall health and productivity of aquaculture systems.

Biotechnology for Biosecurity and Disease Control

biotechnology is a key component of biological biosecurity (Nouri & Chyba, 2008). Biotechnology provides tools and techniques, such as genetic modification, molecular diagnostics, probiotics, immunostimulants, vaccines and nutrigenomics, that help enhance the health and disease resistance of aquatic species. These biotechnological methods contribute to biological biosecurity by improving the immune systems of aquatic animals, preventing the spread of diseases, and ensuring healthier, more sustainable aquaculture practices. Biotechnology plays an essential role in bolstering biosecurity efforts through early detection, pathogen surveillance, and the development of disease-resistant organisms (John *et al.*,2023).

Genetic Resistance and Breeding for Disease Resistance

An important area where biotechnology is applied in aquaculture biosecurity is in the development of disease-resistant strains through genetic manipulation or selective breeding (Bartley, 1998). These efforts aim to reduce the incidence of infections, improve the resilience of farmed species, and reduce dependency on chemical treatments.

Genetically Modified Organisms (GMOs)

Genetically Modified Organisms (GMOs) are organisms that have been altered by incorporating foreign DNA into their genetic material using in vitro genetic engineering techniques (Singh & Rajendran, 2014). The application of genetic technology in aquaculture offers several potential advantages, such as creating fish with enhanced growth rates, better tolerance to temperature variations, and improved resistance to diseases. Through genetic modification, fish can be engineered to grow up to six times faster than their natural counterparts, thrive in colder environments, and possess inherent protection against certain diseases (Singh& Rajendran, 2014). The first documented production of transgenic aquatic species occurred with rainbow trout by Maclean and Talwar in 1984, and with goldfish by Zhu *et al.,* in 1985. GloFish is a good example of a genetically modified organism. GloFish is a brand of genetically modified aquarium fish that are known for their fluorescent colours. This brand is both patented and trademarked. GloFish are a type of aquarium fish that have been genetically modified with fluorescent protein genes originating from Jellyfish or sea coral (Davies, 2014). AquAdvantage salmon is also an example of GMO. In 1989, American company AquaBounty Technologies begun developing a new genetically engineered breed of Atlantic salmon that could grow twice as fast as an Atlantic salmon in the wild.

In salmon, Infectious Salmonid Anaemia Virus (ISAV) is a highly contagious and notifiable disease that represents a significant threat to Atlantic salmon (*Salmo salar*) aquaculture globally. Since there is no completely effective treatment or vaccine available for ISAV, selective breeding to boost resistance to the virus offers a promising approach for preventing the disease. Resistance to Infectious Salmonid Anaemia Virus (ISAV) in salmon involves genetic traits that help fish either prevent infection or reduce disease severity. It is a polygenic trait, meaning it is controlled by multiple genes. Main factors in resistance include an effective immune response, particularly the upregulation of the interferon pathway. Genetic studies, such as RNA sequencing, have identified candidate genes linked to resistance, and quantitative trait locus (QTL) mapping has pinpointed specific chromosomal regions associated with ISAV resistance. Selective breeding, based on these genetic markers, is a primary strategy to improve disease resistance in commercial salmon populations (Gervais *et al.*, 2021).

Research conducted by Weifeng Mao at the Chinese Academy of Sciences has led to the development of a transgenic grass carp with improved immunity to *Aeromonas hydrophila*, an opportunistic pathogen responsible for tail/fin rot, ulcers, and *Hemorrhagic septicemia*. By incorporating specific genes into the grass carp genome, this transgenic fish exhibits enhanced resistance to infections caused by *Aeromonas hydrophila*, which is a major concern in aquaculture due to its ability to cause severe diseases and significant economic losses. The genetically modified grass carp demonstrates a strengthened immune response, offering a promising strategy for disease control and improving the sustainability of grass carp farming.

The introduction of the human lactoferrin gene into grass carp (*Ctenopharyngodon idellus*) aims to enhance their resistance to the *Grass Carp Hemorrhagic Virus* (GCHV), a significant pathogen in aquaculture that causes severe economic losses in grass carp farming. Lactoferrin is a glycoprotein found in human milk and other body fluids that has antimicrobial properties, including antiviral activity (Zhong *et al.*, 2002). By incorporating the human lactoferrin gene into the grass carp genome through genetic engineering techniques, researchers aim to boost the carp's innate immune response and increase its ability to combat GCHV.

GM tilapia have been developed with increased resistance to the *Tilapia Lake Virus* (TiLV), a virus that causes high mortality in tilapia aquaculture. The gene for the production of interferon, a protein involved in the immune response, has been introduced to improve resistance to viral infections (Barría *et al.*, 2020). The tilapia genome is modified to produce higher levels of immune proteins that help them fight off infections, making them more resilient in aquaculture environments.

In some experiments, genetically modified shrimp have been engineered to express the *Bacillus thuringiensis* (Bt) toxin, which is toxic to certain pathogens like *Vibrio* bacteria, a common cause of disease in shrimp farms (Rojas-Avelizapa *et al.*,1999). The Bt gene is inserted into the shrimp's DNA, enabling them to resist infections by specific pathogenic bacteria that cause diseases such as vibriosis and white spot syndrome.

Selective Breeding for Disease Resistance

Selective breeding for disease resistance in aquaculture is a crucial practice aimed at improving the health and productivity of farmed aquatic species. Selective breeding programs hold significant potential to tackle this challenge by gradually enhancing key production traits, such as disease resistance. At present, less than 10% of aquaculture production comes from stocks that have been selectively bred (Gjedrem *et al.*, 2012). Selective breeding to develop populations with enhanced resistance to key pathogens, by utilizing naturally occurring genetic variation (heritability) for resistance in farmed aquaculture species, is becoming a crucial and increasingly significant aspect of disease management (Rakkannan & Priyadarshi, 2023). breeding for disease resistance, this process involves selecting and breeding individuals with desirable genetic traits that enhance their ability to resist diseases (Johnston *et al.*, 2024). The goal is to pass on these beneficial traits to future generations, thereby improving the overall health and resilience of a population, whether in plants, animals, or aquaculture species.

In breeding for disease resistance, the first step is identifying individuals that show natural resistance to specific pathogens. Individual selection is used across all species for traits that can be measured in breeding candidates and progeny testing of males is sometimes used (Gjedrem & Robinson, 2014). This can be done through controlled exposure to diseases or by genetic testing to identify markers linked to disease resistance (Gjedrem, 2005). Once these individuals are identified, they are selectively bred to produce offspring with similar or stronger disease resistance. Over multiple generations, the resistance traits become more widespread in the population, significantly reducing the impact of disease outbreaks.

The benefits of breeding for disease resistance include reduced reliance on antibiotics and chemicals, leading to more sustainable and environmentally friendly farming practices. It also increases farm productivity by reducing mortality rates due to diseases, which are often a major source of loss. However, the process of genetic resistance and breeding for disease resistance can be time-consuming and must be carefully managed to avoid reducing genetic diversity, which could make the population more vulnerable to new diseases. The earliest known selection experiments in aquaculture can be traced back to the 1920s. For brook trout, selection was conducted to reduce mortality from furunculosis (Embody and Hyford, 1925). Similarly, the first selection experiment for common carp also began in the 1920s, leading to the development of two highly productive strains (Kirpichnikov, 1987).

Selective breeding for disease resistance in Atlantic salmon targets improving resistance to sea lice, which can severely affect farmed populations. By identifying naturally resistant fish, breeders selectively mate them to develop salmon strains that survive better in the presence of sea lice (Gharbi *et al.*, 2015). Over time, this approach has led to improved survival rates, less reliance on chemical treatments, and greater sustainability in aquaculture, reducing environmental impacts and supporting healthier fish populations.

In the mid-1990s, outbreaks of Taura syndrome, caused by the Taura syndrome virus (TSV), significantly impacted farmed shrimp (*Litopenaeus vannamei*) populations across the Americas. In response, breeding programs focused on improving survival rates against TSV have provided valuable insights into the quantitative genetics of disease resistance in shrimp. These programs have successfully produced shrimp families with high survival rates following TSV exposure (Moss *et al.*, 2012).

Molecular Diagnostics

Molecular diagnostics plays a crucial role in enhancing disease detection and biosecurity in aquaculture. Molecular diagnostic method offers higher accuracy and quicker results for detecting pathogens compared to conventional techniques. Over the past fifteen years, significant progress has been made in understanding the molecular biology of fish pathogens and their hosts. As a result, molecular biology has become an essential tool in developing better diagnostic methods, controlling fish diseases, and studying the epidemiology of bacterial, viral, and parasitic infections (Altinok & Kurt, 2003). These advanced techniques use the genetic material (DNA or RNA) of pathogens to identify infections in aquatic organisms with high specificity and sensitivity (Mishra *et al.*, 2020). Molecular diagnostics encompasses various methods and techniques.

Polymerase Chain Reaction

PCR is widely used in aquaculture to detect specific pathogens, such as bacteria, viruses, and parasites, by amplifying their genetic material. PCR is one of the most widely used molecular techniques for detecting and identifying pathogens in aquaculture. It provides high sensitivity and specificity, allowing for the detection of low levels of pathogens. Polymerase chain reaction (PCR) is a method used to amplify a specific segment of DNA, defined by two "primers" that mark the starting points for DNA synthesis, which is carried out by a heat-resistant DNA polymerase. The design of primers is crucial for achieving the highest sensitivity and specificity. Primers need to be long enough to enable a high annealing temperature, which helps minimize the chance of nonspecific binding. However, if the primers are too long, they may bind non-specifically to regions of DNA that are not an exact match to the primer sequence. The reaction contains template DNA, which can range from a basic tissue lysate to purified DNA, along with primers, a polymerase enzyme to facilitate the synthesis of new DNA copies, and nucleotides to build the new

DNA strands. In the process, template DNA is denatured, primers bind to their complementary regions, and the polymerase enzyme adds nucleotides to the primers, creating new copies of the target DNA. With each cycle, the amount of product increases geometrically. Quantitative PCR (qPCR) further enhances the capabilities of traditional PCR by allowing for the real-time quantification of pathogen DNA or RNA in a sample (Geraci-Yee *et al.*, 2022). This enables more accurate monitoring of pathogen load and disease progression, which is critical for early intervention and treatment.

Reverse transcriptase polymerase chain reaction (RT-PCR) is used to detect specific mRNA and determine gene expression levels (Koo and Jaykus, 2000). Compared to Northern blot analysis and RNase protection assay, RT-PCR can quantify mRNA from much smaller samples, even as little as a single cell. The development of real-time RT-PCR has made it the preferred method for quantifying gene expression changes and validating results from other techniques. Its sensitivity and specificity make RT-PCR ideal for monitoring covert infections.

In eukaryotes, prokaryotic rRNA genes also contain highly conserved sequences. These conserved regions are valuable for identifying or amplifying rRNA genes, while the more variable regions or spacers can be used to detect or identify bacteria that are challenging or impossible to culture (Barry *et al.*, 1990; Altinok & Kurt, 2003). PCR assays targeting rRNA genes have been utilized for detecting *Renibacterium salmoninarum* (Rhodes *et al.*, 1998), *Aeromonas salmonicida* (Høie *et al.*, 1999), *Yersinia ruckeri* (Altinok *et al.*, 2000) etc.

Multiplex PCR

New developments, such as the design of PCR conditions to detect multiple pathogens in a multiplex reaction, improve the time and cost-efficiency of this methodology, addressing one of the major arguments against its routine adoption (Williams *et al.*, 1999). In multiplex PCR, multiple target sequences can be amplified by including several pairs of primers in the reaction. This technique offers substantial time and effort savings in the laboratory without compromising test utility. Since its introduction, multiplex PCR has been successfully applied in gene deletion analysis, quantitative analysis, and RNA detection. In infectious disease diagnostics, it has proven valuable for identifying viruses, bacteria, fungi, and parasites.

Next-Generation Sequencing (NGS)

Next-generation sequencing (NGS) allows for high-throughput sequencing of entire genomes, enabling the detection of known and novel pathogens in aquaculture systems. NGS technology starts with the preparation of the DNA library, where double-stranded DNA (dsDNA) is broken into fragments of sizes suitable for the specific platform being used. These fragments are then ligated with adaptors, and polymerase chain reaction (PCR) or clonal amplification is performed to amplify the fragments (Swain *et al.*, 2022). Next-generation sequencing (NGS) has the potential to transform pathogen detection in aquaculture. NGS allows for the sequencing of entire genomes or transcriptomes, providing comprehensive information about the genetic makeup of pathogens or farmed organisms. In aquaculture, NGS can be used to identify previously unknown pathogens, track genetic variations within pathogen populations, and understand the genetic basis of disease resistance in farmed species. NGS can be used to identify multiple infections simultaneously. Unlike traditional sequencing methods, which were limited to sequencing a small number of short DNA fragments that were amplified through polymerase chain reaction (PCR), NGS represents a groundbreaking advancement. Its applications have expanded significantly, transitioning from research tools to essential diagnostic techniques (Nafea *et al.*, 2024).

Microarrays

DNA microarrays can be utilized in various methods to detect specific DNA (or RNA) sequences. Microarray technology involves the use of a grid of DNA probes to detect a large number of pathogens simultaneously. One common approach involves fluorescently labelling all DNA sequences in the test sample. The DNA in the sample that binds to a specific spot on the microarray can be identified through fluorescent detection, with the data then analysed using computer programs. A more commonly used method is competitive hybridization, where the test sample competes with a fluorescently labelled competitor oligonucleotide for binding to a tethered oligonucleotide on the chip. If the test DNA perfectly matches the tethered oligonucleotide, it will bind to the chip. However, if the test DNA is not an exact match, the fluorescently labelled competitor oligonucleotide will bind to the tethered oligonucleotide and displace the test DNA. A fluorescent microarray detector and software are then used to assess the presence or absence of the species/strain-specific DNA sequence.

In comparison to traditional nucleic acid hybridization using membranes, microarrays provide several additional benefits, including higher density, greater sensitivity, faster detection, reduced costs, the ability for automation, and minimal background noise (Shalon *et al.*, 1996). Microarrays offer a more effective solution for large-scale diagnostic testing, as they can analyze a sample for multiple sequences at the same time (Southern, 1996). A single microarray chip can detect a wide range of microbes, making microarray-based technology highly effective for identifying fish pathogens in fish populations.

Random Amplified Polymorphic DNA (RAPD)

RAPD is a technique used for genetic fingerprinting and can be used to assess genetic variation in pathogens. The RAPD method has been used to study the crayfish plague fungus *Astacus astaci* (Huang *et al.*, 1994). RAPD employs a single primer in low-stringency polymerase chain reactions, where random binding of primers results in different fragment sizes from nonidentical DNA. This technique has grouped different fungal isolates and aided epidemiological investigations (Oidtmann *et al.*, 1999). RAPD has also been applied to study other fish pathogens, though issues with reproducibility and contamination limit its use as a stand-alone diagnostic tool. However, it can be valuable as an initial step in developing specific primers or probes.

Loop-mediated isothermal amplification

LAMP is an affordable, straightforward, rapid, and highly precise technique for amplifying nucleic acids. This method utilizes six distinct primer sequences and Bst DNA polymerase to target specific genes. The Bst polymerase enzyme exhibits strong 5′–3′ DNA polymerase activity and strand displacement capabilities but does not possess 5′–3′ exonuclease activity (Laatri *et al.*, 2024). This enzyme is essential for the LAMP amplification process. The amplified products are stem-loop DNA structures containing multiple inverted repeats of the target sequence, forming cauliflower-like structures with several loops. This results in the production of PCR products at concentrations greater than 500 mg/ml (Nagamine et al. 2001). LAMP is highly specific and does not need advanced equipment like a thermocycler. Additionally, the accumulation of magnesium pyrophosphate, an insoluble salt, in the final reaction mixture leads to the formation of a visible white precipitate (Farzadnia and Naeemipour 2020). Thus, positive and negative reactions can be easily distinguished by the naked eye. LAMP has been used for the identification of *E. tarda*, *V. anguillarum*, and *S. iniae* (Austin, 2019).

Fluorescent In Situ Hybridization (FISH)

FISH is a molecular technique used for detecting specific nucleic acid sequences in intact cells or tissues. FISH can be applied to identify and localize pathogens in fish tissues, making it useful for detecting intracellular pathogens or parasites. FISH Used to detect Chlamydia-like organisms, Mycobacterium spp., or Salmonella in fish tissues. It Allows for visualization of pathogens within tissues and cells, providing insights into the pathogenesis and distribution of infections.

These are the tools which used for detection of pathogens from aquatic animals. These biotechnology molecular diagnostics tools are very useful in aquaculture for advance biosecurity and proper management. In modern era this technology is very timesaving and improved. Molecular diagnostics is important aspect for biotechnology and aquaculture.

Probiotics

Probiotics play a crucial role in supporting the sustainability of aquaculture. These are beneficial microorganisms that can be provided as single-species or mixed microbial formulations, often in conjugation with prebiotics, to help maintain a healthy gut microbiota. In recent decades, high fish mortality due to fungal, viral, or bacterial infections has been a significant concern (Samuel & Paari, 2022). To prevent disease outbreaks, antibiotics were commonly used. However, this led to the development of antimicrobial resistance in bacteria, as well as the suppression of the host's immune system (Van Doan *et al.*, 2020). The capacity of probiotic strains to enhance disease resistance, improve growth performance, positively affect gut morphology, modulate the gastrointestinal microbiota, and boost feed efficiency by improving nutritional quality and antioxidant enzyme activities has driven the widespread use of probiotics. This has also spurred further research and development in the field. The specificity of new probiotic strains is determined through screening criteria that include phenotype and genotype stability, plasmid stability, carbohydrate utilization patterns, acid and bile tolerance, antimicrobial production, antibiotic resistance, and the ability to adhere to the intestinal mucosa (Samuel & Paari, 2022).

Genetically stable probiotic strains with desirable viability are evaluated for pathogenicity, virulence factors, metabolic activity, and antibiotic resistance as safety criteria for selection. These factors ensure the strains are safe for use in aquaculture and other applications. Metagenomic studies uncover the antagonistic and symbiotic relationships between microbes and aquatic organisms such as fish, crustaceans, and molluscs. The information gained from analysing DNA through sequencing and bioinformatics technologies provides valuable insights into the microbial communities and their roles in the health and disease resistance of these species (Gianoulis *et al.*, 2009; Suttle, 2007). This technology has also been instrumental in evaluating multi-resistant bacteria and their mutations by analysing sequence-based and functional metagenomics. Additionally, metagenomics has been used to identify viral pathogens, helping to understand their mode of action and mechanisms. This information is crucial for developing practical diagnostic tools and treatment methods for host fish (Samuel & Paari, 2022).

Biotechnology plays a significant role in probiotic selection, as probiotics utilis aquatic animals to acquire disease resistance, enhanced growth & various other beneficial characteristics. So, biotechnology is an important factor in biosecurity.

Immunostimulants

Immunostimulants are substances that boost the immune system of farmed animals, helping them to better fight off infections. These can be natural substances like plant extracts, yeast derivatives, and beta-glucans. The use of natural immunostimulants is a distinctive strategy for fish culturists to reduce disease-related losses in their farms. Various polysaccharides from different sources possess the ability to activate the immune system, thereby acting as effective immunostimulants (Wang *et al.*, 2017). Studies have demonstrated that immunostimulants can enhance vaccine effectiveness and improve the general health of fish, reducing the reliance on antibiotics and other medications (Harikrishnan *et al.*, 2022). Immunostimulants are dietary supplements that boost the body's natural (non-specific) defence systems and improve its ability to fight off specific infections (Sakai, 1999). Biotechnology and microbiology offer crucial tools that can enhance both the quality and quantity of products. Through genetic modification, microbial processes, and advanced scientific techniques, these fields enable more efficient production, improved product characteristics, and higher yields. By leveraging microorganisms and biotechnological advancements, industries can boost production output, reduce inefficiencies, and create superior, more sustainable products (Wang *et al.*, 2017).

biotechnological techniques can be used to produce immunostimulants such as cytokines, lectins, or other bioactive compounds through genetic engineering of microorganisms or plants (Wang *et al.*, 2017). Immunostimulants are an important factor in aquaculture. Using biotechnology immunostimulants become upgraded that's why biotechnology is very important.

Vaccines

A vaccine is a biologically derived substance designed to enhance the body's immune response against a particular disease or a group of related diseases. Vaccines are regarded as biological substances that stimulate an immune response to a specific antigen derived from a disease-causing pathogen (Czochor & Turchick, 2014). Vaccination is considered an effective method for preventing a broad range of both bacterial and viral infections (Ma *et al.*, 2019).

Fish vaccines are categorized into two types: modified live vaccines and killed (inactivated) fish vaccines (Mondal & Thomas, 2022). Killed fish vaccines are made from heat-killed or formalin-killed microorganisms. In contrast, modified live vaccines contain live microorganisms that generally offer higher immunogenicity. This is due to their ability to reproduce, easily enter the host, and stimulate a stronger cellular response, activating both innate and adaptive immune systems more effectively than killed vaccines (Levine & Sztein, 2004; Mondal & Thomas, 2022).

Modern vaccine technology focuses on specific components of pathogens, and vaccines developed through these methods may include subunit vaccines, recombinant vaccines, or DNA/RNA particle vaccines. These approaches target particular parts of the pathogen, such as proteins or genetic material, to stimulate an immune response without using the whole pathogen. Subunit vaccines utilize only the antigenic components of a pathogen for vaccination. Since these vaccines do not contain live pathogens, they cannot replicate within the host, eliminating the risk of causing disease in the vaccinated organism (Hansson *et al.*, 2000). DNA vaccines contain an expression plasmid that carries a gene encoding a specific antigenic protein. When this gene is expressed in the host, it triggers a robust immune response, helping the body recognize and defend against the targeted pathogen (Ma *et al.*, 2019). RNA-based vaccines are advancing quickly, and several have shown promising results in both humans and animals. These vaccines use messenger RNA (mRNA) to instruct cells to produce a specific antigen, triggering an immune response without using live pathogens (Pardi *et al.*, 2018). Modern vaccines developed using biotechnology and its very important in aquaculture.

Nutrigenomics

Nutrigenomics is a crucial part in biotechnology. Nutrigenomics is the study of how nutrients affect gene expression and how genetic variations influence an organism's response to nutrients (Vyas *et al.*, 2018). In the context of biosecurity and disease resistance, nutrigenomics offers several advantages. By understanding how different diets or nutrients influence the immune system at a genetic level, nutrigenomics can help develop tailored diets that enhance the natural defence mechanisms of organisms, making them more resistant to diseases (Reen *et al.*, 2015). Certain dietary components like vitamins, trace minerals, and fatty acids can modulate the expression of genes involved in immune function and disease resistance (Wintergerst *et al.*, 2007). Studies have shown that supplementation with specific fatty acids can enhance the expression of immune-related genes in fish, providing greater resistance to infections (Nayak *et al.*, 2018).

Nutrigenomics can identify which genes are involved in the immune response, allowing for the creation of specific nutritional strategies that can activate these genes (Martin *et al.*, 2017). This can be particularly useful in aquaculture and livestock industries to improve the health and disease resistance of animals. With a deeper understanding of how nutrition affects genetic pathways related to disease resistance, nutrigenomics can help prevent outbreaks by enhancing the health of populations, certain amino acids, minerals, or bioactive compounds might help regulate cortisol levels (a stress hormone) and activate cellular pathways that protect against oxidative damage and inflammation (Carlberg *et al.*, 2020). This can also aid in minimizing the use of antibiotics and other chemicals, improving overall biosecurity practices. By linking specific genes with disease resistance traits, nutrigenomics can help develop breeds or strains of animals and plants that are genetically predisposed to resist infections (Pal & Chakravarty, 2019).

Recent research on selenium (Se), an essential micronutrient, has highlighted its impact on immune responses in fish. Selenium exists in different chemical forms, and their bioavailability can vary. Organic selenium, such as the yeast-derived Sel-plex® (produced by Alltec), has been found to be more bioavailable than inorganic selenium forms (Pacitti *et al.*, 2015). This makes organic selenium a suitable option for boosting Se levels in fish diets without causing toxicity.

In a study where rainbow trout were fed Sel-plex-enriched diets, their antiviral responses were assessed after a poly I:C challenge, which simulates a viral infection. Transcriptomic profiling through microarray analysis revealed that the fish fed with selenium-enriched diets had a much stronger immune response compared to those fed a basic diet (Pacitti *et al.*, 2016). The immune response was more pronounced in the head kidney compared to the liver, with a particular increase in the expression of interferon γ and other proteins involved in cell-mediated immunity (Martin *et al.*, 2017). These findings emphasize that a comprehensive understanding of the immune system requires high-throughput transcriptomics, which can assess the broader range of immune responses, rather than relying on the analysis of individual genes.

As the aquaculture industry faces increasing demands for food security and environmental sustainability, nutrigenomics offers exciting opportunities for advancing both the science and the practice of sustainable aquaculture. Because of this nutrigenomics is an important tool of biotechnology.

Conclusion

Biosecurity and disease control in aquaculture are critical for the industry's sustainability and growth, especially as the demand for farmed seafood increases. Implementing effective biosecurity measures, including physical, operational, and biological strategies, is essential to prevent the introduction and spread of pathogens. Biotechnology plays a pivotal role in enhancing biosecurity through genetic resistance, molecular diagnostics, probiotics, immunostimulants, vaccines, and nutrigenomics. These advancements not only improve the health and resilience of aquatic species but also contribute to more sustainable aquaculture practices. By integrating these biotechnological tools, the aquaculture industry can better manage disease risks, ensuring the health of aquatic populations and the overall sustainability of food production systems.

References

Ahmed N, Turchini G.M. Recirculating aquaculture systems (RAS): environmental solution and climate change adaptation. J Clean Prod 297:126604

Altinok, İ., & Kurt, İ. (2003). Molecular diagnosis of fish diseases: a review. Turkish Journal of Fisheries and Aquatic Sciences, 3(2).

Altinok, I., Grizzle, J.M. and Liu, Z. (2000). Detection of Yersinia ruckeri in rainbow trout blood by use of polymerase chain reaction. Dis. Aquat. Org., 44: 29 34.

Aly, S. M., & Fathi, M. (2024). Advancing aquaculture biosecurity: a scientometric analysis and future outlook for disease prevention and environmental sustainability. Aquaculture International, 1-27.

Austin B (2019). Methods for the diagnosis of bacterial fish diseases. Mar Life Sci Technol 1:41–49

Austin B, Lawrence A, Can E, Carboni C, Crockett J, Demirtas N, Schleder D, Adolfo J, Kayis S, Karacalar U, Kizak V, Kop A, Thompson K, Ruiz CAM, Serdar O, Seyhaneyildiz Can S, Watts S, Yucel Gier G. (2022). Selected topics in sustainable aquaculture research: current and future focus. Sustainable Aquatic Research 1(2):74–125.

Barría, A., Trinh, T. Q., Mahmuddin, M., Benzie, J. A., Chadag, V. M., & Houston, R. D. (2020). Genetic parameters for resistance to Tilapia Lake Virus (TiLV) in Nile tilapia (Oreochromis niloticus). Aquaculture, 522, 735126.

Barry, T., Powell, R. and Gannon, F. (1990). A general method to generate DNA probes for microorganisms. Bio/Tech., 8: 233–236.

Bera KK, Karmakar S, Jana P, Das S, Purkait S, Pal S, Haque R. (2018). Biosecurity in aquaculture: an overview. Aqua International 42:44–46

Bondad-Reantaso MG, Arthur JR, Subasinghe R.P. (2012). Improving biosecurity through prudent and responsible use of veterinary medicines in aquatic food production. FAO Fisheries and Aquaculture Technical Paper (547):I

Carlberg, C., Ulven, S. M., & Molnár, F. (2020). Nutrigenomics: how science works. Berlin/ Heidelberg, Germany: Springer.

Czochor, J., & Turchick, A. (2014). Introduction. Vaccines. The Yale Journal of Biology and Medicine, 87(4), 401-402.

Davies, G. (2014). Searching for GloFish®: aesthetics, ethics, and encounters with the neon baroque. Environment and Planning A, 46(11), 2604-2621.

Dong SL, Zhou Y.G. (2023). Health maintenance and welfare of aquatic animals. Aquaculture Ecology. Springer, Singapore, pp 447–472

Embody, G.C. and Hyford, C.D. (1925). The Advantage of Rearing Brook Trout Fingerlings from Selected Breeders.Transactions of the American Fisheries Society, 55, 135-138

FAO. (2016). The state of world fisheries and aquaculture. Contributing to food security and nutrition for all. Rome.

Farzadnia A, Naeemipour M. (2020). Molecular techniques for the detection of bacterial zoonotic patho gens in fish and humans. Aquac Int 28:309–320

Geraci-Yee, S., Allam, B., & Collier, J. L. (2022). Keeping up with advances in qPCR pathogen detection: An example for QPX disease in hard clams. Diseases of Aquatic Organisms, 148, 127-144.

Gervais, O., Barria, A., Papadopoulou, A., Gratacap, R. L., Hillestad, B., Tinch, A. E., ... & Houston, R. D. (2021). Exploring genetic resistance to infectious salmon anaemia virus in Atlantic salmon by genome-wide association and RNA sequencing. BMC genomics, 22(1), 345.

Gianoulis, T. A., Raes, J., Patel, P. V., Bjornson, R., Korbel, J. O., Letunic, I., ... & Gerstein, M. B. (2009). Quantifying environmental adaptation of metabolic pathways in metagenomics. Proceedings of the National Academy of Sciences, 106(5), 1374-1379.

Gharbi, K., Matthews, L., Bron, J., Roberts, R., Tinch, A., & Stear, M. (2015). The control of sea lice in Atlantic salmon by selective breeding. Journal of the Royal Society Interface, 12(110), 20150574.

Gjedrem, T., & Robinson, N. (2014). Advances by selective breeding for aquatic species: a review. Agricultural Sciences, 5(12), 1152.

Gjedrem, T. (2005). Selective Breeding for Disease Resistance in Aquaculture. Aquaculture, 246(1-4), 1-13.

Gjedrem, T.; Robinson, N. and Rye, M. (2012). The importance of selective breeding in aquaculture to meet future demands for animal protein: A review. Aquaculture 350–353:117–129.

Harikrishnan, R., Kim, M.C., Balasundaram, C. (2022). Immunostimulants and their effects on health management of fish. Aquaculture Research, 53(1), 33-50.

Høie, S., Dalsgaard, I., Aase, I.L., Heum, M., Thornton, J.M. and Powell, R. (1999). Polymerase chain reaction (PCR)-based typing analysis of atypical isolates of the fish pathogen Aeromonas salmonicida. Syst. Appl. Microbiol., 22: 403–411.

Johnston, I. A., Kent, M. P., Boudinot, P., Looseley, M., Bargelloni, L., Faggion, S., ... & Lien, S. (2024). Advancing fish breeding in aquaculture through genome functional annotation. Aquaculture, 740589.

Kirpichnikov, V.S. (1987). Selection and New Breeds of Pond Fishes in the USSR. In: Tiewes, K. (Ed.), Selection, Hybridization and Genetic Engineering in Aquaculture, Volume II, 461-473

Koo, K., & Jaykus, L. A. (2000). Selective amplification of bacterial RNA: use of a DNA primer containing mismatched bases near its 3′ terminus to reduce false□positive signals. Letters in applied microbiology, 31(3), 187-192.

Laatri, S., El Khayari, S., & Qriouet, Z. (2024). Exploring the molecular aspect and updating evolutionary approaches to the DNA polymerase enzymes for biotechnological needs: A comprehensive review. International Journal of Biological Macromolecules, 133924.

Levine MM, Sztein M.B. (2004). Vaccine development strategies for improving immunization: the role of modern immunology. Nat Immunol 5(5):460–464

Ma, J., Bruce, T.J., Jones, E.M., & Cain, K.D. (2019). A review of fish vaccine development strategies: conventional methods and modern biotechnological approaches. Microorganisms, 7(11), 569.

Martin, S. A., & Król, E. (2017). Nutrigenomics and immune function in fish: new insights from omics technologies. Developmental & Comparative Immunology, 75, 86-98.

Mishra, S. S., Das, R., Sahoo, S. N., & Swain, P. (2020). Biotechnological tools in diagnosis and control of emerging fish and shellfish diseases. In Genomics and biotechnological advances in veterinary, poultry, and fisheries (pp. 311-360). Academic Press.

Mondal, H., & Thomas, J. (2022). A review on the recent advances and application of vaccines against fish pathogens in aquaculture. Aquaculture international, 30(4), 1971-2000.

Mustapha UF, Alhassan AW, Jiang DN, Li G.L. (2021). Sustainable aquaculture development: a review on the roles of cloud computing, internet of things and artificial intelligence (CIA). Rev Aquac 13(4):2076–2091

Muthu MP, George MR, John R. (2020). Biosecurity strategies in aquaculture for fish health management. J Aquac Tropics 35(1–4):9–26

Nafea, A. M., Wang, Y., Wang, D., Salama, A. M., Aziz, M. A., Xu, S., & Tong, Y. (2024). Application of next-generation sequencing to identify different pathogens. Frontiers in Microbiology, 14, 1329330.

Nagamine K, Watanabe K, Ohtsuka K, Hase T, Notomi T (2001) Loop-mediated isothermal amplification reaction using a nondenatured template. Clin Chem 47(9):1742–3.

Nouri, A., & Chyba, C. F. (2008). Biotechnology and biosecurity. Global catastrophic risks, 450-480.

Hasimuna OJ, Maulu S, Mphande J. (2020). Aquaculture health management practices in Zambia: status, challenges and proposed biosecurity measures. J Aquac Res Dev 11(3):1–6

Honda H, Iwata K. (2016). Personal protective equipment and improving compliance among healthcare workers in high-risk settings. Curr Opin Infect Dis 29(4):400–406

Huang, T.S., Cerenius, L. and Soderhall, K. (1994). Analysis of genetic diversity in the crayfish plague fungus, Aphanomyces astaci, by random amplification of polymorphic DNA. Aquaculture, 126: 1–10.

John, M. A., Bankole, I., Ajayi-Moses, O., Ijila, T., Jeje, O., & Lalit, P. (2023). Relevance of advanced plant disease detection techniques in disease and Pest Management for Ensuring Food Security and Their Implication: A review. American Journal of Plant Sciences, 14(11), 1260-1295.

Maclean, N., & Talwar, S. (1984). Injection of cloned genes into rainbow trout eggs. Journal of Embryology and Experimental Morphology, 82(5), 187-200.

Naylor R, Hindar K, Fleming IA, Goldburg R, Williams S, Volpe J, Mangel M. (2005). Fugitive salmon: assessing the risks of escaped fish from net-pen aquaculture. BioScience, 55(5):427–437

Oidtmann, B., Cerenius, L., Schmid, I., Hoffmann, R. and Soderhall, K. (1999). Crayfish plague epizootics in Germany-classification of two German isolates of the crayfish plague fungus Aphanomyces astaci by random amplification of polymorphic DNA. Dis. Aquat. Org., 35: 235–238.

Oidtmann, B. C., Thrush, M. A., Denham, K. L., & Peeler, E. J. (2011). International and national biosecurity strategies in aquatic animal health. Aquaculture, 320(1-2), 22-33.

Pal, A., & Chakravarty, A. K. (2019). Genetics and Breeding for Disease Resistance of Livestock. Academic press.

Panigrahi, A. (2012). Biosecurity and the essential principles shrimp health management in Litopenaeus vannamei hatcheries and farms. Management Of Emerging Diseases of Shrimp With Special Reference To Pacific White Shrimp, Litopenaeus Vannamei. National Fisheries Development Board, Hyderabad. 10th–14th December.

Pardi, N., Hogan, M. J., Porter, F. W., & Weissman, D. (2018). mRNA vaccines—a new era in vaccinology. Nature reviews Drug discovery, 17(4), 261-279.

Pacitti, D., Lawan, M. M., Feldmann, J., Sweetman, J., Wang, T., Martin, S. A. M., & Secombes, C. J. (2016). Impact of selenium supplementation on fish antiviral responses: a whole transcriptomic analysis in rainbow trout (Oncorhynchus mykiss) fed supranutritional levels of Sel-Plex®. BMC genomics, 17, 1-26.

Rakkannan, G., & Priyadarshi, H. (2023). Selective Breeding for Disease Resistance in Aquaculture: Current Methods & Future Directions. Biotica Research Today, 5(1),45-47.

Reen, J. K., Yadav, A. K., & Singh, J. (2015). Nutrigenomics: concept, advances and applications. Asian Journal of Dairy and Food Research, 34(3), 205-212.

Rhodes, L.D., Rhodes, W.B. and Strom, M.S. (1998). Sensitive detection of Renibacterium salmoninarum in whole fry, blood, and other tissues of Pacific salmon by reverse transcription-polymerase chain reaction. Mol. Mar. Biol. Biotechnol., 7: 270–279.

Rojas-Avelizapa, L. I., Cruz-Camarillo, R., Guerrero, M. I., Rodríguez-Vázquez, R., & Ibarra, J. E. (1999). Selection and characterization of a proteo-chitinolytic strain of Bacillus thuringiensis, able to grow in shrimp waste media. World Journal of Microbiology and Biotechnology, 15, 299-308.

Rurangwa, E., & Verdegem, M. C. (2015). Microorganisms in recirculating aquaculture systems and their management. Reviews in aquaculture, 7(2), 117-130.

Sakai, M. (1999). Current research status offish immunostimulants. Aquaculture,172(1-2),63-92.

Singh, A., Deb, R., & Rajendran, N. (2014). Benefits and risks of genetically modified organisms in aquaculture. International Journal of Pharmaceutical Sciences Review and Research, 27, 79-84.

Shalon, D.D., Smith, S.J. and Brown, P.O. 1996. A DNA microarray system for analyzing complex DNA samples using two-colour fluorescent probe hybridization. Genome Res., 6: 639–645.

Schreier, H. J., Mirzoyan, N., & Saito, K. (2010). Microbial diversity of biological filters in recirculating aquaculture systems. Current opinion in biotechnology, 21(3), 318-325.

Snieszko, S. F. (1970). Immunization of fishes: a review. Journal of Wildlife Diseases, 6(1), 24-30.

Southern, E.M. (1996). DNA chips: analysing sequence by hybridisation to oligonucleotides on a large scale. TIG, 12: 110–115.

Suttle, Curtis A. (2007). Marine Viruses--Major Players in the Global Ecosystem. Nature Reviews. Microbiology 5 (10): 801–12.

Swain, P. P., Sahoo, L., Kumar, R., & Sundaray, J. K. (2022). Applications of next-generation sequencing in aquaculture and fisheries. In Advances in Fisheries Biotechnology (pp. 41-64). Singapore: Springer Nature Singapore.

Ubina, N. A., Cheng, S. C., Chen, H. Y., Chang, C. C., & Lan, H. Y. (2021). A visual aquaculture system using a cloud-based autonomous drones. Drones, 5(4), 109.

Vyas, P., Singh, D., Singh, N., Kumar, V., & Dhaliwal, H. S. (2018). Nutrigenomics: advances, opportunities and challenges in understanding the nutrient-gene interactions. Current Nutrition & Food Science, 14(2), 104-115.

Wang, W., Sun, J., Liu, C., & Xue, Z. (2017). Application of immunostimulants in aquaculture: current knowledge and future perspectives. Aquaculture Research, 48(1), 1-23.

Williams, K., Blake, S., Sweeney, A., Singer, J.T. and Nicholson, B.L. (1999). Multiplex reverse transcriptase PCR assay for simultaneous detection of three fish viruses. J. Clin. Microbiol., 37: 4139–4141.

Wintergerst, E. S., Maggini, S., & Hornig, D. H. (2007). Contribution of selected vitamins and trace elements to immune function. Annals of nutrition and metabolism, 51(4), 301-323.

White PG, Shipton TA, Bueno PB, Hasan MR (2018). Better management practices for feed production and management of Nile tilapia and milkfish in the Philippines. FAO Fish Aquacult Techl Paper, (614):I-79

Zhong, J., Wang, Y., & Zhu, Z. (2002). Introduction of the human lactoferrin gene into grass carp (Ctenopharyngodon idellus) to increase resistance against GCH virus. Aquaculture, 214(1-4), 93-101.

Zhu, Z., He, L., & Chen, S. (1985). Novel gene transfer into the fertilized eggs of goldfish (Carassius auratus L. 1758). Journal of Applied Ichthyology, 1(1), 31-34.

5

Genetic Improvement in Aquaculture Integrating Traditional and Modern Breeding Techniques

Trisha Tiwari*[*] *and Madhumita Mishra

Department of Zoology, Institute of Science, Banaras Hindu University, Varanasi Uttar Pradesh (221 005), India

Abstract

Aquaculture, farming aquatic organisms such as fish, mollusks, and crustaceans, plays a critical role in meeting the growing global demand for high-quality protein. Various genetic improvement strategies have been employed to enhance aquaculture sustainability and productivity, including traditional and modern techniques. This chapter explores key approaches such as selective breeding, integration breeding (intra- and inter-specific crossbreeding), chromosome set manipulation (artificial androgenesis, gynogenesis, and polyploid breeding), and transgenic breeding technologies. Chromosome manipulation, in particular, has proven to be a powerful tool for enhancing the genetic potential of farmed species, contributing to improvements in growth, disease resistance, and reproductive performance. Over the last few decades, advances in these techniques have led to the development of robust strains of fish and invertebrates, improving aquaculture output and ensuring the sustainability of the industry. Moreover, genetically modified organisms (GMOs) have applications beyond aquaculture, serving as valuable models in medical research, pharmacology, and environmental toxicity studies. This chapter delves into the application of chromosome manipulation and genetic modification in aquaculture, highlighting their potential to increase farmed stock productivity, improve marketability, and conserve natural resources. The use of genetic enhancement techniques in aquaculture promises to meet the demands of an ever-growing global population while ensuring sustainable practices.

Keywords: *Chromosome manipulation, Genetic enhancement, Selective breeding, Hybridization, Sustainable aquaculture.*

Introduction

Chromosome manipulation, the targeted alteration of chromosome quantity and combination, represents a cornerstone of genetic enhancement in aquaculture. Initially rooted in early twentieth-century studies focused on fundamental biology—particularly in amphibian models—this field has significantly expanded to enhance desirable traits through methods like polyploidy and uniparental inheritance (Arai, 2001; Benfey, 1999; Benfey, 2011; Cherfas, 1975; Devlin & Nagahama, 2002; Felip *et al.,* 2001; Guo *et al.,* 2009; Hulata, 2001; Komen & Thorgaard, 2007; Nagy, *et al.,* 1979; Nagy, *et al.,* 1978; Neyfakh, 1956; Ojima & Makino 1978). Over time, researchers worldwide have meticulously refined these techniques, transforming chromosome manipulation into a powerful tool for developing aquaculture strains with improved performance traits. By the late 20th century, aquaculture research began leveraging chromosomal modifications in finfish and aquatic invertebrates to achieve increased growth rates, disease resistance, and enhanced survival, with such techniques becoming extensively applied in commercial aquaculture operations (Pandian & Koteeswaran, 1998; Piferrer *et al.,* 2009; Purdom, 1972). Over the past 30 years, a wide body of research has focused on optimizing treatment conditions specific to each target species, yielding robust evaluations of various aquaculture performance metrics. These metrics, including survival, growth, reproductive maturation, and disease resistance, have proven essential in assessing the effectiveness of polyploid and uniparental (gynogenetic and androgenetic) progeny (Purdom, 1976; Refstie *et al.,* 1977; Romashov *et al.,* 1960; Romashov & Belyaeva 1964; Stanley 1976; Stanley, 1976; T. J. Pandian, 2011; Devlin & Nagahama 2002; Guo, *et al.,* 2009; Hulata, 2001; Komen & Thorgaard, 2007). As aquaculture has grown into the fastest-expanding sector of animal food production globally, the demand for sustainably produced, high-quality protein continues to rise. This surge in demand is driven, in part, by the decline of global fisheries due to overfishing, pollution, and habitat degradation. Consequently, the aquaculture industry faces the challenge of increasing output sustainably, necessitating the development of improved fish strains that support the resilience and efficiency of farmed stocks. To meet these demands, a range of genetic modification methods has been employed, aiming to produce fish with desirable traits such as rapid growth, superior meat quality, and enhanced resistance to disease (Xu *et al.,* 2015). Chromosome manipulation has thus emerged as a critical tool, facilitating precise genetic interventions to enhance farmed species' overall performance and adaptability.

This chapter focuses on the applications and advancements in genetic improvement technologies in aquaculture, encompassing both classical and modern genetic techniques across diverse aquatic species. These technologies include:

- **Selective Breeding:** Traditional techniques that capitalize on natural variation within species.
- **Integration Breeding:** Intra-specific and inter-specific crossbreeding to introduce and stabilize desirable traits.
- **Chromosome Set Manipulation:** Advanced methods of altering chromosome structure and number, such as artificial androgenesis, gynogenesis, and polyploid breeding.
- **Modification Breeding:** Contemporary approaches, including transgenic breeding, to introduce specific gene variants conferring beneficial traits.

The utilization of these genetic enhancement strategies has opened new pathways for aquaculture sustainability, with chromosome manipulation playing a particularly transformative role. These approaches offer the potential to increase production efficiency, improve marketability, and support the conservation of natural resources (Omole, 2017). This chapter will explore the technical foundations and practical implications of chromosome manipulation, emphasizing its role in meeting future food production demands while addressing the environmental and economic challenges facing global aquaculture.

Selective Breeding Technique

Selective breeding, one of the earliest genetic improvement strategies, is a process aimed at enhancing economically significant traits in aquaculture species, including growth rate, disease resistance, feed conversion efficiency, and reproductive fitness (Gjedrem and Robinson, 2014; Gjedrem and Baranski, 2010; Lind *et al.,* 2012). This method involves intentionally selecting individuals with desirable genetic traits, ensuring that these traits become prevalent in subsequent generations. Selective breeding has long been successfully applied in various animal species to improve attributes that offer economic or adaptive advantages. The history of selective breeding in aquaculture dates back to the 1920s in the United States, where one of the first reported trials focused on developing disease resistance in brook trout (*Salvelinus fontinalis*) against furunculosis. Embody and Hyford, (1925) conducted these studies, selectively breeding survivors from a population affected by endemic furunculosis. Through three generations of selection, survival rates improved dramatically from 2% in the initial population to 69%, setting a precedent for disease resistance breeding in aquaculture. Following these early successes, significant progress was made with the development of the "Donaldson" strain of rainbow trout (*Oncorhynchus mykiss*). In 1932, Donaldson initiated a selective breeding program that spanned 35 years, yielding substantial improvements in the fertility and size of rainbow trout through individual selection. The enhanced strain became widely disseminated and valued among trout culturists. Despite these advancements, few selective breeding programs were conducted on aquaculture species before the 1970s. In the mid-1960s, a mass selection program was initiated for common carp (*Cyprinus carpio*), focusing on growth rate improvement, but this initial effort yielded limited results (Moav *et al.,* 1976). However, a subsequent breeding program targeting disease resistance and growth rate proved successful. By incorporating wild carp strains from the Amur River, this effort successfully improved resistance to dropsy, leading to the development of three common carp stocks. These stocks were later crossbred and commercially produced in the Krasnodar region (Kirpichnikov *et al.,* 1993). Selective breeding efforts in tilapia (*Oreochromis niloticus*) initially faced challenges. Early studies focusing on individual selection for growth rate in tilapia, conducted by Teichert-Coddington 1983, Hulata *et al.,* 1986, and Huang & Liao, 1990, showed limited success. However, in 1988, a transformative breeding initiative began with the establishment of the GIFT (Genetically Improved Farmed Tilapia) program by ICLARM in the Philippines. This project utilized four wild African strains and four cultured Filipino strains, with initial comparisons revealing faster growth rates among the African strains. Subsequent family-based selection led to a robust growth response, averaging a 17% improvement per generation over five selection cycles (Eknath *et al.,* 1998). Other Nile tilapia breeding programs focusing on growth rate selection have since reported similarly impressive outcomes (Bolivar and Newkirk, 2002). Today, some of these early selective breeding initiatives have evolved into large-scale industry programs. For instance, Atlantic salmon and Nile tilapia breeding have integrated extensively into commercial aquaculture, while rainbow trout and channel catfish breeding projects have transitioned into private enterprises. Selective breeding programs for species such as Atlantic salmon, rainbow trout, and Nile tilapia have significantly enhanced the economic viability of the aquaculture industry (Hulata, 1995). These efforts improve production efficiency and support sustainable practices.

Integration Breeding Technologies

Intra-Specific Crossbreeding

Intra-specific crossbreeding, which involves crossing distinct strains within a species, has shown the potential to enhance lucrative traits in aquaculture species. However, the manifestation of heterosis—improved traits in the progeny compared to the parents—is not always achieved. Nonetheless, several commercially farmed species have benefited from crossbreeding. For instance, common carp, channel catfish, rainbow trout, and Pacific oysters have demonstrated varying degrees of heterosis, particularly for growth rate, survival, disease resistance, and reproductive features. In common carp, the development of crossbreeds with positive heterosis for growth rate has significantly impacted aquaculture in regions such as Vietnam, China, Israel, and Hungary (Wohlfarth, 1993; Gorda *et al.,* 1995). A prime example is the crossbreeding program in Szarvas, Hungary, which has led to over 100 experimental crosses since 1965. Three specific crosses were selected for culture purposes based on a 20% increase in growth rate and other qualitative traits compared to the parent and control lines. These Szarvas crossbreeds now account for approximately 80% of common carp production in Hungary. The crossbreeding effort also extended to the development of gynogenetic female lines and gynogenetic sex-reversed inbred male lines from the common carp, selected for their high combining capacity. In Indonesia, the creation of ten inbred common carp lines using artificial gynogenesis and sex reversal set the foundation for successful crossbreeding efforts (Sumantadinata, 1995). In Vietnam, significant heterosis was observed in the first generation of crosses between indigenous Vietnamese common carp and foreign strains from Ukraine, the Czech Republic, Indonesia, and Hungary. These hybrids contributed to the initiation of Vietnam's carp selective breeding program (Thien & Trong, 1995). Similarly, in China, intra-specific crossbreeds such as Feng carp, Heyuan carp, Yue carp, and Tri-crossed carp were developed with desirable traits, including increased growth rate and improved feed conversion efficiency, making them dominant strains in Chinese aquaculture (Gupta, *et al.,* 2005). Channel catfish crossbreeding, particularly among strains like Marion x Kansas and Auburn x Kansas, also demonstrated increased growth rates of 8-13% over pure strains (Dunham & Smitherman, 1983). These crossbreeds were praised for their enhanced growth, reproductive performance, and greater resistance to *E. ictaluri*. In addition, rainbow trout and Pacific oysters have shown improvements in growth and disease resistance due to crossbreeding (Gall, 1975; Ayles & Baker, 1983; Hedgecock, *et al.,* 1989). However, crossbreeding between seabream and Atlantic salmon strains yielded negligible heterosis (Gjerde & Refstie, 1984; Knibb, *et al.,* 2000).

Inter-Specific Crossbreeding

Interspecific crossbreeding, or hybridization, has proven effective in several fish and shellfish species, particularly for increasing growth rates, manipulating sex ratios, improving flesh quality, producing sterile animals, and enhancing disease resistance. Although hybrids often do not express the same level of heterosis as intra-specific crosses, they can still offer a desirable combination of beneficial traits from both parent species. While the majority of interspecific hybrids have not been commercially successful, certain hybrids in freshwater and marine environments have demonstrated significant benefits. Among cyprinid hybrids, the crosses between silver carp (*Hypophthalmichthys molitrix*) and bighead carp (*Aristichthys nobilis*) have shown superior survival and production traits, leading to commercialization in the 1970s and 1980s (Isa *et al.,* 1986). In India, interspecific crosses between common carp and indigenous carps such as rohu, mrigal, and catla have produced hybrids with improved growth rates and better performance. These hybrids are often sterile, triploid, and exhibit increased meat content with reduced escapability, enhancing their suitability for aquaculture (Reddy *et al.,* 1990). Another notable example is the hybrid between female grass carp (*Ctenopharyngodon idella*) and male bighead carp (*Aristichthys nobilis*), which produced sterile triploid offspring, reducing the chances of unwanted reproduction. This hybrid was also valued for its herbivorous properties, making it an effective tool for controlling aquatic vegetation (Sutton *et al.,* 1981). In the United States, the hybrid between female channel catfish (*Ictalurus punctatus*) and male blue catfish (*Ictalurus furcatus*) combined the desirable culture traits of the channel catfish with the improved harvestability and disease resistance of the blue catfish, proving effective in culture and outperforming purebred channel catfish (Dunham and Smitherman, 1985). Similarly, hybridization of African catfish species, such as *Clarias gariepinus* and *Heterobranchus longifilis*, has resulted in faster growth rates, enhancing their viability for aquaculture in Thailand. Catfish hybrids of *Pseudo platystoma corruscans* and *Pseudo platystoma reticulatum* have also been developed in Brazilian hatcheries to select individuals with optimal culture characteristics for fish farming (Pardo *et al.,* 2011). Tilapia hybrids have been widely utilized, especially for producing 100% male populations, thus preventing unwanted reproduction. Crosses between Nile tilapia (*Oreochromis niloticus*) and blue tilapia (*Oreochromis aureus*) achieved this through differences in sex determination systems, resulting in all-male populations. Other tilapia hybrids, such as those between Nile tilapia and Wami tilapia, and between Mozambique tilapia and Wami tilapia, also produced entirely male offspring (Wohlfarth, 1994). Additionally, hybrids of various tilapia species have demonstrated improved environmental tolerance, including cold and salinity tolerance, expanding their range for aquaculture in diverse climates (Cnaani *et al.,* 2000; Lahav & Lahav, 1990). The sunshine bass, a hybrid of white bass (*Morone chrysops*) female, and striped bass (*Morone saxatilis*) male, has been successfully cultivated due to its superior growth and overall culture characteristics. This hybrid has been commercialized in the United States and internationally in Israel and Taiwan. Similarly, hybridization between striped bass and yellow bass (*Morone mississippiensis*) produced only female populations, which can be advantageous in aquaculture systems where sex-based growth variation is a concern (Wolters & Demay, 1996). Hybridization in salmonids, such as between coho salmon (*Oncorhynchus kisutch*) and rainbow trout, has led to increased resistance to salmonid viruses, although overall viability was lower. Viability improved through triploidization, a technique also applied to rainbow trout and char hybrids to enhance disease resistance. This approach has been widely adopted in the salmon industry for genetic improvement (Dorson *et al.,* 1991; Seeb *et al.,* 1993). Despite the challenges with viability, triploid hybrids of Pacific salmon species demonstrated earlier saltwater acclimatization, which can be beneficial for aquaculture (Galbreath & Thorgaard, 1996).

Chromosome Set Manipulation Technologies

Over recent decades, chromosome manipulation technologies have seen widespread application across various aquatic species. Androgenesis, gynogenesis, and induced polyploidy have proven instrumental in managing uncontrolled reproduction and enhancing sex culture performance. These methods are pivotal in generating mono-sex gametes, which can be utilized to produce mono-sex diploid or triploid populations, a strategy with significant implications for aquaculture and fisheries management.

Gynogenesis

Gynogenesis is a reproductive process that results in offspring with genetic material exclusively inherited from the mother. This occurs when sperm from a male is used to fertilize an egg, but the male's genetic contribution is nullified. This is typically achieved through the irradiation of sperm, which either prevents the extrusion of the second polar body (meiotic gynogenesis) or inhibits the first cleavage of the fertilized egg (mitotic gynogenesis) (Arai & Mukaino, 1997; Pandian & Koteeswaran, 1998). The resulting offspring, depending on the method used, may be diploid and all-female (meiogynogenesis) or also triploid and sterile (mitogynogenesis). The application of gynogenesis spans various fish species, such as common carp, tilapia, rainbow trout, Atlantic halibut, European seabass, and others (Komen & Thorgaard, 2007; Liu & Yang, 2009; Felip, 2001). Gynogenetic lines have been utilized commercially in species like rainbow trout, red seabream, and olive flounder (Komen, *et al.,* 1992; Fopp-Bayat, 2007). In salmonids, gynogenesis is often used to generate all-female populations, while triploidization, achieved through pressure or shock treatments, produces sterile female populations (Donaldson, 1996). The combination of gynogenesis with other techniques, such as hormonal sex-reversal, allows for the production of either neomales or neofemales, or triploid gynogenetic females that are sterile. Over the past few decades, these methods have found economic applications across various aquaculture species globally. For instance, during the 1980s, many trout farms relied heavily on all-female diploid stocks, with sterile triploid trout becoming increasingly popular, particularly for sport fisheries (Hulata, 2001). Similarly, in Japan, the commercial production of all-female diploid or triploid salmonids was once a common practice, although the government later restricted it (Yamamoto *et al.,* 1999). Triploid salmonids, created either by triploidizing gynogenetic females or by crossing tetraploid XXXX females with diploid XX males, were also used in the cultivation of non-native salmonid species, particularly in Canada. Despite the economic potential, the performance of triploid salmonids was often less optimal compared to diploids (Davis & Hetzel, 2000). Meanwhile, triploidy in grass carp species has been widely adopted in the U.S. to control aquatic plants. Additionally, methods such as gynogenesis combined with hormonal sex-reversal have been developed to produce completely infertile, all-female triploid black and grass carp (Rothbard *et al.,* 1997; Zhang *et al.,* 2010).

Research has suggested that using sperm from distantly related fish species can improve the survival rates of gynogenetic offspring, while also simplifying their identification. For example, gynogenetic crucian carp produced using sperm from blunt snout bream exhibited a higher survival rate compared to those created using common carp sperm (Sun *et al.,* 2007). Furthermore, gynogenetic strains have been developed from a variety of species, such as natural red crucian carp, goldfish, common carp, grass carp, and even hybrid strains (Xiao *et al.,* 2011; Xu *et al.,* 2015), all of which show increased survival and easy identification.

Androgenesis

Androgenesis is a process in which offspring are produced solely from the genetic contribution of the male, with no input from the female. This is achieved by fertilizing inactivated eggs with active sperm. The resulting offspring will have only paternal genetic material, and in some cases, androgenesis can produce all-male populations, which are beneficial in aquaculture (Nwokwa, 2012). This technique can also be used for restoring lost genotypes from preserved sperm or for producing homozygous fish lines (Xu *et al.,* 2015).

Several species, including cyprinids, cichlids, and salmonids, have been the subject of androgenesis studies. A notable application of androgenesis in aquaculture is in tilapia. By producing super-male individuals (YY) and crossing them with regular females (XX), it is possible to generate all-male tilapia (XY), which are commercially valuable (Xu *et al.,* 2015). These genetically male tilapia (GMT) have been extensively used in countries such as the Philippines, Thailand, China, and the U.S. (Hulata, 2001). While the use of haploid sperm to induce androgenesis has been limited in other species, its success in tilapia demonstrates its potential for widespread application. Although androgenesis can be challenging due to difficulties in inactivating the egg's DNA and the low survival rate of androgenetic individuals, recent advancements have led to improved methods. The use of tetraploid fish and diploid sperm has been explored to generate diploid androgenetic offspring without the need for chromosomal doubling treatment. This approach has shown higher survival rates than those achieved using haploid sperm (Thorgaard *et al.,* 1990; Arai *et al.,* 1995; Fujimoto *et al.,* 2010).

Polyploidy

Polyploidy involves the presence of multiple sets of chromosomes in an organism and has been applied commercially to improve the breeding of farmed fish species. The most common types of polyploidies are triploidy (three sets of chromosomes) and tetraploidy (four sets of chromosomes), which are created through various physical, biological, or chemical methods (Pandian & Koteeswaran, 1998). Polyploidy offers significant advantages in fish breeding, such as providing a means of gonadal sterilization, facilitating sex control, and enabling cloning (Lakran & Ayyappan, 2003). Polyploidy can be induced through several methods, including physical approaches (e.g., temperature shock, hydrostatic pressure), biological techniques (e.g., hybridization, nuclear transfer), and chemical treatments (e.g., cytochalasin, polyethylene glycol, colchicine) (Song *et al.,* 2012). The physical and chemical methods prevent the exclusion of the second polar body or inhibit the first cleavage of the zygote, leading to polyploidy.

Triploidy

Triploidy is the most commonly employed method for producing sterile fish for aquaculture and fisheries management. Triploid fish are sterile due to the presence of an odd number of chromosome sets, which disrupts normal reproduction. Triploids can be generated by preventing the extrusion of the second polar body after fertilization, using heat shock or high hydrostatic pressure (Pandian & Koteeswaran, 1998). Triploids offer several benefits, including rapid growth, improved meat quality, enhanced disease resistance, and increased survival. Moreover, triploid fish are typically sterile, which reduces the risk of interbreeding with wild populations if farmed fish escape into the environment (Benfey, 1999). Methods for inducing triploidy include temperature shock (both cold and hot), hydrostatic pressure shock, and chemical treatments, such as colchicine or cytochalasin-B. Hybridization between tetraploid females and diploid males is another way to produce triploid offspring. For instance, in some species, hybridization between Cyprinus carpio females and Carassius auratus males has been used to create triploid carp (Wu *et al.,* 1993).

Tetraploidy

Tetraploidy, which involves the presence of four sets of chromosomes, can be induced in fish using techniques such as hydrostatic pressure or temperature shock. Tetraploid individuals are fertile and can be used to produce sterile triploid populations through hybridization. The creation of tetraploids is critical in the commercial production of sterile triploids, as hybridization between tetraploid females and diploid males results in sterile triploid offspring (Guo and Harrison, 1996). The success of tetraploid induction depends on various factors, including the timing and intensity of the shock, the duration of exposure, and the quality of the gametes. Heat shock, hydrostatic pressure, and cold shock have been successfully employed to induce tetraploidy in species such as rainbow trout, channel catfish, and loach (Pandian & Koteeswaran, 1998; Myers, 1986).

Modification Breeding

Modification breeding refers to the deliberate manipulation of genetic traits in aquatic species to enhance specific characteristics, such as growth rate, disease resistance, stress tolerance, or reproductive performance. These breeding techniques can include transgenesis and genome editing, among others. It is closely related to chromosome manipulation and plays an essential role in sustainable aquaculture practices.

Transgenesis

Transgenesis, a powerful tool that involves the introduction of foreign genes into an organism's genome to create genetically modified organisms (GMOs), has emerged as a transformative approach in aquaculture. This technology allows for the incorporation of beneficial traits from other species or the enhancement of existing characteristics, such as growth rates, disease resistance, and the ability to withstand environmental stress. Recombinant growth hormone and gene transfer technologies have played a key role in advancing aquaculture by facilitating substantial growth enhancements in transgenic fish, mollusks, and crustaceans (Chen, 1995). A prime example is the AquAdvantage Salmon, approved by the FDA in 2015, which incorporates a growth hormone gene from Pacific chinook salmon. This modification accelerates growth, reduces production time, and lowers associated costs, making the species more commercially viable. Similarly, the development of cold-resistant Tilapia, engineered with antifreeze genes, allows this traditionally tropical species to thrive in colder climates, extending its farming range and providing new opportunities for aquaculture in temperate zones. Disease resistance in aquaculture has also been improved, as demonstrated by disease-resistant catfish modified with antimicrobial genes via CRISPR technology. These genetically modified fish show enhanced resilience to pathogens, reducing the need for antibiotics and improving overall farm productivity. In the ornamental fish industry, GloFish, created by inserting fluorescent protein genes from marine organisms into zebrafish, has gained popularity due to its bright, inheritable colours. These genetically modified fish not only add aesthetic value to aquariums but also offer a new market for ornamental fisheries. Collectively, these advancements demonstrate the significant potential of transgenesis to enhance aquaculture efficiency, biodiversity, and sustainability (Danve *et al.,* 2024). Despite these successes, continued refinement of gene editing techniques, as well as increased consumer education, are essential for broadening the acceptance and market integration of transgenic fish (Aydin *et al.,* 2020).

Genome Editing (CRISPR-Cas9)

Genome editing technologies, particularly CRISPR-Cas9, have emerged as powerful tools for targeted genetic modifications in aquatic species. Unlike transgenesis, which typically involves inserting foreign genes, genome editing allows for precise changes to be made in the organism's own genome. This can be used to knock out genes, add beneficial traits, or correct undesirable mutations. CRISPR/Cas9 genome editing has been successfully applied to a variety of aquaculture species, enhancing traits such as growth, reproduction, immunity, and disease resistance. In Atlantic salmon (*Salmo salar*), edits to the tyr/slc45a2 genes have altered pigmentation, while modifications to elov-2 have improved omega-3 metabolism. Tilapia (*Oreochromis niloticus*) has seen significant advances with edits to genes such as dmrt1, gsdf, and sf-1, leading to improvements in reproduction and successful germline transmission. Sea bream (*Sparus aurata*) has been modified via CRISPR to target the mstn gene for growth enhancement, resulting in increased muscle development. In Channel catfish (*Ictalurus punctatus*), CRISPR editing of the mstn gene has led to greater growth and successful germline transmission, making this species more viable for aquaculture. Common carp (*Cyprinus carpio*) has undergone CRISPR-based editing of genes like sp7a/sp7b and mstn(ba) to improve muscle development, boosting growth rates and aquaculture productivity (Gratacap *et al.*, 2019).

CRISPR/Cas9 genome editing offers significant potential for advancing disease control and molecular breeding in aquaculture, surpassing traditional transgenesis by enabling specific, targeted modifications to the genome without introducing foreign DNA, thus addressing public safety concerns related to GMOs (Okoli *et al.*, 2021). The system's flexibility allows for the editing of multiple genes at once through multiplexing, increasing the likelihood of achieving desired genetic changes. Unlike selective breeding, which involves random genetic variations over many generations, CRISPR/Cas9 accelerates breeding by precisely targeting genes, often establishing desired traits within just 1–3 generations. Key applications of this technology in aquaculture include fixing alleles at quantitative trait loci (QTLs), targeted introgression to introduce beneficial alleles from different species, and creating new variants based on genetic knowledge (Gratacap *et al.*, 2019). Aquaculture species, with their high fecundity and short generation times, are particularly well-suited for genome editing compared to terrestrial animals, making CRISPR/Cas9 an ideal tool for improving production and sustainability. Genome editing experiments in farmed fish typically involve selecting a target gene, designing sgRNA, and delivering the CRISPR/Cas9 complex to fertilized eggs using methods such as microinjection or electroporation at the one-cell stage to ensure heritable modifications (Okoli *et al.*, 2021; Yang *et al.*, 2022). These efficient and precise methods have paved the way for rapid genetic advancements in aquaculture species, enhancing both productivity and sustainability.

Conclusion

Chromosome manipulation techniques are pivotal in advancing sustainable aquaculture by improving growth rates, disease resistance, and reproductive efficiency in farmed species. Methods such as gynogenesis, androgenesis, and polyploidy have significantly enhanced breeding efficiency, while hybridization strategies further boost productivity and minimize environmental impact. Triploidy, which produces sterile fish, and advances in genome duplication, including tetraploids and double haploids, offer new opportunities for strain improvement and cloning. Cold-shock-induced androgenesis and distant hybridization have simplified techniques, broadening their applicability across species.

These genetic advancements in species like carp, trout, and salmon are crucial for meeting the growing demand for aquatic protein. As research progresses, chromosome manipulation will remain central to the sustainable development of aquaculture, supporting both economic viability and ecological conservation.

With the integration of transgenic technologies, these innovations will help address challenges like overfishing and food security, ensuring a resilient future for the aquaculture industry.

References

Arai, K. (2001). Genetic improvement of aquaculture finfish species by chromosome manipulation techniques in Japan. Aquaculture 197, 205–228.

Arai, K., & Mukaino, M. (1997). Clonal nature of gynogenetically induced progeny of triploid (diploid× tetraploid) loach, Misgurnus anguillicaudatus (Pisces: Cobitididae). Journal of Experimental Zoology, 278(6), 412-421.

Aydin, I., Goddard, A., & Cebeci, A. (2020). Bigger, stronger, better: Fish transgenesis applications and methods.

Ayles, G. B., & Baker, R. F. (1983). Genetic differences in growth and survival between strains and hybrids of rainbow trout (Salmo gairdneri) stocked in aquaculture lakes in the Canadian prairies. Aquaculture, 33(1-4), 269-280.

Benfey, T. J. (1999). The physiology and behavior of triploid fishes. Review in Fisheries Science 7(1), 39–67.

Benfey, T. J. (2011). Physiology of triploid fish. In: A. P. Farrell (ed). Encyclopedia of Fish Physiology: From Genome to Environment, vol. 3. Academic Press, San Diego, pp. 2009–2015.

Bolivar, R. B., & Newkirk, G. F. (2002). Response to within family selection for body weight in Nile tilapia (Oreochromis niloticus) using a single-trait animal model. Aquaculture, 204(3-4), 371-381.

Chen, T. T. (1995). Transgenic fish and aquaculture. In Towards sustainable aquaculture in Southeast Asia and Japan: Proceedings of the Seminar-Workshop on Aquaculture Development in Southeast Asia, Iloilo City, Philippines, 26-28 July, 1994 (pp. 81-89). Aquaculture Department, Southeast Asian Fisheries Development Center.

Cherfas, N. (1975). Studies on diploid gynogenesis in the carp (Cyprinus carpio L.). I. Experiments on the mass production of diploid gynogenetic offspring. Genetika 11, 78–86 (in Russian).

Cnaani, A., Gall, G. A. E., & Hulata, G. (2000). Cold tolerance of tilapia species and hybrids. Aquaculture International, 8, 289-298.

Davis, G. P., & Hetzel, D. S. (2000). Integrating molecular genetic technology with traditional approaches for genetic improvement in aquaculture species. Aquaculture research, 31(1), 3-10.

Devlin, R. H. and Nagahama, Y. (2002). Sex determination and sex differentiation in fish: an overview of genetic, physiological, and environmental influences. Aquaculture 208, 191–364.

Donaldson, E. M. (1996). Manipulation of reproduction in farmed fish. Animal Reproduction Science, 42(1-4), 381-392.

Dorson, M., Chevassus, B., & Torhy, C. (1991). Comparative susceptibility of three species of char and of rainbow trout× char triploid hybrids to several pathogenic salmonid viruses.

Dunham, R. A., & Smitherman, R. O. (1985). Improved growth rate, reproductive performance, and disease resistance of crossbred and selected catfish from AU-M and AU-K lines.

Dunham, R. A., Smitherman, R. O., & Webber, C. (1983). Relative tolerance of channel x blue hybrid and channel catfish to low oxygen concentrations. The Progressive Fish-Culturist, 45(1), 55-57.

Eknath, A. E., Dey, M. M., Rye, M., Gjerde, B., Abella, T. A., Sevilleja, R., ... & Bentsen, H. B. (1998). Selective breeding of Nile tilapia for Asia. In 6th World Congress on Genetics Applied to Livestock Production (Vol. 27, pp. 89-96). University of New England Armidale Australia.

Embody, G. C., & Hayford, C. O. (1925). The advantage of rearing brook trout fingerlings from selected breeders. Transactions of the American Fisheries Society, 55(1), 135-148.

Felip, A., Zanuy, S., Carrillo, M., & Piferrer, F. (2001). Induction of triploidy and gynogenesis in teleost fish with emphasis on marine species. Genetica, 111, 175-195.

Fopp-Bayat, D. (2007). Verification of meiotic gynogenesis in Siberian sturgeon (Acipenser baeri Brandt) using microsatellite DNA and cytogenetical markers. Journal of Fish Biology, 71, 478-485.

Gall, G. A. (1975). Genetics of reproduction in domesticated rainbow trout. Journal of Animal Science, 40(1), 19-28.

Gjedrem, T., & Baranski, M. (2010). Selective breeding in aquaculture: an introduction (Vol. 10). Springer Science & Business Media.

Gjedrem, T., & Robinson, N. (2014). Advances by selective breeding for aquatic species: a review. Agricultural Sciences, 5(12), 1152.

Gjerde, B., & Refstie, T. (1984). Complete diallel cross between five strains of Atlantic salmon. Livestock Production Science, 11(2), 207-226.

Gorda, S., Bakos, J., Liska, J., & Kakuk, C. (1995). Live gene bank of common carp strains at the Fish Culture Research Institute, Szarvas. Aquaculture, 129(1-4), 199-202.

Gratacap, R. L., Regan, T., Dehler, C. E., Martin, S. A., Boudinot, P., Collet, B., & Houston, R. D. (2019). Efficient CRISPR/Cas9 genome editing in a salmonid fish cell line using a lentivirus delivery system. bioRxiv, 734442.

Guo, M., Harrison, P. J., & Taylor, F. J. R. (1996). Fish kills related to Prymnesium parvum N. Carter (Haptophyta) in the People's Republic of China. Journal of Applied Phycology, 8, 111-117.

Guo, X., Wang, Y., Xu, Z., & Yang, H. (2009). Chromosome set manipulation in shellfish. In New technologies in aquaculture (pp. 165-194). Woodhead Publishing.

Gupta, M. V., Dey, M. M., & Penman, D. (2005). Importance of Carp genetic resources. Carp genetic resources for aquaculture in Asia. World Fish Center Technical Report, 65, 1-5.

Hedgecock, D., Hutchinson, E. S., Li, G., Sly, F. L., & Nelson, K. (1989). Genetic and morphometric variation in the Pacific sardine, Sardinops sagax caerulea: comparisons and contrasts with historical data and with variability in the northern anchovy, Engraulis mordax. Fish. Bull, 87(3), 653-671.

Huang, C. M., & Liao, I. C. (1990). Response to mass selection for growth rate in Oreochromis niloticus. Aquaculture, 85(1-4), 199-205.

Hulata, G. (1995). A review of genetic improvement of the common carp (Cyprinus carpio L.) and other cyprinids by crossbreeding, hybridization and selection. Aquaculture, 129(1-4), 143-155.

Hulata, G. (2001). Genetic manipulations in aquaculture: a review of stock improvement by classical and modern technologies. Genetica 111(1–3), 155–173.

Hulata, G., Wohlfarth, G. W., & Halevy, A. (1986). Mass selection for growth rate in the Nile tilapia (Oreochromis niloticus). Aquaculture, 57(1-4), 177-184.

Isa, Y. (1986). An electron microscope study on the mineralization of the skeleton of the staghorn coral Acropora hebes. Marine Biology, 93, 91-101.

Kirpichnikov, V. S., Ilyasov, I., Shart, L. A., Vikhman, A. A., Ganchenko, M. V., Ostashevsky, A. L., ... & Tjurin, V. V. (1993). Selection of Krasnodar common carp (Cyprinus carpio L.) for resistance to dropsy: principal results and prospects. In Genetics in Aquaculture (pp. 7-20). Elsevier.

Knibb, W. (2000). Genetic improvement of marine fish–which method for industry? Aquaculture Research, 31(1), 11-23.

Komen, H. and Thorgaard, G. H. (2007). Androgenesis, gynogenesis and the production of clones in fishes: a review. Aquaculture 269, 150–173.

Lahav, M., & Lahav, E. (1990). The development of all-male tilapia hybrids in Nir David.

Lind, C. E., Ponzoni, R. W., Nguyen, N. H., & Khaw, H. L. (2012). Selective breeding in fish and conservation of genetic resources for aquaculture. Reproduction in domestic animals, 47, 255-263.

Moav, R., Brody, T., Wohlfarth, G., & Hulata, G. (1976). Applications of electrophoretic genetic markers to fish breeding. I. Advantages and methods. Aquaculture, 9, 217-228.

Nagy, A., Rajki, K., Bakos, J., & Csanyi, V. (1979). Genetic analysis in carp (Cyprinus carpio) using gynogenesis. Heredity, 43(1), 35-40.

Nagy, A., Rajki, K., Horvárth, L., & Csárnyi, V. (1978). Investigation on carp, Cyprinus carpio L. gynogenesis. Journal of Fish Biology, 13(2), 215-224.

Neyfakh, A. (1956). The effect of ionizing radiation on gametes of the loach (Misgurnus fossilis L.). Doklady Academy Nauka SSSR (DAN) 111, 585–588 (in Russian).

Ojima, Y. and Makino, S. (1978). Triploidy induced by cold shock in fertilized eggs of the carp. A preliminary study. Proceedings of the Japan Academy, Ser. B. Physical and Biological Sciences 54, 359–362.

Okoli, A. S., Blix, T., Myhr, A. I., Xu, W., & Xu, X. (2022). Sustainable use of CRISPR/Cas in fish aquaculture: the biosafety perspective. Transgenic Research, 31(1), 1-21.

Omole, I. A. (2017). Concentration of heavy metals in fish (Synodontis clarias) of Benin river, Delta state, Nigeria. Journal of Energy, Environmental & Chemical Engineering, 2(1), 6-9.

Pandian, T. J. (2011). Sex determination in fish. Science Publishers, CRC Press, New York.

Pandian, T. J. and Koteeswaran, R. (1998). Ploidy induction and sex control in fish. Hydrobiologia 384, 167–243.

Pardo, L. M., González, K., Fuentes, J. P., Paschke, K., & Chaparro, O. R. (2011). Survival and behavioral responses of juvenile crabs of Cancer edwardsii to severe hyposalinity events triggered by increased runoff at an estuarine nursery ground. Journal of Experimental Marine Biology and Ecology, 404(1-2), 33-39.

Piferrer, F., Beaumont, A., Falguière, J. C., Flajšhans, M., Haffray, P., & Colombo, L. (2009). Polyploid fish and shellfish: production, biology and applications to aquaculture for performance improvement and genetic containment. Aquaculture, 293(3-4), 125-156.

Purdom, C. E. (1969). Radiation-induced gynogenesis and androgenesis in fish. Heredity, 24(3), 431-444.

Purdom, C. E. (1972). Induced polyploidy in plaice (Pleuronectes platessa) and its hybrid with flounder (Platichthys flesus). Heredity 29, 11–24.

Purdom, C. E. (1976). Genetic techniques in flatfish culture. Journal of the Fisheries Research Board of Canada 33, 1088–1093.

Reddy, P. V. G. K., Kowtal, G. V., & Tantia, M. S. (1990). Preliminary observations on induced polyploidy in Indian major carps, Labeo rohita (Ham.) and Catla catla (Ham.). Aquaculture, 87(3-4), 279-287.

Refstie, T., Vassvik, V., Gjedrem, T. (1977). Induction of polyploidy in salmonids by cytochalasin B. Aquaculture 10, 65–74.

Romashov, D. D. and Belyaeva V. N. (1964). Cytology of the radiation induced gynogenesis and androgenesis in the loach (Misgurnus fossilis L.). Doklady Academy Nauka SSSR (DAN) 157(4), 964–967 (in Russian).

Romashov, D. D., Golovinskaia K., Beliaeva, V., Bkulina E., Pokrovskaia, G. & Cherfas, N. B. (1960). On radiation-induced lipoid hypogenesis in fish. Biofizika, 5, 461-467.

Seeb, J. E., Thorgaard, G. H., & Tynan, T. (1993). Triploid hybrids between chum salmon female× chinook salmon male have early sea-water tolerance. Aquaculture, 117(1-2), 37-45.

Stanley, J. G. (1976). Female homogamety in grass carp (Ctenopharyngodon idella) determined by gynogenesis. Journal of the Fisheries Research Board of Canada 33, 1373–1374.

Stanley, J. G. (1976). Production of hybrid, androgenetic and gynogenetic grass carp and carp. Transactions of the American Fisheries Society, 105, 10–16.

Sumantadinata, K. (1995). Present state of common carp (Cyprinus carpio L.) stocks in Indonesia. Aquaculture, 129(1-4), 205-209.

Sutton, D. L., Stanley, J. G., & Miley, W. W. (1981). Grass carp hybridization and observations of a grass carp X bighead hybrid.

T. J. Pandian (1998). Ploidy induction and sex control in fish. Hydrobiologia 384, 167–243.

Teichert-Coddington, D. R., & Smitherman, R. O. (1988). Lack of response by Tilapia nilotica to mass selection for rapid early growth. Transactions of the American Fisheries Society, 117(3), 297-300.

Thien, T. M., & Trong, T. D. (1995). Genetic resources of common carp in Vietnam. Aquaculture, 129(1-4), 216-216.

Wohlfarth, G. W. (1993). Genetic management of natural fish populations. In Genetic Conservation of Salmonid Fishes (pp. 227-230). Boston, MA: Springer US.

Wolters, W. R., & DeMay, R. (1996). Production characteristics of striped bass× white bass and striped bass× yellow bass hybrids. Journal of the World Aquaculture Society, 27(2), 202-207.

Xu, D. H., Shoemaker, C. A., & Zhang, D. (2015). Treatment of Trichodina sp reduced load of Flavobacterium columnare and improved survival of hybrid tilapia. Aquaculture Reports, 2, 126-131.

Xu, M., Fang, Y. J., Chen, Y. M., Lu, M. S., Pan, Z. Z., Yan, B., ... & Zhang, C. X. (2015). Higher freshwater fish and sea fish intake is inversely associated with colorectal cancer risk among Chinese population: a case-control study. Scientific reports, 5(1), 12976.

Yamamoto, N., Yoshimoto, M., Albert, J. S., Sawai, N., & Ito, H. (1999). Tectal fiber connections in a non-teleost actinopterygian fish, the sturgeon Acipenser. Brain behavior and evolution, 53(3), 142-155.

Yang, H., Wang, H., Huang, M., Cao, G., Tao, F., Shen, Q., ... & Yang, H. (2022). Repurposing fish waste into gelatin as a potential alternative for mammalian sources: A review. Comprehensive Reviews in Food Science and Food Safety, 21(2), 942-963.

6

Gene Editing in Aquaculture: Strategies for Accelerating Genetic Improvement

Maharshi Limbola, Sonal Suman and Vatsalya Bhatt

ICAR-Central Institutes of Fisheries Education, Mumbai, Maharastra (400061), India

Abstract

As demand for seafood increases globally, aquaculture becomes the primary means to ensure food security and nutrition. However, traditional breeding cannot overcome the limitations associated with disease susceptibility, slow growth, and environmental stress. Gene editing with CRISPR-Cas9 in combination with TALENs and ZFNs brings on the transformative potential to hasten genetic improvement in aquaculture. This chapter deals with gene editing in terms of applications in improving resistance to diseases, growth rates, feed efficiency, environmental tolerance, reproductive control, and flesh quality. Ethical concerns, technical challenges, and regulatory barriers of gene editing technology in aquaculture are also addressed. The convergence of gene editing with traditional breeding and genomic selection approaches is also discussed as a means to achieve maximal genetic gain. In fact, by focusing on the ecological, regulatory, and societal implications of these technologies, aquaculture can develop sustainable, resilient, and efficient production systems to meet the demand for global food security in the best possible manner with due concern for environmental and ethical considerations. Finally, the future directions in gene editing technologies such as prime editing are going to revolutionize aquaculture practices with precise, ethical, and sustainable genetic modifications.

Keywords: *Gene editing, aquaculture, CRISPR-Cas9, sustainable aquaculture, genetic improvement.*

Introduction

As global demand for seafood and high-protein foods grows, aquaculture has become a vital, fast-growing sector in food production. Fish farming supports food security, livelihoods, and nutrition worldwide but faces challenges like disease outbreaks, low growth rates, inefficient feed conversion, and limited genetic diversity. To address these issues, genetic improvement has become a key focus, enhancing farmed species' resilience, productivity, and sustainability. Traditional breeding methods have driven genetic advancements, yet modern biotechnologies, particularly gene editing, now offer transformative potential (Gjedrem *et al.*, 2012). Gene editing enables precise DNA modifications, enhancing desired traits without adding foreign genes. CRISPR-Cas9 TALENs (Transcription Activator-Like Effector Nucleases) and ZFNs (Zinc Finger Nucleases) are vital tools. Among these, CRISPR-Cas9 has gained particular prominence due to its simplicity, affordability, and high specificity (Doudna & Charpentier, 2014). In aquaculture, gene editing can increase disease resistance, accelerate growth rates, and improve feed conversion efficiency—each vital for production efficiency and environmental sustainability (Yáñez *et al.*, 2020). Despite its promise, gene editing in aquaculture faces technical, ethical, and regulatory challenges. Issues such as off-target effects, public acceptance, regulatory diversity, and animal welfare concerns impact the adoption of gene-edited fish in food production (Van Eenennaam & Young, 2017). This chapter explores gene editing's transformative role in aquaculture, assessing strategies to enhance disease resistance, growth, feed efficiency, and regulatory and ethical considerations. By understanding gene editing's current landscape, stakeholders can evaluate its potential to support sustainable aquaculture and food security worldwide.

Overview of Genetic Improvement in Aquaculture

Genetic improvement has been a crucial component of aquaculture, focusing on enhancing the productivity, resilience, and quality of farmed aquatic species. As aquaculture is expected to meet the world's increasing food demands, genetic improvement in traits like growth rate, disease resistance, reproductive efficiency, and feed conversion becomes increasingly critical. Advances in biotechnology have introduced novel and more precise genetic manipulation techniques, including gene editing, which can accelerate and further refine the process of genetic improvement in aquaculture (Gjedrem & Rye, 2018). This section thus, describes both traditional breeding approaches and modern gene-editing approaches that have shaped and continue to shape genetic improvement in aquaculture.

Traditional Breeding Approaches

Traditional breeding techniques, including selective breeding and hybridisation, have been the basis of aquaculture for centuries. Selective breeding is choosing individuals with desired traits for improvement, such as a high growth rate or disease resistance. These selected individuals are then bred to produce offspring with enhanced characteristics. Improvement by this technique can accumulate over many generations. This practice has been successfully adopted in several aquaculture species, for instance, Atlantic salmon (*Salmo salar*), tilapia (*Oreochromis niloticus*), and common carp (*Cyprinus carpio*) (Gjedrem *et al.*, 2012). However, a high genetic gain has already resulted from traditional breeding in aquaculture, but at a cost. Long generations ago, characterised selective breeding, and quite a long way still stands before the expected genetic changes. Besides this, many desired traits tend to be polygenic; they depend upon numerous genes. Therefore, unpredictable results often arise through selection due to the inability of such procedures to manipulate directly implicated specific genes for those characteristics. This unpredictability, for instance, makes achieving consistent genetic advances by breeding challenging and very restricted in what can be achieved with these programs (Lind *et al.*, 2019). Hybridisation, another means of genetic improvement, is the development of hybrids by inter-breeding individuals from distinct breeds or species to form more desirable hybrids. Although hybridisation could produce high-performing fish, it can also have the adverse effects of reducing genetic diversity and hybrid sterility. Thus, some aquaculture systems often limit their application (Fujiwara *et al.*, 2021).

Genetic Engineering and Gene Editing

Genetic engineering has opened the way to direct manipulation of desirable traits in aquaculture species. Genetic engineering allows foreign DNA to be introduced into the organism by enhancing its traits, particularly disease resistance and growth potential. Transgenic fish results, such as AquAdvantage salmon, are products of genetically modified Atlantic salmon that grows faster through adding a growth hormone gene extracted from Chinook salmon, *Oncorhynchus tshawytscha* (Van Eenennaam & Muir, 2011). Despite the benefits of these kinds of genetic engineering, there are regulatory and public acceptance constraints because of environmental pressure and the perception of GMOs as “unnatural”. Recently, gene editing tools like CRISPR-Cas9, TALENs, and ZFNs have enabled precise genetic alterations without foreign DNA, addressing many transgenic regulatory concerns. Gene editing allows specific genes to be “knocked out” or enhanced, achieving desired traits within a single generation (Doudna & Charpentier, 2014). For instance, CRISPR-Cas9 introduced disease resistance in catfish and increased growth rate by editing genes regulating metabolism (Yáñez *et al.*, 2020). Gene editing has several significant advantages over traditional methods. It develops traits faster, targets specific genes, and is considered more “natural” because it acts within the species’ genome, with the possibility of fewer regulatory hurdles than transgenics (Devlin *et al.*, 2015).

Comparative Analysis of Traditional Breeding and Gene Editing in Aquaculture

Traditionally, breeding and gene editing could be estimated based on their relative impacts on genetic gain, cycle length, and accuracy. Traditional breeding works well but takes time, and predictability uncertainty exists in each breeding cycle. Gene editing dramatically compresses breeding cycles since changes can directly be effectuated at the genetic levels. For example, in an Atlantic salmon breeding program, it may take years to develop measurable differences in growth rates, whereas gene editing could have a similar effect in one generation (Wargelius *et al.*, 2016). Furthermore, the natural genetic variation within a species' population could constrain the range of genetic changes achieved by traditional breeding. On the other hand, gene editing does not have such a limitation because genetic variation already exists and can immediately alter a gene that might control an attribute of interest, be it immune response or metabolic efficiency, even if that particular gene might not have natural variability available within the population (Zhang *et al.*, 2020). This ability makes gene editing exceptionally valuable for improving complex traits controlled by numerous genes or enhancing a trait with slight natural variation. While gene editing can offer some advantages, aquaculture still relies strongly on traditional breeding, at least in many cases, more so as a secondary tool. In most instances, the joint use of selective breeding alongside gene editing maximises genetic gain achieved. The other advantage is selective breeding for managing genetic diversity and upgrading polygenic traits, though gene editing focuses on specific traits with accurate precision. Thus, these combined approaches would result in robust and resilient fish populations adapted for commercial production and responding well to environmental challenges (Lind *et al.*, 2019).

Gene Editing Technologies for Aquaculture

It is a revolution in genetics, as it applied gene editing technologies, providing unprecedented precision in changing an organism's genome. Aquaculture opened new avenues for improving the growth rate, disease resistance, and environmental tolerance determinants for the productivity and sustainability of aquatic animals. The most common tools used for gene editing in aquaculture are CRISPR-Cas9, TALENs, and ZFNs. Each of these is summarised by its mechanism, aquaculture applications, and advantages and disadvantages.

CRISPR-Cas9

Mechanism and Function

CRISPR-Cas9 is the acronym for Clustered Regularly Interspaced Short Palindromic Repeats-associated protein 9. It has become one of the most widely used and versatile gene-editing technologies. Derived from a natural defence mechanism in bacteria, it helps protect the cell from viral attacks by cutting DNA from invading viruses. Scientists utilise CRISPR-Cas9 in the laboratory by introducing cuts into the genome at specified points. It has two main components: Cas9 protein, like molecular scissors, and a guide RNA (gRNA), which guides Cas9 to the genome site where one would like to modify (Doudna & Charpentier, 2014).

Applications in Aquaculture

CRISPR-Cas9 has been widely applied to enhance economically essential traits of fish and other aquaculture organisms. The most frequently targeted area is disease resistance, a prominent challenge for farmed fish populations. For example, CRISPR-Cas9 has been documented in deleting or modulating genes that make it susceptible to viral or bacterial infection in catfish and salmon. Thus far, research has aimed at producing fish resistant to pathogens with an improved immunity response by disrupting or editing such susceptibility genes (Yáñez *et al.*, 2020). Besides improving disease resistance, CRISPR-Cas9 technology has also shown potential in improving growth rates and feed efficiency. For example, growth hormone-regulating genes' editing results in faster-growing fish reaching market size in fewer days, which means higher productivity and reduced feed uptake. CRISPR-Cas9 has also been used to investigate and manipulate genes involved in environmental tolerance, including salinity and temperature tolerance, for farmed fish to tolerate changing or extreme environmental conditions.

Advantages and Challenges

The simplicity, efficiency, and cost-effectiveness make it a powerful tool in handling CRISPR-Cas9. Compared with TALENs and ZFNs, CRISPR is more straightforward to design and use because it needs just one gRNA for a particular gene. However, there are challenges because it risks off-target effects when unintended parts of the genome are edited, particularly on complex traits governed by several genes. Such ethics and regulations surrounding using CRISPR-Cas9 in aquaculture are ongoing. It may raise several issues among some parties concerned about the effects when they enter the wild environment (Hsu *et al.*, 2014).

TALENs (Transcription Activator-Like Effector Nucleases)

Mechanism and Function

TALENs are synthetic enzymes capable of identifying and cutting DNA sequences. TALES derive from plant pathogenic bacteria, and the two primary parts are DNA-binding, which will recognise and bind to the target DNA sequence, and FokI nuclease, which generates a double-strand break in the DNA. Because the binding domain can be engineered to complement the target site, TALENs can target any DNA sequence. Therefore, TALENs are a versatile tool for targeted gene editing (Boch & Bonas, 2010).

Applications in Aquaculture

Other aquaculture species, such as tilapia and carp, can also be engineered using TALENs to introduce traits of alteration in growth, fertility, and tolerance to stress. For instance, sterile fish can be produced by the interruption of genes essential for the development of reproduction. It would be beneficial if farmed fish accidentally escaped into the wild so they do not mate with the wild population. TALENs have also been used in tilapia to knock out specific genes, which is aiding scientists in understanding how some changes in genetics can enhance growth and disease-resistance traits (Liu *et al.*, 2013).

Advantages and Challenges

TALENs are, therefore, well suited for high-precision targeting with minimal off-target effects and, thus, ideal for complex traits. Additionally, TALENs can delete more prominent genes or edit genes in multiples simultaneously, which is essential for aquaculture. Complex multi-gene traits are often of interest when it comes to aquaculture. However, the production of TALENs can be technically more challenging and more expensive than CRISPR-Cas9, leading to some limitations in making TALENs accessible for some researchers. TALENs are labour-intensive in design and construction against specific genes and may extend the experimental timeline in large projects.

ZFNs (Zinc Finger Nucleases)

Mechanism and Function

ZFNs are among the earliest gene-editing technologies and comprise two main components: a DNA-binding domain consisting of zinc finger motifs and a FokI nuclease domain. Each zinc finger motif can recognise a specific DNA triplet, and multiple zinc fingers can be combined to recognise a more extended DNA sequence. When two ZFN constructs bind to opposite strands of a DNA target, FokI domains form a double-strand break that enables the exact DNA editing at that particular site (Kim *et al.*, 1996).

Applications in Aquaculture

ZFNs have been applied in aquaculture, although less frequently than CRISPR-Cas9 or TALENs, primarily because of the complexity of their design and cost. Recent studies used ZFNs in zebrafish (*Danio rerio*) to produce knockouts that can be studied to identify gene function and traits in growth, pigmentation, and immune responses. Less often applied, ZFNs have been used in aquaculture as an entree into how targeted gene modifications could offer specific desired features in a research and developmental scenario (Ekker, 2008).

Advantages and Challenges

ZFNs have high specificity and can make well-suited edits for targeted genetic changes. However, the expertise in developing effective ZFNs is very complex, labour-intensive, and costly compared with CRISPR-Cas9. Furthermore, off-target effects are likely challenging with ZFNs because zinc fingers can bind to the DNA sequences outside the targeted sequence, introducing uncharacterised genetic changes. Hence, ZFNs are less popular than gene-editing tools in aquaculture than other novel techniques.

Applications of Gene Editing in Aquaculture

The ability to gene editing in aquaculture brought revolutionary capabilities, such as precision and efficient modification of genes with various traits, including resisting diseases, growing, adapting to the environment, and controlling reproduction. Researchers could use CRISPR-Cas9, TALENs, and ZFNs to modify specific genes in farmed fish and shellfish and amplify desirable traits or suppress negative ones that impede the production process. It identifies the main applications of gene editing in aquaculture and gives its scope to solving some problems within this industry by making its application more efficient.

Enhancing Disease Resistance

A critical application of gene editing in aquaculture is enhancing disease resistance in farmed species. Infectious diseases pose significant risks in fish and shellfish farming, and treatments like antibiotics can harm sustainability and species' health. Gene editing enables precise modifications of immunity-related genes, building innate resistance in farmed animals. For example, CRISPR-Cas9 has been used to knock out genes linked to pathogen susceptibility in catfish and salmon and edit immune regulation genes to reduce bacterial infection mortality (Yáñez *et al.*, 2020). Similar efforts have targeted immunity genes in shrimp to combat the white spot syndrome virus (Jaroenram & Owens, 2014). Creating disease-resistant lines minimises antibiotic reliance, promoting sustainable aquaculture.

Accelerating Growth and Improving Feed Efficiency

Optimising growth rates and feed efficiency are significant priorities in aquaculture: higher growth rates and better feed conversion lead to higher productivity. Direct modification of growth-related genes would bypass the time-consuming cycles associated with generational generations from selective breeding. Growth hormone-regulating genes have been applied in gene editing in various fish species. For example, CRISPR-Cas9 can introduce a gene knockout in tilapia, a gene that, by its natural function, is a repressor of growth hormone activity. This editing provided rapid growth and feed efficiency; fish could achieve commercial size faster and incur lower-cost feeds (Dong *et al.*, 2019). Other edited genes that impact the metabolic process can help correctly use nutrients in the body, hence efficient feed use. This application of gene editing reduces the environmental footprint of fish farming by minimising feed waste and lowering the demand for wild-caught fish as feed ingredients.

Increasing Environmental Tolerance

Climate change and environmental variability are still significant challenges for aquaculture because temperature fluctuations, changes in salinity, and reduced water oxygen levels have stressed the industry. Gene editing allows the researcher to edit genes linked to physiological responses to stress caused by the environment so that it is possible to produce fish and shellfish with a higher resistance towards diverse and changing conditions. For instance, CRISPR-Cas9 has been employed to edit genes related to thermal tolerance in rainbow trout and improve survival at higher water temperatures (Laghari *et al.*, 2020). Similarly, editing genes controlling osmoregulation- the ability of fish like tilapia to maintain salt and water balance and the tolerance of fish to salty environments opens up new opportunities for aquaculture. This would reduce the climatic factor risk and, conversely, offer possibilities where, otherwise, aquaculture is not possible because of environmental reasons.

Enhancing Reproductive Control and Sterility

Gene editing is also beneficial in reproductive control in aquaculture. Reproductive control will prevent interbreeding between farmed populations and wild populations, and this is especially crucial with genetically modified or gene-edited fish that can impact the gene pool of the wild species. Sterility can be induced using gene editing, so farmed fish do not reproduce if they escape into the natural ecosystem. Sterility can be obtained by gene editing. These comprise removing the genes responsible for the development of the reproductive system. CRISPR-Cas9 is utilised in knockout fertility-related genes for salmon, raising sterile fish with average size but not fertile enough to reproduce (Wargelius *et al.*, 2016). There is an avoidance of adverse ecological consequences, and the energy that would otherwise have been used for reproduction can now be used for growth, resulting in bigger sizes and more yield. It also allows selective breeding programs to focus on growth and health traits without regard for reproductive capability in modified fish.

Improving Flesh Quality and Nutritional Value

Flesh quality is a concern, particularly about fat content, texture, and nutritional profile, directly relating to consumer preference and, therefore, to the value in the marketplace. Gene editing can directly manipulate metabolic pathways that control flesh composition, so aspects of meat quality and nutritional value can be improved, like its content of omega-3 fatty acids. For example, gene editing has been recently used to insert Atlantic salmon for the modification of lipid metabolism in increasing healthy fats, such as omega-3 fatty acids, that are good for the health of humans (Zohar *et al.*, 2022). By adjusting genes controlling the synthesis and deposition of fatty acids, scientists can produce fish with a more advantageous nutritional profile. This application can, therefore, increase the health benefits associated with aquaculture products, thus making them more appealing to the health-conscious consumer

Enhancing Immune Response and Reducing Stress Sensitivity

Some exciting gene editing applications in aquaculture are improved immune response and stress-sensitivity reduction. Fish are subjected to stressful conditions, such as handling, high-density environments, and sudden environmental changes, which weaken their immune system and make them more susceptible to diseases. Modifying genes controlling stress response and immunity will make the fish resistant to such conditions. For example, research on tilapia and carp used CRISPR-Cas9 to knock out the genes of the stress response pathway. The result was decreased cortisol levels and increased resistance to stress (Dai *et al.*, 2021). Gene editing in the form of improvement of immune and stress responses makes fish populations healthier and more robust, thus enhancing animal welfare and reducing outbreaks in cases of disease in the environment of high-density farming.

Functional Genomics and Basic Research

Besides the application of gene editing for the production of market-ready fish, it is applied more broadly in functional genomics research, which allows scientific study and elucidation of the roles of specific genes in aquaculture species. Gene knocking out or modifying enables researchers to analyse genes' functions and effects on complex traits such as growth, reproduction, and immunity. Danio rerio, or zebrafish, has been extensively used as a model organism in aquaculture research in gene function studies with the help of the CRISPR-Cas9 technique. The studies benefit genetic improvement programs and further breeding strategies. Further, gene editing research in functional genomics has enabled the breakthrough toward identifying crucial genes for disease resistance, tolerance for stress, and nutrient utilisation toward a greater understanding of molecular mechanisms for important aquaculture traits (Jin *et al.*, 2018).

Challenges and Ethical Considerations

Gene editing in aquaculture a transformative technology, but it also presents notable challenges and ethical concerns. Key issues include technical limitations, regulatory hurdles, ecological risks, public acceptance, and ethical considerations related to animal welfare and biodiversity.

Technical Limitations and Off-Target Effects

This way, off-target effects influence the health and behaviour of the fish. Therefore, the reasons for precise gene editing arise in this context. A safer tool with reduced risks, such as CRISPR-Cas12a, is not tantamount to technical accuracy regarding safe application (Oakes *et al.*, 2019).

Regulatory and Policy Challenges

Gene editing adoption is hindered by regulatory uncertainty worldwide. For instance, the U.S. and EU regulate gene-edited fish under GMO frameworks requiring extensive testing. Harmonised, clear policies could facilitate innovation and responsible use (van Eenennaam, 2018).

Ecological Risks and Biodiversity Concerns

Released gene-edited fish can potentially harm wild populations by altering gene pools or disrupting the balance of ecosystems. Sterility solutions like CRISPR-induced sterility in salmon minimise ecological impacts, but containment is required to prevent unintended interbreeding (Wargelius *et al.*, 2016).

Public Acceptance and Consumer Perception

Consumer perception about GMOs also reflects public scepticism toward genetically modified fish. The Major elements determining consumer acceptance are transparency in labelling and effective communication of the safety and benefits (Cunningham & Graham, 2021).

Ethical Concerns: Animal Welfare and Intrinsic Value

Rapid growth changes or disease resistance raise ethical concerns about fish welfare and altering natural life forms. These issues can be addressed with humane treatment protocols and balanced benefits against ethical values (Kokko & Koenig, 2020).

Social Equity and Access to Technology

Gene-editing technology will likely favour large-scale aquaculture, increasing disparities with smaller farms. Training and public-private initiatives for equitable access can help grow the industry inclusively (Gjedrem & Rye, 2018).

Potential Long-Term Implications and Unknowns

Gene editing technology is new, and long-term ecological and health impacts are unknown. Continuous monitoring, adaptive policies, and proper research are necessary to mitigate unknown consequences (van Eenennaam, 2018).

Future Directions and Conclusion

Future research in gene editing for aquaculture appears promising as new technologies emerge to help overcome the existing limitations and maximise the benefits. Future research will likely focus more on precision in gene-editing tools to reduce the off-target effects, increasing safety and ethical standards. Innovations of the CRISPR toolkit will include prime editing and base editing, which should open up more targeted and less-invasive methods for the fish genome to almost indefinitely control the specific genetic outcome (Anzalone *et al.*, 2019). Other possible applications involve using gene-editing technology to enhance the survival of the fish in a given environment or nutritional profile by alteration of vital functional genes. This approach to integrating the editing gene into other more popular aquaculture-related technologies like selective breeding or genomic selection will likely accelerate further genetic gain and enhance aquaculture-product availability to respond to fast-rising demands. Meeting this potential demand will surely rely on effective interdisciplinary consultation and coordination involving genetics, ecologists, ethicists, and policymakers; after all, innovation would only succeed by balancing that against sustainability and ethics concerns. End. This means gene editing is a transforming technology if communicated correctly to the public, and by doing further research that provides ecological safety and respects the welfare of animals, a sustainable approach to a very responsible food supply chain would satisfy global demands.

References

Anzalone, A. V., Randolph, P. B., Davis, J. R., Sousa, A. A., Koblan, L. W., Levy, J. M., ... & Liu, D. R. (2019). Search-and-replace genome editing without double-strand breaks or donor DNA. Nature, 576(7785), 149-157.

Boch, J., & Bonas, U. (2010). Xanthomonas AvrBs3 family-type III effectors: discovery and function. Annual review of phytopathology, 48(1), 419-436.

Cunningham, F., & Graham, M. (2021). Public perceptions of gene-edited fish: Communication strategies for consumer acceptance. Aquaculture International, 29(3), 201-214.

Dai, X., et al. (2021). Targeted gene editing for reduced stress response in farmed fish: Applications and impacts. Aquaculture Research, 52, 102-115.

Leggatt, R. A., Sundström, L. F., Woodward, K., & Devlin, R. H. (2017). Growth-Enhanced Transgenic Coho Salmon (Oncorhynchus kisutch) Strains Have Varied Success in

Simulated Streams: Implications for Risk Assessment. PloS one, 12(1), e0169991. https://doi.org/10.1371/journal.pone.0169991

Devlin, R. H. (2015). Potential risks of genetically modified fish escaping into wild populations. Reviews in Fish Biology and Fisheries, 25(1), 1-19.

Dong, Z. (2019). "CRISPR-Cas9 mediated knockout of growth suppressor genes in tilapia enhances growth performance." Aquaculture Reports, 14, 100209.

Doudna, J. A., & Charpentier, E. (2014). The new frontier of genome engineering with CRISPR-Cas9. Science. https://doi.org/1258096

Ekker, S. C. (2008). Zebrafish genome project: Current and future goals. Nature Reviews Genetics, 9(1), 62-68.

Fujiwara, K., (2021). Hybrid vigor and genetic diversity: Opportunities and limitations for improving aquaculture species. Aquaculture Research, 52, 1020-1034.

Gaudelli, N. M., Komor, A. C., Rees, H. A., Packer, M. S., Badran, A. H., Bryson, D. I., & Liu, D. R. (2017). Programmable base editing of A• T to G• C in genomic DNA without DNA cleavage. Nature, 551(7681), 464-471.

Gjedrem, T., & Rye, M. (2018). Selection and breeding programs in aquaculture. Reviews in Aquaculture, 10(2), 318-328.

Gjedrem, T., & Rye, M. (2018). Selection and breeding programs in aquaculture. Reviews in Aquaculture, 10(1), 5-21.

Gjedrem, T. (2012). Genetic improvement for the development of efficient global aquaculture: a personal opinion review. Aquaculture, 344, 12-22.

Hsu, P. D., Lander, E. S., & Zhang, F. (2014). Development and applications of CRISPR-Cas9 for genome engineering. Cell, 157(6), 1262-1278.

Jaroenram, W., & Owens, L. (2014). Improving disease resistance in shrimp through CRISPR-Cas9: Applications and insights. Fish & Shellfish Immunology, 38(1), 271-280.

Jin, Y., (2018). "Functional genomics in aquaculture research: Zebrafish as a model for gene editing." Marine Genomics, 40, 20-30.

Kokko, K., & Koenig, B. (2020). Ethics of rapid growth gene editing in fish: Implications for animal welfare.Animal Welfare Science, 14(2), 110-125.

Laghari, Z. H., (2020). CRISPR-Cas9 technology for thermal tolerance improvement in rainbow trout. BMC Genomics, 21, 52.

Lind, C. E., (2019). Genetic improvement of aquaculture species in developing countries: Prospects and challenges." Frontiers in Genetics, 10, 533.

Oakes, B. L., (2019). Advances in gene editing for precision aquaculture: Reducing off-target effects. Nature Biotechnology, 37(6), 620-625.

Van Eenennaam, A. L. (2018). Regulatory implications of gene editing in aquaculture. Journal of Applied Aquaculture, 30(1), 43-55.

Van Eenennaam, A. L., & Muir, W. M. (2011). Transgenic salmon production and ecological impacts. Nature Biotechnology, 29, 1155-1162.

Wargelius, A., (2016). CRISPR-Cas9 mediated knockout of reproductive genes in Atlantic salmon induces sterility. Scientific Reports, 6, 24272.

Yáñez, J. M., (2020). Opportunities and challenges in applying genome editing for disease resistance in aquaculture. Reviews in Aquaculture, 12(1), 255-273.

Zhang, F., Wen, Y., & Guo, X. (2014). CRISPR/Cas9 for genome editing: progress, implications and challenges. Human molecular genetics, 23(R1), R40-R46.

Zohar, Y., (2022). Improving omega-3 content in farmed salmon through gene editing of lipid metabolism pathways. Aquaculture Nutrition, 28, 384-392.

7

Harnessing Biotechnology for Climate-Smart Aquaculture Adaptation and Mitigation Strategies

Khoob Singh[1*], C.P. Singh[2], Narendra Kumar Maurya[1], Shiwam Dubey[1] and Khusbu Samal[1]

[1]*College of Fisheries, Mangaluru, Veterinary, Animal and Fisheries Sciences University, Bidar, Karnataka (575 002), India*

[2]*College of Fisheries, Kumarganj, Acharya Narendra Deva University of Agricultural and Technology, Ayodhya, Uttar Pradesh (224229), India*

Abstract

Biotechnology has played a very important role in increasing production in the aquaculture sector. The biotechnology field has provided new opportunities to improve the ecological conditions in aquaculture systems, which are negatively impacted by several challenges. Hence, aquatic plants and animals depend on the environment for survival and growth. so, it is very necessary to monitor the environment, because day by day the degradation of the environment is due to pollution and harmful contaminants. The unique role of environmental biotechnology in the aquaculture industry is focused on recognizing chances to provide new technology and ideas for cleaning up polluted areas, reducing future waste discharge, and developing alternatives for control of pollution. Professionals have found that sustainable aquaculture must regulate negative effects on the natural and social environment, as well as minimize the formation of adverse impacts. Aquaculture development must be environmentally sustainable to protect the environment.

Keywords: *Aquaculture, Biotechnology, Climate change mitigation, Sustainable development.*

Introduction

Aquaculture is the farming of aquatic organisms, which includes fish, mollusks, crustaceans, and aquatic plants. Over the past forty years, aquaculture has played a significant role in increasing the global supply of fish for human consumption, leading to a marked rise in worldwide fish production. Nearly half (49.3%) of the world's food fish are currently farmed for aquaculture (FAO, 2023). The proportion of inland fisheries and aquaculture in India's overall fish production has increased from 46% in the 1980s to over 85% in recent years, resulting in the country with the second-largest aquaculture production. Fishing is a thriving industry with a wide range of resources and enormous potential in India. It is also a highly significant economic activity. Aquaculture may be the solution to ensuring global food security because fish can produce significantly more biomass per unit surface area than terrestrial animals. Aquaculture has a significant impact on world food security and is also considered a potential factor for the pollution of vast marine resources. The application of eco-friendly aquaculture methods can help the aquaculture industry thrive sustainably by reducing its negative effects on the environment. In the twenty-first century, biotechnology is becoming a crucial technology for long-term sustainability management and protection (Cantor, 2000).

One of the components of biotechnology that is developing in the context of environmental protection is environmental biotechnology, as the clean environment has been challenged by contemporary lifestyles and increased industrialisation. Chemical engineering gave birth to environmental biotechnology, but other fields such as biochemistry, ecological engineering, environmental microbiology, biology, and ecology have also contributed to its advancement (Hashim and Ujang, 2004).

Role of Biotechnology in Aquaculture

According FAO (2005), biotechnology is the use of biological systems or living organisms in the production process and has a variety of beneficial uses in aquaculture and fisheries. The rising global fish production reached about 185.4 million tonnes in 2024, with aquaculture representing about 49 per cent of the total. Aquaculture has been responsible for the impressive growth in the supply of fish for human consumption (SOFIA, 2018). India, the second-largest country in aquaculture production after China, has a share of inland fisheries and aquaculture in total fish production of about 13.11 million tonnes (Handbook of Fisheries Statistics, 2023). Over 95 per cent of worldwide aquaculture production comes from freshwater aquaculture. With an increment in demand for fish products, capture fisheries harvest reaches to lag phase or may decline as the expansion of the human population is soaring.

The use of biotechnological tools like polymerase chain reaction (PCR) and gene probes is demonstrating its significant potential in fish health analysis and stabilisation. Gene probes and PCR-based diagnostic techniques have been developed for many infections affecting prawns and fish. (Karunasagar and Karunasagar, 1999). Environmental biotechnology is one of the branches of biotechnology that is emerging in the context of environmental conservation. This permits the use of fewer pesticides and antibiotics in the environment. Significant contributions have been made in the past decade in the development of vaccines for fish (Thompson *et al.*, 2004) in addition to diagnostic probes and tests (Cunningham, 2004). There are currently a lot of advanced technologies available to help improve fish health and are beneficial to the aquaculture sector. Advances in biotechnology aimed at preserving and safeguarding the environment have proven advantageous.

Fish, Food Security and Livelihoods

At the global and local scales, fisheries and aquaculture play a significant role in the production of food, also in food security, and provides livelihoods to the people. Due to the high nutritional content of aquatic foods, they are among the most exported and traded food items from many developing nations. For about 4 billion people, the majority of which live in developing nations, they provide at least 15% of their animal protein (FAO, 2012). In certain nations and communities, this figure is significantly greater, and in many small island states, it can make up about 50% of total animal protein. Up to 200 million people are employed in the fishing and aquaculture-related supply, marketing, distribution, and processing sectors, which employ 43.5 million people overall, many of whom are employed in poor nations (Cochrane *et al.*, 2009).

Climate Change Impacts

Acidification, variations in sea temperatures and circulation patterns, the frequency and intensity of extreme events, sea level rise, and related ecological changes are some of the essential factors that impact fisheries and aquaculture. Both direct and indirect effects include effects on the range and productivity of targeted populations, habitats, and food webs, as well as effects on the productivity and expenses of fisheries and aquaculture, as well as the livelihoods and safety of fishing communities (Daw *et al.*, 2009).

Changes in their physical surroundings will have an indirect impact on growth, mortality, and reproduction while feeding, migratory, and breeding behaviour will be directly impacted (Brander, 2010). Additionally, fisheries sustainability may have uncertain effects on the species and ecosystems on which fish depend. Fish species likely shift their distribution, as warm-water species and cold-water species are expected to move towards the warmer and colder poles (Beare *et al.*, 2004a, 2004b). While human activities will hurt fish production, a rise in sea level, storm surges, and also floods can have both good as well as bad effects. However, it will be impacted by the salinization of coastal waterways, water scarcity, and temperature changes.

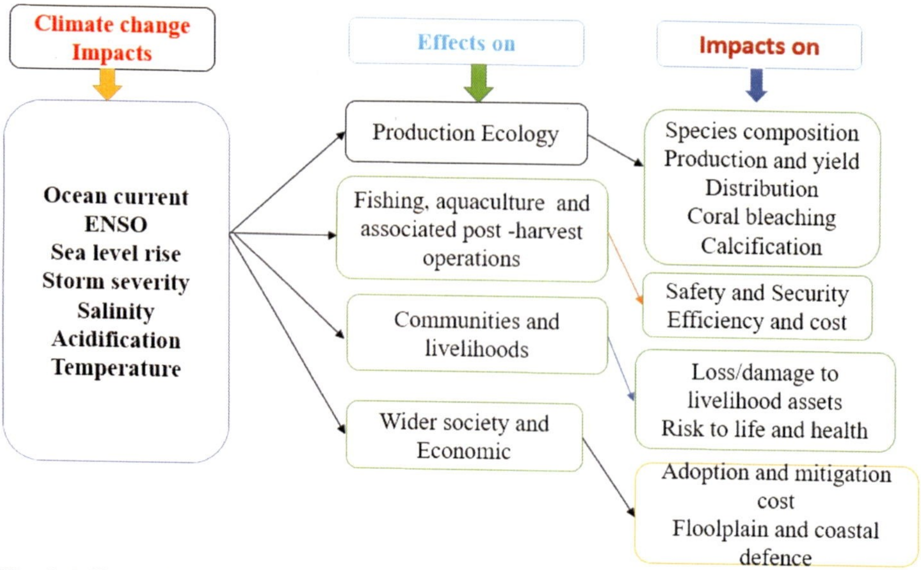

Fig. 1: Effects of climate change on aquaculture and fisheries pathways, ***Source:*** (Badjeck *et al.*, 2010)

Adaptation of Fisheries to Climate Change

Climate Change Adaptation is the process of changing ecological, social, or economic systems in response to real or anticipated changes in climatic stimuli and their impacts to minimize the negative effects of change or take advantage of new opportunities.

To combat climate change, actions must be taken to reduce the CO_2 and other greenhouse gases in the atmosphere. The effectiveness of mitigation and adaptation efforts is dependent upon the context of the activity, the action goals, and the temporal and spatial scale of impacts (Coulthard, 2009). It is thought that proactive responses to the immediate effects of extreme events on communities and fisheries infrastructure, as part of long-term integrated coastal and disaster risk management planning are more effective. The pattern of adaptation might be helpful in conceptualising local fisheries that react to environmental change (Berkes *et al.*, 2001).

Examples of Adaptation in Fisheries

Role of Institutions in Adaptation

The role of institutions mainly the informal types, in promoting or restricting adaptation may be underestimated by a technological approach to adaptation. In some cases, traditional methods or other sources of income can be used to adjust to decreasing fish stocks (Coulthard, 2009).

Around the world, fishermen and their communities are continuously adapting to different types of changes (Coulthard, 2009). Climate change impact pathways are addressed by the following projects and activities, and the following symbols represent the pathways:

Fig. 2: Impact of Climate change

Source: (FAO information products-www.fao.org/publications)

Biotechnological Tools in Climate Change Adaptation and Mitigation

Biotechnological tools play a crucial role in addressing climate change, particularly in the areas of adaptation and mitigation. These tools leverage biological processes and organisms to develop solutions that can reduce greenhouse gas emissions, and enhance ecosystem resilience. some key biotechnological approaches for climate change adaptation and mitigation:

Microbial Solutions for Climate Change Mitigation

Carbon sequestration: It is the absorption of carbon, especially carbon dioxide, from the atmosphere. It aids in absorbing CO_2 from the atmosphere and raising the amount of organic carbon in the soil, which implies that more soil carbon storage slows down global warming. Therefore, sequestering carbon dioxide from the atmosphere is one of the best ways to reduce the effects of climate change.

Algal biofixation: Algae can photosynthesize and absorb CO_2 from the atmosphere, and genetically engineered algae strains may be able to absorb more CO_2 or grow faster, enhancing their potential for carbon capture.

Soil microbes and carbon storage: Certain soil bacteria and fungi can enhance soil's ability to store carbon. Biotechnology can optimize these processes, creating more efficient natural carbon sinks.

Restoration and Conservation of Ecosystems

Bioremediation: Certain microbes can be used to degrade or neutralize pollutants in soils and water. By cleaning up polluted environments (such as oil spills or contaminated soils), biotechnological solutions can restore ecosystems that are vulnerable to climate impacts.

Coral reef restoration: Climate change poses a major threat to coral reefs through warming waters and ocean acidification. Biotechnology is being used to develop heat-tolerant and acid-resistant coral species, which could help preserve these critical ecosystems.

Biochar and Soil Carbon Sequestration

Biochar: Pyrolysis is the process of heating organic material in a low-oxygen environment to produce biochar, a type of charcoal. It can be used as a soil amendment to improve soil fertility and increase the soil's capacity to sequester carbon. Biotechnology can help optimize the production of biochar or enhance its carbon retention capacity.

Genomic Tools for Biodiversity Conservation

Gene Editing for Conservation: Techniques like CRISPR can be used to preserve endangered species by enhancing their resilience to disease, environmental stress, or inbreeding. For example, genetic tools can help save coral reefs by developing more heat-resistant coral species, improving biodiversity conservation efforts.

De-Extinction of Species: Though it is controversial, some scientists are exploring the possibility of bringing back extinct species that could help restore ecological balance in ecosystems affected by climate change. These species could play important roles in biodiversity restoration.

Climate-Smart Aquaculture

Genetic Selection for Resilient Fish: Genetic tools can improve aquaculture practices by producing fish that are more resistant to diseases, tolerate higher temperatures, or grow more efficiently. This can reduce the environmental impact of fish farming and ensure sustainable fish production even as ocean conditions change.

Ocean Algae and Aquaculture Integration: Biotechnology can enhance the integration of algae cultivation with aquaculture. This can provide a source of food for marine species, while also capturing excess nutrients from aquaculture to prevent eutrophication in marine environments.

Environmental monitoring: Advances in biotechnology also enable more precise environmental monitoring, allowing for better understanding and management of ecosystems and carbon stocks. Genetically engineered biosensors can be used to detect pollutants, toxins, or greenhouse gases, providing real-time data for climate mitigation efforts.

Fisheries and Climate Change Mitigation

The Role of Fisheries in Greenhouse Gas Emissions

GHG emissions from fishing operations during the capture process and the transportation, processing, and storage of fish, so are the main factors that contribute to human-induced climate change. As compared to other sectors, fisheries and aquaculture do not produce significant amounts of greenhouse gases. Fisheries and aquaculture don't produce an excessive amount of greenhouse gases in comparison to other sectors. Emissions from fishing activities could be decreased by increasing fuel economy by switching to more efficient equipment or boats, switching to sails, or changing fishing techniques (FAO, 2007).

Fuel costs would decrease even when switching to more fuel-efficient boats or equipment could only result in a 20% decrease in fuel use. Certain coastal ecosystems can store carbon in addition to reducing emissions. Fuel costs would decrease even when switching to more fuel-efficient boats or equipment could only result in a 20% decrease in fuel use. Some of the aquatic ecosystems such as mangroves and salt marshes also provide more benefits to the communities through flood management, buffering coastlines from storms, water quality, and provide habitat for many juvenile fish.

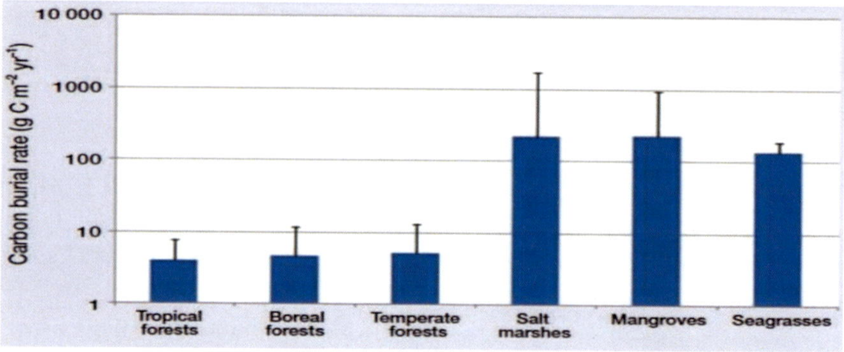

Fig. 3: The estimated quantity of carbon stored in major terrestrial forest systems and "blue carbon" storage regions, *Source:* (McLeod *et al.*, 2011)

Potential contributions from fisheries to mitigation

Utilizing fish processing waste materials to produce biodiesel has been the subject of a little early study. In many cases where there are abundant supplies of fish fats, this could provide substitutes for fossil fuels or terrestrial biodiesels. In common capture fisheries, such uses are unlikely to be acceptable due to the nutritional value of fish, unless there are comparable vast quantities of fish products that would be considered trash.

Sustainable Aquaculture

Gandhi Ji said - "The earth provides every human being's need, but not every human being's greed".

The purpose of sustainable development is to ensure that human needs are fulfilled and will continue to be satisfied for both current and future generations by regulating and preserving the base of renewable resources and guiding institutional and technical progress (FAO, 1995).

Only when environmental conditions—including ecological, socio-anthropological, and economic aspects of the environment—are suitable and sustained can sustainability be achieved.

Criteria of Ecological Sustainability

Like other food production systems, aquaculture's ecological sustainability is endangered by a number of factors. Aquaculture poses the following risks: habitat destruction or modification, excessive freshwater use, organic contamination and inorganic contamination and spread with infected organisms and depletion of wild fish stocks (Leung and Dudgeon, 2008). This was determined by calculating the number of hectares required to produce one unit of fish. Currently, ecological sustainability may be measured by the use of land, the wild fish required to culture the fish, the amount of water consumed that influence on soil and groundwater, and the environmental impact on biodiversity.

Biodiversity

The area of land use, the use of alien or genetically modified species and the utilization of captured fish to feed cultured fish have an impact on biodiversity. Genetically modified organisms (GMOs), hybrids, and non-indigenous species are prohibited by several certification standards. The alteration of aquatic ecosystems by using wild fish, escapees, and water quality ultimately causes adverse effects on biodiversity. The supply of locally pleasant and marketable species can be diminished by these alien species, which can be highly invasive (Casal, 2006). Genetic alteration and selection may intensify the effect (Hallerman and Kapuscinski, 1995).

As available land for nature decreases, the extent of land used for various purposes indirectly affects biodiversity. The density of fish and the depth of the water body influence how much land is needed for effective fish farming. Recommended stocking densities depend on the species' characteristics, such as their tendency for cannibalism, as well as social and welfare considerations. The practice of using fish from capture fisheries to produce farmed fish poses challenges to sustainable aquaculture. Additionally, the quality of protein and fatty acids in fish diets significantly impacts the health benefits associated with consuming fish (Cheung and Sumaila, 2008).

Fig. 4. Biodiversity flourishing in our ecosystem
Source: (https://easydrawingguides.com/how-to-draw-biodiversity)

Implementation

The ability of the country that establish a program and effectively manage aquatic resources is the basis for the aquatic resource management program's success. Monitoring and assessment of the use of land, their selections as well as modifications to aquatic resources, should be established. In coastal areas, they support sustainable aquaculture development and environmental protection by offering continuous monitoring and management. Environmental groups, authorities, and aquaculture and other coastal sectors have acknowledged the importance of BMPs. Site planning and ICZM practice BMPs offer chances for early intervention and cooperative evaluation of novel initiatives.

Environmental Impact Assessment

It is necessary to give stakeholders and decision-makers comprehensive information regarding the environmental effects of aquaculture. Every stage of aquaculture development programs should include impact evaluations. Additionally, it must offer unambiguous choices for impact reduction and ecologically responsible management.

Policy framework and regulatory measures

This is an essential tool for education and training, as well as for involving the local community in decision-making. A state or country needs to assess and control aquaculture practices that affect the sustainable use of renewable resources and should be mentioned in the policy statement (PAP/RAC, 1996). The policy framework is going to transform into an important tool for educating and training stakeholders, scientists, resource managers, decision-makers, and the general public.

Recommendations

According to national and international initiatives, it encourages sustainable aquaculture. The following important rules and guidelines need to be considered and put into effect:

- Developing integrated plans for rural and community management along the coast (IUCN, 2000). Creating aquaculture suitability criteria and indicators such as using benthic fauna.
- Establishing systems for monitoring and evaluation (M&E).
- Low trophic level fish should be farmed using more broad and responsible methods by the rapidly expanding finfish aquaculture sector (Stickney and McVey, 2002)
- Government incentives for ecologically harmful firms should be eliminated, and laws protecting coastal and ocean ecosystems should be implemented (fish and agriculture should be no exception).

For aquaculture to flourish sustainably, environmental site suitability indicators must be developed and put into use. Indicators, which might be physical, biological, chemical, social, or economic, are important variables that indicate change. They can be computed from measurements of various data sets, directly measurable, derived from many sources (derived indicators) (NRC, 2000).

For environmental monitoring of impact and carrying capacity modelling at farm sites, surveys should make the best methods and technologies available (Frankic, 1998). The latest techniques and technology should be used in environmental monitoring surveys for carrying capacity modelling and environmental monitoring of impact at farm locations.

References

Badjeck, M-C., Allison, E. Halls, A. & Dulvy, N. (2010). Impacts of climate variability and change on fishery based livelihoods. Marine Policy, 34: 375–383.

Berkes, F., Mahon, R., McConney, P., Pollnac, R. & Pomeroy, R. (2001). Managing small scale fisheries: alternative directions and methods. Ottawa, International Development Research Centre.

Casal, C.M.V., (2006). Global documentation of fish introductions: the growing crisis and recommendations for action. Biological Invasions 8, 3–11.

Cheung, W.W.L., Sumaila, U.R., (2008). Trade-offs between conservation and socio-economic objectives in managing a tropical marine ecosystem. Ecological Economics 66, 193–210.

Cochrane, K., De Young, C., Soto, D., & Bahri, T. (2009). Climate change implications for fisheries and aquaculture. FAO Fisheries and aquaculture technical paper, 530, 212.

Coulthard, S. (2009). Adaptation and conflict within fisheries: insights for living with climate change. In W.N. Adger, I. Lorenzoni, & K. O'Brien, eds. (2009 in press) Adapting to climate change: thresholds, values, governance. pp 255-268. Cambridge, UK, Cambridge University Press

Cunningham, C.O. (2004). Use of molecular diagnostic tests in disease control: making the leap from laboratory to field application. In Molecular Aspects of Fish and Marine Biology: Current Trends in the Study of Bacterial and Viral Fish and Shrimp Diseases, 3: 292–312.

Daw, T., Adger, W.N., Brown, K. & Badjeck, M.-C. (2009). Climate change and capture fisheries: potential impacts, adaptation and mitigation. In K. Cochrane, C. De Young, D. Soto & T. Bahri, eds. Climate change implications for fisheries and aquaculture: overview of current scientific knowledge, pp. 107–150. FAO Fisheries and Aquaculture Technical Paper No. 530. Rome, FAO. 212 pp. (also available at www.fao.org/docrep/012/i0994e/i0994e00.htm).

FAO (1995). Code of Conduct for Responsible Fisheries. Food and Agriculture Organization of the United Nations, Rome, 41 p.

FAO (2012). The State of World Fisheries and Aquaculture – 2012. Rome. 209 pp. (also available at http://www.fao.org/fishery/sofia/en).

FAO (2024). The state of world fisheries and Aquaculture 2024. Fisheries and Aquaculture Department of the Food and Agriculture Organization (FAO) of the United Nations, Rome.

Frankic A. (1998). A framework for planning sustainable development in coastal regions: an island pilot project in Croatia. Dissertation. Virginia institute of Marine Science, College of William & Mary. http://www.vims.edu/library/Frankic/Frankic.pdf.

Hallerman, E.M., Kapuscinski, A.R., (1995). Incorporating risk assessment and risk management into public policies on genetically modified finfish and shellfish. Aquaculture 137, 9–17

Handbook of Fisheries Statistics. 2018. Published by Department of fisheries, Ministry of Fisheries, Animal Husbandry and Dairying, Government of India.

Hashim, M.A. and Uijang, Z. (2004). Environmental biotechnology: its relevance and prospects for developing countries. In: Vjang Z, Menze M (Eds) Environmental Biotechnology IWA Publishing, 7-12.

IUCN (2000). Coastal and Marine Biodiversity: policy recommendations. Fifth Meeting of the Subsidiary Body on Scientific, Technical and Technological Advice. Montreal, Canada, 31 January–4 February.

Karunasagar, I. and Karunasagar, I. (1999). Diagnosis treatment and prevention of microbial diseases of fish and shellfish. Curr. Sci., 76: 387-399.

Leung, K.M.Y., Dudgeon, D., (2008). Ecological risk assessment and management of exotic organisms with aquaculture activities. In: BondaReantaso, M.G., Arthur, J.R., Subasinghe, R.P. (Eds.), Understanding and Applying Risk Analysis in Aquaculture. FAO Fisheries and Aquaculture TP 519, pp. 67–100.

Nicholls, R.J., Wong, P.P., Burkett, V.R., Codignotto, J.O., Hay, J.E., McLean, R.F., Ragoonaden, S. & Woodroffe, C.D. (2007). Coastal systems and low-lying areas. In M.L. Parry, O.F. Canziani, J.P. Palutikof, P.J. v. d. Linden & C.E. Hanson, eds. Climate Change 2007: impacts, adaptation and vulnerability, pp. 315–356. Contribution of working group II to the Fourth Assessment Report of the Intergovernmental Panel on Climate Change. Cambridge, UK, Cambridge University Press.

NRC (2000). Clean Coastal Waters: Understanding and Reducing the Effects of Nutrient Pollution. National Academy Press, Washington, DC, 75 p

PAP/RAC (1996). Approaches for zoning of coastal areas with reference to Mediterranean aquaculture. PAP-10/EAM/GL.1. Split, Croatia.

Phillips, M.J., Subasinghe, R.P., (2008). Application of risk analysis to environmental issues in aquaculture. In: Bonda-Reantaso, M.G., Arthur, J.R., Subasinghe, R.P. (Eds.), Understanding and Applying Risk Analysis in Aquaculture. FAO Fisheries and Aquaculture TP 519, pp. 101–119.

Stickney R.R. and McVey J.P. (eds) (2002). Responsible Marine Aquaculture. CAB International, Wallingford, UK.

Thompson, K.D. and Adams, A. (2004). Current trends in immunotherapy and vaccine developme

8

Molecular Breeding in Fisheries and Aquaculture

Pragati Inwati[1*], Vaijnath Aitwar[1], Rohini Swami[1], Sweksha Chauhan[2] and Durgesh Kumar Verma[3]

[1]College of Fisheries, Mangalore, Karnataka Veterinary, Animal and Fisheries University, Bidar, Karnataka (575 002), India

[2]College of Fisheries, Ratnagiri, Dr. B.S. Konkan Krishi Vidyapeeth, Dapoli Maharashtra (415 712), India

[3]ICAR-Central Inland Fisheries Research Institute, Regional Centre, Prayagraj Uttar Pradesh (211 020), India

Abstract

Applying molecular breeding techniques and relevant indicators in aquaculture and fisheries is vital in advancing genetic enhancement and fostering sustainability within these industries. Genetic variation is essential for species adaptation and survival in changing environments, facilitating differentiation across taxonomic levels. Such diversity yields valuable data for research in evolution, conservation, resource management, and genetic enhancement initiatives. The advent of molecular genetic markers has significantly advanced genetic studies at both individual and population levels. Researchers can more effectively analyze genetic diversity by combining these markers with sophisticated statistical methods. Numerous genetic markers are often used in the fisheries and aquaculture industries to monitor and control aquatic populations. These indicators include randomly amplified polymorphic DNA (RAPD), microsatellites, single nucleotide polymorphisms (SNPs), and mitochondrial DNA (mtDNA). Their uses are numerous, from determining variation in genetics and population structure to assisting with selective breeding and conservation initiatives. They support critical applications such as species identification, genetic variation studies, comparisons between wild and hatchery populations, demographic bottleneck assessments, and rehabilitation efforts. This review emphasizes the foundational concepts of molecular genetics, frequently utilized markers, and their uses and constraints in the discipline.

Keywords: *Molecular markers, SNPs, Mitochondrial DNA, RAPD, Microsatellite DNA*

Introduction

Aquaculture has emerged as the fastest-growing sector in global agriculture. The total fish production worldwide includes both wild catches and aquaculture outputs. In recent years, fish production has consistently risen, reaching around 130.9 million tons, which accounts for 52% of the fish available for human consumption (FAO, 2024). Molecular breeding in fisheries and aquaculture utilizes cutting-edge genetic technologies to enhance favourable traits in aquatic species. This approach encompasses the application of molecular markers, genome-wide association studies (GWAS), and gene editing techniques to improve attributes such as growth rates, disease resistance, and resilience to environmental stressors. For example, marker-assisted selection (MAS) facilitates the identification and selection of superior broodstock by leveraging genetic markers associated with advantageous traits.

Furthermore, whole-genome sequencing (WGS) and the development of high-density genetic maps are employed to elucidate the genetic underpinnings of economically significant traits. These innovations not only enhance production efficiency but also promote the sustainability and profitability of the aquaculture sector. The distinct biological characteristics of each species influence the fitness and survival of organisms within specific environments. Genetic variation within species and populations is crucial for adapting to constantly changing environmental conditions. A comprehensive understanding of genetic and morphological diversity is vital for preserving ocean resources, studying patterns of adaptive evolution, and managing marine ecosystems effectively. Successful fisheries management relies on accurately identifying fish species to protect biodiversity. Understanding stock structures is key to implementing targeted management approaches. Recent advancements in molecular techniques have facilitated the precise identification and mapping of stock structures, empowering conservationists to develop more effective strategies for management and conservation (Sukumaran, 2023). Fish mutants arise from regular cellular processes and environmental interactions, resulting in genetic variation (polymorphism). This genomic variation enhances a species' ability to adapt to changing conditions, which is critical to its survival and overall fitness.

Genetic diversity and evolutionary processes like genetic drift and natural selection contribute to the differences observed among individuals, populations, and broader taxonomic groups. Molecular genetic markers are key for identifying genetic variations across individuals, populations, or species, significantly enhancing our understanding of genetic diversity (Hillis *et al.*, 1996). The development and statistical analysis of molecular markers have transformed genetic diversity research. Various molecular markers, including proteins, mitochondrial DNA (mt-DNA), and nuclear DNA markers like microsatellites, SNPs, and RAPD, are widely used in fisheries studies. These markers support various essential scientific investigations related to aquaculture, such as research on genetic diversity and population dynamics in wild populations, identifying species, evaluating demographic bottlenecks in natural populations, contrasts between wild and hatchery-raised populations, and assistance for rehabilitation programs that rely on propagation (Haldar, 2020). Progress in genetic modification and molecular biology has made creating enhanced fish species with favourable characteristics easier. The phrase "molecular breeding" describes breeding methods that utilize genetic engineering techniques. Compared to traditional selection methods, molecular breeding offers significant advantages for aquaculturists, particularly for traits with low heritability, such as growth rate in fish. This approach aids in identifying existing genetic variations in broodstock, which can be leveraged to enhance the target species' desirable characteristics. By integrating traditional selection with molecular tools, improving performance traits becomes more precise and efficient, enabling a deeper understanding of the genetic mechanisms that govern these traits (Haldar, 2020).

Classification of Molecular Markers

Molecular markers are classified into two types: Type I and Type II. Type I markers are linked to genes with known functions, while Type II markers are associated with genes whose functions are not fully understood. Allozyme markers fall under Type I as they are related to proteins with specific functions. In contrast, microsatellites and other neutral markers are considered Type II unless they are linked to genes with known functions (O'Brien, 1991).

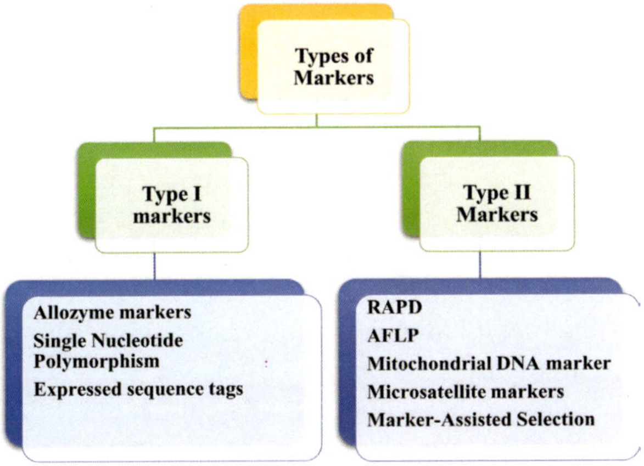

Fig. 1: Different Types of Molecular Markers used in fisheries

Type I Markers

Allozyme markers

Allozyme markers belong to the Type I category since they are linked to genes with clearly defined roles. These markers originate from DNA, and variations in enzyme activity can be observed through allozyme electrophoresis, a method that reveals genetic differences at the enzyme level. Allozymes are protein variants caused by differences in the electric charge of alleles, which can be detected through electrophoresis. According to Mendelian inheritance, these markers are codominant, revealing different traits in heterozygous individuals. Allozyme analysis provides valuable insights into genetic variation at specific loci, helping explore intraspecific and interspecific diversity in fish populations. They are extracted from tissues and analyzed using electrophoresis on acrylamide or cellulose acetate gels to identify allozyme differences. A single band indicates homozygosity, while two bands indicate heterozygosity (Chauhan & Rajiv, 2010). Allozymes have been widely employed in numerous studies because they are straightforward and economical, allowing for the analysis of any soluble protein. Moreover, various loci can be examined simultaneously utilizing allozyme markers. Nonetheless, allozyme analysis comes with notable constraints, including requiring a considerable quantity of tissue, which poses difficulties when applied to smaller species, such as larvae. The process of sampling tissue involves an invasive procedure, necessitating the euthanization of the fish and the later cryogenic preservation of the samples. Additionally, protein electrophoresis cannot identify point mutations in nucleotide sequences since these mutations might not lead to changes in the amino acid structure. Although there are limitations, allozymes remain an important resource in fishery biology, aiding research in fields like fish systematics, genetic structure of populations, conservation genetics, and forensic studies.

Single Nucleotide Polymorphism (SNP)

(SNPs) exhibit polymorphism due to point mutations resulting from adding or deleting a single base at a specific nucleotide position within a locus. SNPs have gained prominence in the development of molecular markers because they are abundantly found throughout the genome of any organism, encompassing both coding and non-coding regions. This widespread presence allows SNPs to uncover hidden polymorphisms often not detectable using other markers and methods (Rasal *et al.*, 2016). SNPs are abundantly found in the non-coding regions of the genome. SNPs can be classified as either non-synonymous, leading to an amino acid change, or silent within coding regions. This does not alter the amino acid sequence but may influence phenotypic variations by affecting mRNA splicing (Richard & Beckman, 1995). SNPs play a crucial role in developing genetic linkage maps and identifying quantitative trait loci (QTL) associated with important traits such as growth, body weight, grilse development, thermal tolerance, low oxygen tolerance, and disease resistance. They are also important for locating sex-determination loci, recognizing offspring in the selection and chromosome manipulation studies, and evaluating genomic and marker-assisted selection in aquaculture. Genome-wide association studies (GWAS) allow for the discovery of links between SNPs and particular traits in related and unrelated individuals.

Nevertheless, numerous characteristics are intricate and can be affected by various QTLs. Genotyping using reduced complexity sequencing has become a valuable method for genomic selection. Identifying genes, sequences, and nucleotide variations that directly affect phenotypic traits significantly enhances the potential for marker-assisted selection in aquaculture (Wenne 2018).

Expressed sequence tags (ESTs)

Expressed Sequence Tags (ESTs) are single-stranded sequences obtained by randomly sequencing complementary DNA (cDNA) clones (Adams *et al.*, 1991). They are used to identify genes and study their expression through expression profiling. They facilitate quick and informative evaluations of genes expressed in specific tissues, under certain physiological states, or at various stages of development. ESTs are also crucial for developing cDNA microarrays, which facilitate the study of genes with varying expression levels (Wang *et al.*, 1999) and play a key role in genome mapping (Boguski & Schuler, 1995). In genome mapping, ESTs are vital for both linkage and physical mapping in animal genomics, particularly in cattle and swine, where radiation hybrid panels map non-polymorphic DNA markers (Cox *et al.*, 1990). These panels consist of hybrid cell lines containing segments of irradiated chromosomes from the studied species, created by exposing the target species' cells to radiation to produce small chromosome fragments.

ESTs are instruments for identifying genes and analyzing their expression. They facilitate swift and precise evaluations of genes that are expressed in specific tissues, under certain physiological states, or at different stages of development. Moreover, cDNA microarrays can be employed to methodically detect genes with differential expression. ESTs are significant for linkage mapping in genomic research.

Type II Markers

Random amplified polymorphic DNA (RAPD) markers

RAPD markers originate from the amplified products of regions in the genome that are less critical in function and do not substantially impact phenotypic selection. These areas of DNA can accumulate more nucleotide mutations, making them valuable for evaluating genetic variation between populations (Mamuris *et al.*, 2002). Utilizing random nucleotide sequence primers for the PCR amplification of genomic DNA, RAPD can demonstrate significant levels of DNA polymorphism (Welsh & McClelland, 1990). This method identifies both coding and non-coding DNA sequences, with many informative polymorphic sequences arising from the genome's repetitive (non-coding) areas (Williams *et al.*, 1990). Considering that around 90% of the nuclear genome in vertebrates consists of non-coding sequences, it is probable that most amplified loci do not undergo selective pressure. RAPD loci follow Mendelian inheritance and are evaluated as being either present or absent. They provide all the benefits of PCR-based markers, such as the accessibility of commercially available primers, without needing a prior understanding of the target DNA sequence or genome organization. RAPDs also offer the advantage of assessing numerous loci and individuals simultaneously. Nonetheless, this type of marker has limitations, including difficulties in showcasing Mendelian inheritance and distinguishing between homozygotes and heterozygotes. The assessment relies on the premise that the populations under investigation adhere to Hardy-Weinberg equilibrium. Additionally, challenges like the occurrence of paralogous PCR products (various DNA regions that can be mistaken for a single locus due to their similar lengths) and inconsistent results stemming from the low annealing temperatures utilized during PCR amplification have limited the use of RAPDs in fisheries science (Wirgin & Waldman, 1994).

Amplified fragment length polymorphism (AFLP)

Amplified Fragment Length Polymorphism (AFLP) analysis is a powerful, PCR-based DNA fingerprinting method that provides an efficient, reliable, and cost-effective way to study population genetics. In contrast to environmental and mitochondrial markers, AFLPs are nuclear DNA markers that follow Mendelian inheritance patterns. Compared to other nuclear markers like RFLP and RAPD, AFLPs exhibit significantly higher polymorphism and broader genomic coverage. Furthermore, AFLP has the potential to surpass microsatellites in population genetic research because it can examine hundreds of loci at once. Recognizing that AFLP markers are passed down as dominant markers is crucial, and care must be taken when sharing information between different laboratories. The need for specialized apparatus, including sequencers, could also restrict its extensive application. Despite this, the benefits of the strong multilocus AFLP analysis—such as significant polymorphism and broadband sharing—support the determination of relationships between populations. Significantly, AFLP (similar to RAPD) does not necessitate a prior understanding of the genome, which makes it appropriate for conducting population genetic studies on any species. Considering these advantages, the use of AFLP in studies of fish population genetics is becoming more prevalent, and its broader acceptance in this area is expected to grow (Liu, 2005).

Mitochondrial DNA marker

Mitochondrial DNA (mtDNA) can be found in the cytoplasm in structures called mitochondria, which categorizes it as a type of non-nuclear DNA. It is classified as a haploid genome passed down from the mother and encoded as a single entity. Mitochondrial DNA (mtDNA) does not undergo recombination, is neutral in selection, and is present in several copies within every cell. Isolating mitochondrial DNA from tissue or blood specimens is relatively easy, as it is physically separate from the nuclear DNA of the cell (Boore, 1999). Using mtDNA helps identify population bottlenecks and hybridization events, as its adequate population size is smaller than nuclear DNA due to maternal inheritance. Molecular taxonomic methods based on mtDNA use RFLP-based length polymorphisms or sequence variations to analyze genetic diversity. RFLP analysis involves digesting mitochondrial DNA with restriction enzymes to produce length polymorphisms, which are then visualized through electrophoresis. This technique has been widely applied to study species-specific trends in marine fish (Cronin *et al.*, 1993).

Advancements in sequencing technologies have made mitochondrial DNA sequencing increasingly popular. Universal primers designed from conserved regions allow for use across different tissue types. Slowly evolving gene regions are used for interspecific comparisons, while rapidly evolving regions are preferred for intraspecific studies. The Cytochrome C Oxidase 1 gene (about 600 bp) is a common species identifier due to its high conservation across taxa. The D-loop region, a non-coding segment, evolves rapidly and is helpful for intraspecific and population-level comparisons. Other regions, such as cytochrome b, ND-1, and ND-5/6, are also used for intraspecific analysis. Mitochondrial DNA plays a key role in fish systematics and population genetics studies.

Microsatellite markers

Microsatellites consist of numerous repetitions of simple sequence repeats (SSRs) arranged in tandem, generally varying in length from 1 to 6 base pairs, like ACA or GATA (Litt and Luty, 1989). These patterns are commonly found in all species studied up to this point, with estimates indicating they appear as often as once every 10 kb in fish (Wright, 1993). Microsatellites are typically found throughout the genome and on every chromosome; however, recent whole genome sequencing data has uncovered some intricacies regarding this distribution. They may be found in regions that code for genes (Liu *et al.*, 2001), within introns, and in non-coding sequences. Most microsatellite loci are small, comprising only a few to several hundred repeat units. Changes in the quantity of these repeat units can result in considerable variation in the alleles present at each microsatellite locus within a population. The inheritance of microsatellites exhibits a Mendelian pattern, and they function as codominant markers. The usefulness of these markers has been especially significant in species that show minimal variation compared to other markers. However, using microsatellite markers requires a significant initial outlay of funds since to create efficient PCR primers, each locus must be located and its surrounding areas sequenced. Furthermore, tiny variations (as low as two bp in di-nucleotide repeats) may occur between alleles due to polymerase slippage during replication. As a result, microsatellites are now a prevalent marker type in many genetic research projects (Askari *et al.*, 2013).

Marker-Assisted Selection (MAS)

Marker-assisted selection (MAS) is an indirect method for identifying superior breeding individuals. MAS is particularly advantageous for traits that are challenging or costly to measure and those with low heritability or recessive characteristics. This approach leverages genetic diversity within breeding populations to enhance desirable traits in livestock. MAS relies on establishing a link between genetic markers and Quantitative Trait Loci (QTL), where the distance between the marker and the target traits influences their association. Selective breeding programs can pick brooders with better genetic potential for the desired characteristics by using markers linked to QTL after they have been found. MAS accelerates the improvement of performance traits, providing a more precise method and deeper insights into the genetic mechanisms that influence these traits (Eze, 2019). In fisheries and aquaculture, microsatellite markers are valuable instruments for several purposes, such as phylogenetic analysis, biodiversity conservation, and the evaluation of genetic population structure. They are also essential for phylogeographic research and comprehending how hybridization and stocking affect fish populations. Microsatellites may also map genomes, identify family and behavioural patterns, and forensically identify individuals. Because of their adaptability, they are crucial for improving aquatic environment research and management techniques.

Importance of Marker-Assisted Selection (MAS) in Fish Breeding

Candidate gene association studies examine the relationship between genetic variations in specific target genes and performance traits in fish. This approach aims to boost aquaculture yields, enhance farmer incomes, and improve food security. While numerous selective breeding programs have been implemented in fish, optimizing desired traits can be time-consuming (Haldar, 2018). Molecular marker-assisted selection involves identifying DNA markers associated with desirable traits in fish. Pathogen resistance, growth rates, meat quality, and other qualitative and quantitative features may be included in this category. The ability of fish breeders to choose suitable markers for certain features, allowing for early identification and selection before breeding, is a significant benefit of employing molecular markers.

Quantitative Trait Loci (QTL)

The genetic gains achievable through marker-assisted selection (MAS) can be enhanced by continuously identifying new quantitative trait loci (QTL) associated with targeted traits. Nevertheless, it can be challenging to forecast the pace of new QTL discovery, and the extra genetic benefits from MAS often fade quickly in subsequent generations of selection for the same QTL. Creating selection criteria that connect the genetic data obtained from identified QTL to phenotypic variance is crucial. MAS efficiently identifies major QTLs for characteristics that are costly to assess phenotypically. QTLs for sex determination, cold and salinity tolerance, disease resistance, and growth have all been mapped in different fish species thus far. Such studies are helpful for marker-assisted selection in selective breeding programs and offer important insights into evolutionary processes. In aquaculture, mapping genes and QTLs is a fundamental and practical problem. QTL mapping aims to make selective breeding easier by clarifying how particular genes affect and contribute to different characteristics. More than 20 aquaculture species, such as finfish, mussels, and crustaceans, have undergone QTL analyses for important traits like bacterial cold-water disease, infectious hematopoietic necrosis, the development of cells stress response, tolerance to salinity, spawning timing, embryonic development, thermal tolerance, and spawning timing (Haldar, 2020).

Table 1. Classification of Quantitative Trait in aquaculture species

Quantitative Traits	**Species**	**Reference**
Body size	Prawn	(Li *et al.*, 2006)
Growth rate	Arctic charr	(Tao and Boulding, 2003)
Fat percentage and distribution	Atlantic salmon	(Derayat *et al.* 2007)
Fork length	Rainbow trout	(Perry *et al.*, 2005)
Body length	Tilapia	(Cnaani *et al.*, 2003)
Embryonic developmental rate	Rainbow trout	(Robison *et al.*, 2001)
Body weight	Tilapia	(Cnaani *et al.*, 2003)
Caudal length	Asian sea bass	(Wang *et al.*, 2011)
Fecundity	Japanese flounder	(He *et al.*, 2008)
Qualitative Traits		
Body color	Tilapia	(Lee *et al.*, 2005)
Numbers of vertebrae, lateral line, fin rays and gill rakers	Rainbow trout	(Nichols *et al.*, 2004)
Sex maturation	Arctic charr	(Moghadam *et al.* 2007
Flesh color	Atlantic salmon	(Derayat, 2009)
Skin pigmentation	Atlantic salmon	(Boulding *et al.*, 2008)

Emerging Molecular Markers in Fisheries and Aquaculture

Numerous DNA markers have been created, including single nucleotide polymorphisms (SNPs), mitochondrial DNA (mt-DNA), RAPDs, microsatellites, and allozymes. These markers have shown significant genetic diversity throughout the genomes and among fish populations. Recent initiatives aim to enhance the development of DNA markers, genome mapping, and species identification. Notable advancements have been achieved in developing DNA barcoding for several aquaculture species.

DNA barcoding

The preservation and control of biodiversity is the primary goal of conservation biology. Nonetheless, there are several important obstacles in this subject. Accurately evaluating biodiversity to find and rank regions with high species richness is a significant challenge (Dobson *et al.*, 1997). Determining which evolutionary lineages are particularly valuable or urgently need conservation work is another (Moritz, 2002). Particularly vulnerable to extinction are species belonging to understudied taxonomic groupings, as information gaps might result in what is sometimes called a "conservation penalty" (McKinney, 1995). Moreover, the difficulty of international conservation efforts is increased by the enormous number of species that are yet unknown and unidentified.

DNA barcoding is a quick, effective, and economical way to inventory biodiversity. It entails sequencing a section of the mitochondrial gene cytochrome oxidase I (COI) comprising around 600 base pairs. This technique employs universal polymerase chain reaction (PCR) primers to amplify and sequence the COI gene fragment. The resulting sequences are then compared to an existing database of known sequences from previously identified specimens using distance-based algorithms. Although DNA barcoding focuses on a small portion of the mitochondrial genome, it offers a promising approach to assessing at least a basic level of biodiversity. In well-studied groups, particularly birds and mammals, molecular analyses based on these barcoded sequences have uncovered cryptic DNA lineages and provided valuable insights (Hebert *et al.*, 2004).

DNA microarrays or DNA chips

A DNA microarray is a tool made of glass slides, silicon chips, or nylon membranes, each containing a grid of immobilized DNA fragments arranged in a specific pattern. This platform allows for comparing DNA from a sample with a known reporter probe sequence. It can be used to identify species by incorporating species-specific DNA sequences. A fluorescent marker is applied to the target sample's DNA, which is then hybridized to the microarray. The presence of a fluorescent signal, detected through scanning or imaging, indicates successful hybridization (Sukumaran and Gopalakrishnan, 2019).

Table 2. Major attributes of molecular markers commonly used in fisheries and aquaculture genetics, *Source:* (Park and Moran 1995; O'Connell and Wright 1997; Parker *et al.,* 1998; Sunnucks 2000)

Marker	General ease	PCR	Expression	Loci	Genome	No. loci readily available
Multilocus mini/ or microsatellite 'fingerprint nuclear RELPs'	Extraction, digestion, electrophoresis, gel blot hybridization; analysis problematic extraction, digestion, electrophoresis, library screening, gel blot hybridization, analysis straightforward	No No	Dominant Co-dominant	Multilocus Single	Nuclear Nuclear	Many Many
RFLPs	extraction, digestion, electrophoresis, gel blot hybridization; analysis straightforward	No	Varies / Codominant	Single	Organellar	Single
mtDNA sequence	extraction, PCR, prepare DNA template, electrophoresis; analysis straightforward	Yes	Varies / Codominant	Single	Organellar	Single
Allozymes	electrophoresis; analysis straightforward	No, protein	Co-dominant	Single	~Nuclear	Moderate
RAPDs/AFLP	extraction, PCR; analysis problematic but developing	Yes	Dominant	Multilocus	~Nuclear	Many
VNTR: minisatellite VNTR: microsatellite	Mostly similar to microsatellites Primer development may be long extraction, PCR, electrophoresis; analysis straightforward	Few Yes	Co-dominant Co-dominant	Single Single	Nuclear Nuclear	Moderate Many
Anonymous scn	Similar microsatellite	No	Co-dominant	Single	Nuclear	Many

DNA Markers: Applications in Fisheries and Aquaculture

Inter-specific and intra-specific variations

Molecular genetic markers serve as a valuable complementary tool, enhancing the precision and accuracy of taxonomic studies. Molecular evolution varies considerably among different taxa, and the degree of DNA or gene divergence can be effectively used to distinguish between species. Morphological and ecological traits often evolve at a faster rate than genetic variations at neutral loci, leading to a weak correlation between physical characteristics and genetic divergence. Molecular marker proves particularly valuable in identifying species within mixed catches, especially when morphological identification poses challenges. This is especially true for processed fish products such as fillets, smoked, or salted types—in addition to the early phases of fish life, such as planktonic eggs and larvae, which are often difficult to distinguish based on physical characteristics. In these cases, molecular markers serve as essential tools for accurate species identification. Molecular markers have proven effective for sub-species identification as well. Certain highly variable loci exhibit greater divergence within species, making them valuable for detecting intra-specific variations. These markers can thus facilitate a deeper understanding of genetic diversity within species.

Population Structure: Intra- and Inter-Population Dynamics

Understanding stock structure is crucial for effective fisheries management and conservation. Stocks refer to reproductively isolated subgroups within a species that exhibit distinct physiological and behavioral traits. These morphological and meristic traits are influenced by a combination of genetic and environmental factors, making it essential to integrate these characteristics with genetic data for a comprehensive analysis (Guo *et al.*, 2015). By utilizing a combination of approaches, researchers can identify actual genetic variations while assessing the substantial influence of environmental factors on phenotypic traits. Microsatellite DNA is widely regarded as the marker of choice for investigating stock structure due to its high evolutionary rate. In addition, mixed fisheries, consisting of subgroups from various populations, life stages, or distinct stocks, require detailed information from microsatellite markers, allozymes, and mitochondrial DNA markers to trace the origins of stock components within these mixed-stock fisheries (Norris *et al.*, 1999).

Genetic tagging/marking

Marking individual fish is important for tracking migration, estimating population size, and evaluating stock contributions in mixed fisheries. Unlike physical tags, which are non-heritable, genetic marking provides a more effective and enduring solution. By identifying rare alleles and monitoring their presence over generations, researchers can gain insights into hatchery programs' effects, stocked fish's role in target populations, and the identification of migrant fish from their source populations.

Forensic Investigations in Fisheries

When morphological identification is not possible, such as with dead, stranded, preserved, or processed fish, molecular markers offer a reliable method for identification and certification. Forensic techniques, which use scientific methods to investigate past events, are increasingly essential for verifying fishery products and detecting illegal trade. Additionally, molecular tools are key in monitoring intentional and unintentional releases into natural water bodies (Allendorf *et al.*, 2010).

Investigating Trophic Relationships in Ecosystems

A key component of ecological research is understanding trophic relationships, with dietary composition being essential. Identifying species' diets can be difficult, as digestive processes often obscure morphological features. When solid remains are absent, molecular techniques provide a helpful alternative, allowing DNA to be extracted from partially digested samples for diet analysis.

Ancient DNA Analysis: Revealing Historical Ecologies

Advances in extracting DNA from preserved samples, such as those in museums, fossils, and archaeological artefacts, have made it increasingly possible. These studies have greatly improved our understanding of evolutionary relationships among taxa, offering valuable insights into biodiversity and evolutionary history.

Aquaculture Genetics: Fish Breeding & Disease Management

Molecular markers play a crucial role in aquaculture by enabling the genetic identification and differentiation of hatchery stocks. They are instrumental in detecting inbreeding, assigning offspring to their parentages through genetic tagging, and identifying quantitative trait loci (QTLs) linked to commercially important traits. These markers facilitate marker-assisted selection in breeding programs, assess the effects of polyploidy induction and gynogenesis, and provide insights into genetic variability among stocks. Furthermore, they allow for tracking parental contributions during mass spawning events and support the mapping of genomes to enhance breeding strategies. In addition to their role in breeding, molecular markers significantly improve disease diagnosis in aquaculture. Cost-effective and user-friendly PCR assays have been developed for detecting various pathogens, such as white spot syndrome virus (WSSV) and viral hemorrhagic septicemia virus (VHSV). Moreover, identifying genes associated with pathogen resistance, like Major Histocompatibility Complex (MHC) genes, empowers breeders to create resilient strains of finfish and shellfish, ultimately enhancing production and sustainability in the industry.

Conclusion

In summary, molecular breeding is a pivotal advancement in fisheries and aquaculture, providing essential tools to enhance genetic improvements and tackle sustainability challenges. Using molecular markers and advanced sequencing techniques, researchers can make targeted breeding decisions promoting desirable traits such as growth, disease resistance, and environmental adaptability. This approach accelerates the breeding process and helps maintain genetic diversity, leading to more resilient aquaculture systems. The collaboration between molecular biologists, geneticists, and aquaculture practitioners is crucial for identifying traits that align with ecological and economic objectives. As we continue to leverage these molecular tools, we can enhance food security and support sustainable seafood production, ensuring the long-term health of aquatic ecosystems.

References

Adams, M. D., Kelley, J. M., Gocayne, J. D., Dubnick, M., Polymeropoulos, M. H., Xiao, H., ... & Venter, J. C. (1991). Complementary DNA sequencing: expressed sequence tags and human genome project. Science, 252(5013), 1651-1656.

Allendorf, F. W., Hohenlohe, P. A., & Luikart, G. (2010). Genomics and the future of conservation genetics. Nature reviews genetics, 11(10), 697-709.

Askari, G. H., Shabani, A., & Miandare, H. K. (2013). Application of molecular markers in fisheries and aquaculture.

Boguski, M. S., & Schuler, G. D. (1995). Establishing a human transcript map. Nature genetics, 10(4), 369-371.

Boore, J. L. (1999). Animal mitochondrial genomes. Nucleic acids research, 27(8), 1767-1780.

Boulding, E. G., Culling, M., Glebe, B., Berg, P. R., Lien, S., & Moen, T. (2008). Conservation genomics of Atlantic salmon: SNPs associated with QTLs for adaptive traits in parr from four trans-Atlantic backcrosses. Heredity, 101(4), 381-391.

Chauhan, T., & Rajiv, K. (2010). Molecular markers and their applications in fisheries and aquaculture. Advances in Bioscience and Biotechnology, 1(04), 281-291.

Cnaani, A., Hallerman, E. M., Ron, M., Weller, J. I., Indelman, M., Kashi, Y., & Hulata, G. (2003). Detection of a chromosomal region with two quantitative trait loci, affecting cold tolerance and fish size, in an F2 tilapia hybrid. Aquaculture, 223(1-4), 117-128.

Cox, D. R., Burmeister, M., Price, E. R., Kim, S., & Myers, R. M. (1990). Radiation hybrid mapping: a somatic cell genetic method for constructing high-resolution maps of mammalian chromosomes. Science, 250(4978), 245-250.

Cronin, M. A., Spearman, W. J., Wilmot, R. L., Patton, J. C., & Bickham, J. W. (1993). Mitochondrial DNA variation in chinook (Oncorhynchus tshawytscha) and chum salmon (O. keta) detected by restriction enzyme analysis of polymerase chain reaction (PCR) products. Canadian Journal of Fisheries and Aquatic Sciences, 50(4), 708-715.

Derayat, A. (2009). Detection of QTL affecting flesh quality traits (body lipid percentage and flesh colour) using molecular markers (microsatellites and AFLP markers) in Atlantic salmon (Salmo salar L.).

Derayat, A., Houston, R. D., Guy, D. R., Hamilton, A., Ralph, J., Spreckley, N., & Haley, C. S. (2007). Mapping QTL affecting body lipid percentage in Atlantic salmon (Salmo salar). Aquaculture, 272, S250-S251.

Dobson, A. P., Rodriguez, J. P., Roberts, W. M., & Wilcove, D. S. (1997). Geographic distribution of endangered species in the United States. Science, 275(5299), 550-553.

FAO. (2024) The State of World Fisheries and Aquaculture 2024 – Blue Transformation in action. Rome. https://doi.org/10.4060/cd0683en

Guo, B., DeFaveri, J., Sotelo, G., Nair, A., & Merilä, J. (2015). Population genomic evidence for adaptive differentiation in Baltic Sea three-spined sticklebacks. BMC biology, 13, 1-18.

Haldar, C. (2018). Progress and promises of candidate gene association studies for improvement of fish complex traits. International Journal of Fisheries and Aquatic Studies. 2018a, 6(1), 162-165.

Haldar, C. (2020). Molecular marker and its application in aquaculture: New opportunities and challenges. International Journal of Fauna and Biological Studies, 7(2): 41-45.

Hallerman, E. M. (2006). Use of molecular tools for research and improvement of aquaculture stocks. Israeli Journal of Aquaculture-Bamidgeh, 58.

He, F., Wen, H. S., Dong, S. L., Shi, B., Chen, C. F., Wang, L. S. & Zhou, Y. G. (2008). Identification of single nucleotide polymorphism cytochrome P450-c19a and its relation to reproductive traits in Japanese flounder (Paralichthys olivaceus). Aquaculture, 279(1-4), 177-181.

Hebert, P. D. N., Stoeckle, M. Y., Zemlak, T. S., & Francis, C. M. (2004). Identification of birds through DNA barcodes. PLoS biology, 2(10), e312.

Hillis, D. M. (1996). Applications of molecular systematics: The status of the field and a look to the future. Molecular systematics.

Lee, B. Y., Hulata, G., & Kocher, T. D. (2004). Two unlinked loci controlling the sex of blue tilapia (Oreochromis aureus). Heredity, 92(6), 543-549.

Lee, B. Y., Lee, W. J., Streelman, J. T., Carleton, K. L., Howe, A. E., Hulata, G., & Kocher, T. D. (2005). A second-generation genetic linkage map of tilapia (Oreochromis spp.). Genetics, 170(1), 237-244.

Li, Y., Dierens, L., Byrne, K., Miggiano, E., Lehnert, S., Preston, N., & Lyons, R. (2006). QTL detection of production traits for the Kuruma prawn Penaeus japonicus (Bate) using AFLP markers. Aquaculture, 258(1-4), 198-210.

Litt, M., & Luty, J. A. (1989). A hypervariable microsatellite revealed by in vitro amplification of a dinucleotide repeat within the cardiac muscle actin gene. American journal of human genetics, 44(3), 397.

Liu, Z. J. (2005). Amplified fragment length polymorphism (AFLP). In Stock Identification Methods (pp. 389-411). Academic Press.

Liu, Z., Li, P., Kocabas, A., Karsi, A., & Ju, Z. (2001). Microsatellite-containing genes from the channel catfish brain: evidence of trinucleotide repeat expansion in the coding region of nucleotide excision repair gene RAD23B. Biochemical and biophysical research communications, 289(2), 317-324.

Mamuris, Z., Sfougaris, A. I., Stamatis, C., & Suchentrunk, F. (2002). Assessment of genetic structure of Greek brown hare (Lepus europaeus) populations based on variation in random amplified polymorphic DNA (RAPD). Biochemical Genetics, 40, 323-338.

McKinney, M. L. (1999). High rates of extinction and threat in poorly studied taxa. Conservation biology, 13(6), 1273-1281.

Moghadam, H. K., Ferguson, M. M., & Danzmann, R. G. (2007). Linkage variation at the sex□ determining locus within Fraser strain Arctic charr Salvelinus alpinus. Journal of Fish Biology, 71, 294-301.

Moritz, C. (2002). Strategies to protect biological diversity and the evolutionary processes that sustain it. Systematic biology, 51(2), 238-254.

Nichols, K. M., Wheeler, P. A., & Thorgaard, G. H. (2004). Quantitative trait loci analyses for meristic traits in Oncorhynchus mykiss. Environmental Biology of Fishes, 69, 317-331.

Norris, A. T., Bradley, D. G., & Cunningham, E. P. (1999). Microsatellite genetic variation between and within farmed and wild Atlantic salmon (Salmo salar) populations. Aquaculture, 180(3-4), 247-264.

O'connell, M., & Wright, J. M. (1997). Microsatellite DNA in fishes. Reviews in fish biology and fisheries, 7, 331-363.

Park, L. K., & Moran, P. (1994). Developments in molecular genetic techniques in fisheries. Reviews in fish biology and fisheries, 4, 272-299.

Parker, P. G., Snow, A. A., Schug, M. D., Booton, G. C., & Fuerst, P. A. (1998). What molecules can tell us about populations: choosing andusing a molecular marker. Ecology, 79(2), 361-382.

Perry, G. M., Danzmann, R. G., Ferguson, M. M., & Gibson, J. P. (2001). Quantitative trait loci for upper thermal tolerance inoutbred strains of rainbow trout (Oncorhynchus mykiss). Heredity, 86(3), 333-341.

Robison, B. D., Wheeler, P. A., Sundin, K., Sikka, P., & Thorgaard, G. H. (2001). Composite interval mapping reveals a major locus influencing embryonic development rate in rainbow trout (Oncorhynchus mykiss). Journal of Heredity, 92(1), 16-22.

Sukumaran, S. (2023). Genetic and genomic tools in fisheries and aquaculture.

Sukumaran, S., & Gopalakrishnan, A. (2019). Applications of Molecular Markers in Fisheries and Aquaculture.

Sunnucks, P. (2000). Efficient genetic markers for population biology. Trends in ecology & evolution, 15(5), 199-203.

Tao, W. J., & Boulding, E. G. (2003). Associations between single nucleotide polymorphisms in candidate genes and growth rate in Arctic charr (Salvelinus alpinus L.). Heredity, 91(1), 60-69.

Wang, C. M., Lo, L. C., Zhu, Z. Y., Pang, H. Y., Liu, H. M., Tan, J., & Yue, G. H. (2011). Mapping QTL for an adaptive trait: the length of caudal fin in Lates calcarifer. Marine Biotechnology, 13, 74-82.

Wang, K., Gan, L., Jeffery, E., Gayle, M., Gown, A. M., Skelly, M., & Mulligan, J. (1999). Monitoring gene expression profile changes in ovarian carcinomas using cDNA microarray. Gene, 229(1-2), 101-108.

Wenne, R. (2018). Single nucleotide polymorphism markers with applications in aquaculture and assessment of its impact on natural populations. Aquatic Living Resources, 31, 2.

Williams, J. G., Kubelik, A. R., Livak, K. J., Rafalski, J. A., & Tingey, S. V. (1990). DNA polymorphisms amplified by arbitrary primers are useful as genetic markers. Nucleic acids research, 18(22), 6531-6535.

Wirgin, I. I., & Waldman, J. R. (1994). What DNA can do for you. Fisheries, 19(7), 16-27.

Wright, J. M. (1993). DNA fingerprinting of fishes.

9

Genome Sequencing in Aquaculture

Foram D. Vala*, Vishal K. Solanki, Nilesh H. Joshi and Ridhdhisa R. Barad

College of Fisheries Science, Veraval, Kamdhenu University, Gandhinagar, Gujarat (362 265), India

Abstract

Genome sequencing has transformed aquaculture by enabling precise genetic insights into key species. Sequencing aids in selective breeding initiatives meant to increase sustainability and productivity by locating genes associated with characteristics like growth, disease resistance, and stress tolerance. Moreover, genome sequencing improves resistance through marker-assisted selection and helps comprehend disease causes. It also monitors genetic variety, which is crucial for conservation. Programs for selective breeding that aim to increase sustainability and productivity can be developed due to these insights. The biological sciences have been transformed by the quick development of sequencing technology, which has given aquaculture the means to overcome these challenges and preserve its sustainability and economic viability. Recent advances in next-generation sequencing technologies have made genome sequencing more accessible and affordable. We anticipate that the genetic development of farmed fish and the sustainable use of fisheries resources will be significantly aided in the coming decades by applying growing intelligence systems and genomic techniques in aquaculture and fisheries.

Keywords: *Aquaculture, Genome, Sequencing, Selective breeding, Marker-assisted selection*

Introduction

Teleost fish, the largest group of vertebrates, comprise about 30,000 species on Earth, while shellfish include over a million species (Nelson, 2006). About 400 of them have been cultivated, according to Gjedrem and Robinson (2014). One of the most environmentally friendly ways that people may get protein is through fish farming. Over the past few decades, aquaculture has grown rapidly among all agricultural sectors. Fish aquaculture, however, is confronted with several significant obstacles. Most cultivated species have not undergone genetic modification through breeding and are still in their natural state. Although there are a few species-specific feeds, most farmed species do not have tailored diets. Pathogens such as bacteria, parasites, and viruses lead to various fish diseases, resulting in significant financial losses in aquaculture (Meyer and Schnick 1989). Diseases caused aquaculture to lose billions of dollars annually in revenue worldwide. Aquaculture has had significant issues due to aquatic toxicants (Dunier and Siwicki 1993), which has led to environmental problems in the aquaculture sector (Cao *et al.*, 2007). Innovative solutions and technology are needed to address these issues to make aquaculture advantageous and sustainable.

Some of these issues can be resolved with the help of sequencing technology and genomics advancements, which can also help aquaculture and sustainable fisheries (Liu 2010; Bernatchez *et al.,* 2017; Yue and Wang 2017; Canario 2019). Because NGS technologies make it possible to quickly and effectively identify genetic variation inside and between individuals, biological science has revolutionized. The study of fish biology is increasingly using NGS. Recent applications of next-generation sequencing (NGS) technologies on fish species have aimed to address various aquaculture challenges. Significant advancements include the identification of genetic markers linked to commercially desirable traits (Yue, 2014), the exploration of molecular mechanisms underlying numerous physiological processes (Xia *et al.*, 2011; Yasuike *et al.*, 2016), and the investigation of biodiversity in environmental samples, including both prokaryotic and eukaryotic organisms (Hajibabaei *et al.*, 2011). Additionally, researchers have characterized the innate immune system (Ao *et al.*, 2015), identified beneficial probiotics and antimicrobial peptides (Ghanbari *et al.*, 2015), and enhanced our understanding of the relationships between genomes and genes. Furthermore, studies have contributed to our knowledge of how dietary components influence metabolism (Bonacic *et al.*, 2016). Nevertheless, aquaculture applications of NGS technologies lag far behind those in cattle and model organisms because of the variety of fish species grown and the scarcity of financing for aquaculture species.

Evolutionary Process of Sequencing

Finding the precise arrangement of nucleotides within a nucleic acid polymer (DNA or RNA) is accomplished through sequencing. Sanger sequencing, which was developed in the 1970s, is still commonly used today. The subsequent advances to the Sanger method resulted in the creation of ABI DNA sequencing technologies (MacKenzie and Jentoft 2016). Ever since the publication of the human draft genome (Lander *et al.,* 2001), DNA sequencing technology has evolved drastically and rapidly. For more than two decades, sanger's method of DNA sequencing was the preferred approach, it was used to sequence whole genomes of various creatures, such as humans. However, this technology has inherent constraints such as slow speed, high costs, and labour intensity (Heather and Chain 2016). NGS technology has transformed biological research since 2005 by increasing yield by several orders of magnitude and reducing expenses per base. (Margulies *et al.*, 2005; Shendure *et al.*, 2005; Metzker 2010). As the cost of NGS decreases and pipelines for bioinformatics analysis improve, whole genome sequencing is being used more and more in aquaculture.

Tens of millions of nucleotides can now be sequenced in a single run due to NGS technologies, which were developed to overcome the limitations of first-generation sequencing technologies (Sanger sequencing) (Heather and Chain 2016; Shendure *et al.,* 2017). Sequencing methods of the second generation comprise several novel approaches, including pyrosequencing, which was introduced by 454 Life Sciences in 2005 (Margulies *et al.,* 2005); bridge amplification-based sequencing, which was developed by Solexa and subsequently purchased by Illumina (Ruparel *et al.,* 2005; Seo *et al.,* 2005); and sequencing by oligonucleotide ligation and detection (SOLiD), which was developed by Life Technologies and subsequently included in the portfolio of Thermo Fisher Scientific (Heather and Chain 2016). The so-called "genomics revolution," which has been sparked by second-generation sequencing technologies, has greatly advanced research in several disciplines, including aquaculture and fisheries. Third-generation sequencing methods were created to get over their inherent limitations, which include low read lengths and lengthy sequencing runtimes. SMRT (Single Molecule Real-Time) sequencing by PacBio (Pacific Biosciences) and Oxford Nanopore Technologies (ONT) are recent developments in sequencing technology. With comparatively short runtimes, these platforms allow for the creation of lengthy reads (Astier *et al.,* 2006; Eid *et al.,* 2009). Notwithstanding these benefits, third-generation sequencing methods' primary disadvantage is their poor accuracy.

Nevertheless, this problem might be solved by increased sequencing coverage. Combining second- and third-generation sequencing technologies, including PacBio and Illumina, appears to be a viable approach for sequencing non-model species (Wang *et al.,* 2019). Hi-C technology and other fourth-generation sequencing technologies have emerged, allowing for the direct sequencing of nucleic acids in tissue and cells by combining NGS technologies with conventional imaging analysis methods (Mignardi and Nilsson 2014; Ke *et al.*, 2016). One example of this technology is in situ sequencing. The yellow catfish, a type of freshwater fish raised for commercial purposes in Asia, particularly in Southern China, has had its chromosomal-level genome assembled through the application of Illumina, PacBio, and Hi-C technologies (Gong *et al.*, 2018) the terms high-throughput and massively parallel sequencing are also frequently used to refer to technologies that can process millions of sequence reads at once in short periods, while the process of clonally amplifying target DNA to provide a strong enough signal for detection during sequencing is known as clonal sequencing (Kumar and Kocour, 2017).

History of Genome Sequencing in Fisheries and Aquaculture

Fugu rubripes, a Japanese pufferfish, was the first fish whose entire genome was sequenced. China and Japan have both engaged in the commercial cultivation of this food fish (Froese and Pauly, 2010). ABI DNA sequencing systems have been used to sequence the genomes of humans and several model fish species (MacKenzie and Jentoft 2016), such as the spotted green pufferfish (*Tetraodon nigroviridis)* (Jaillon *et al.*, 2004), the torafugu (*Takifugu rubripes*) (Aparicio *et al.*, 2002), the medaka or Japanese rice fish (*Oryzias latipes*) (Kasahara *et al.*, 2007), or zebrafish (*Danio rerio*) (Howe *et al.*, 2013). Since the advent of massively parallel sequencing methods Over 200 fish genomes were sequenced in 2005, and are currently available for public access through repositories (Heather and Chain 2016). The NCBI Genome database has over 270 completed fish genomes as of December 21, 2019. Significant progress has been made in basic fields like comparative genomics, evolutionary biology, and systematics due to these genomic resources. They have also been used practically in aquaculture and fisheries management (Roest Crollius and Weissenbach 2005; MacKenzie and Jentoft 2016; Hughes *et al.*, 2018; Bian *et al.*, 2019).

Sequencing Projects

- A team of biologists and genomic experts launched the genomic 10K Project in 2009 intending to fully sequence 10,000 vertebrate species' genomes, including over 4,000 fish species. This project aims to improve our knowledge of vertebrate evolution and help save many endangered species (Haussler *et al.*, 2009; Bernardi *et al.*, 2012; Koepfli *et al.*, 2015).
- To advance biodiversity study and conservation, the EBP (Earth BioGenome Project) plans to sequence all eukaryotic species' genomes within ten years, including the more than 34k fish species listed in FishBase (Lewin *et al.*, 2018).
- The ICSASG (International Cooperation to Sequence the Atlantic Salmon Genome) is a group of scientists working on the Atlantic

Salmon Genome Sequencing Project, which Canada, Norway, and Chile support.

- A Norwegian group has sequenced the genome of Atlantic cod, which is around a third of the genome's size of Atlantic salmon. The Roche 454 FLX Titanium platform, a second-generation sequencing technology, was used to do whole genome shotgun sequencing in conjunction with paired-end sequencing.
- The Beijing Genomics Institute, the Institute of Oceanology (Chinese Academy of Sciences), and the international Oyster Genome Consortium, including American scientists, are working together to perform genome sequencing on the Crassostrea gigas (Pacific oyster).
- The tilapia genome sequencing project funded by the NIH (National Institute of Health), is being conducted by the broad institute of MIT (Massachusetts Institute of Technology) and Harvard University.
- Shrimp genome sequencing is presently being carried out by the International Shrimp Genome Sequencing Consortium (Iqbal *et al.*, 2023).

An Intelligent and Integrated Aquaculture and Fisheries System

Due to the rapid advancement of technologies for next-generation sequencing and the increased interest in sequencing the genomes of species important to the economy, ecology, or evolution, Draft genomes of hundreds of fish are now accessible in public repositories. There are 265 fish species with genomic data, 64 of which have chromosome-level genome sequences, and the bulk of fish genomes are kept in the U.S. NCBI (National Centre for Biotechnology Information) genome database. Ensembl is another significant genome resource that has made over 60 fish genomes publicly available (https://uswest.ensembl.org/index.html). Since the NCBI and Ensembl genome databases are created to serve a wide variety of research communities, they might not adequately meet the unique requirements of the fisheries and aquaculture industries. This gap has been filled by the development of several species-specific genome resources that offer access to gene expression data, linkage mapping, and genomic data. Notable examples are the Salmonid species molecular data database SalmoBase (Samy *et al.,* 2017) and the Grass carp Genome Database (GCGB) (Chen *et al.,* 2017).

There is currently no database system for fish that combines information about phenotype, genetics, and environment. A comprehensive big data platform encompassing the main species involved in fisheries and aquaculture operations is essential (Chen *et al.*, 2019). A database system, known as Ag-BioData, is used in agriculture and has the potential to be implemented in fisheries and aquaculture. Standardizing protocols and practices will improve the results of genomics, genetics, and breeding research (Harper *et al.,* 2018). As data science and AI (artificial intelligence) advance, the aquaculture and fisheries communities must work together more closely. These industries should collaborate to create cloud-based initiatives that address intricate problems like data integration, exchange, and utilization in aquaculture. They should also support technological advancements like the IoT (Internet of Things) for Aquaculture 4.0 for (Dupont *et al.*, 2018) (Figure 1).

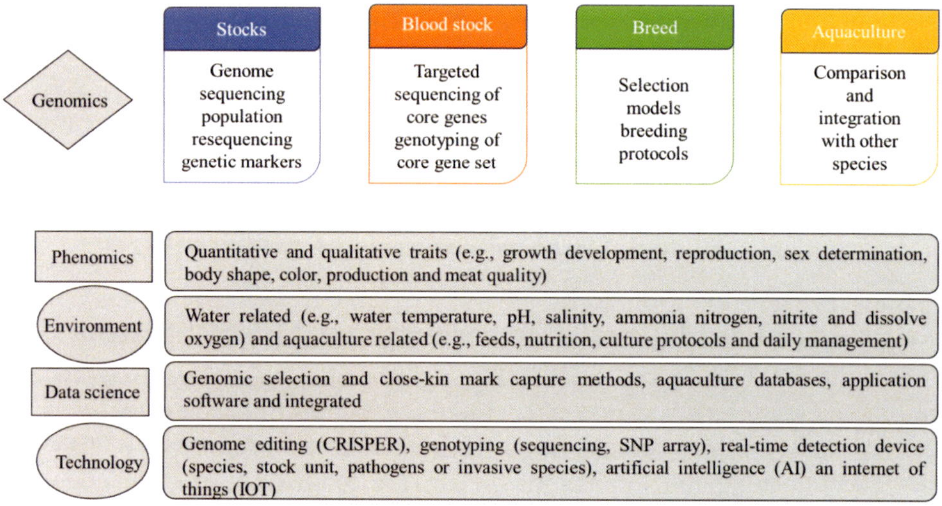

Fig. 1: A suggested pipeline that incorporates phenomics, genetics, and other fields for selective breeding of fish, *Source:* (Lu and Luo, 2020).

NGS Applications

The identification and characterization of genetic markers, including microsatellites and SNP (Single nucleotide polymorphisms), throughout the genome and transcriptome of non-model fish species with limited genomic resources, has been completely transformed by the introduction of NGS technologies. Several freshwater fish species (Carvalho *et al.,* 2011; Luo *et al.,* 2012; Sahu *et al.,* 2012; Wang *et al.,* 2012; Xu *et al.,* 2013; Sahu *et al.*, 2014; Yu *et al.,* 2014; Villanova *et al.,* 2015) and marine fish species (e.g., Reid *et al.,* 2012; Slattery *et al.,* 2012; Yin *et al.,* 2012; Fernandez-Silva *et al.,* 2013; Schultz *et al.,* 2013; Taguchi *et al.*, 2013; Barnes *et al.,* 2014; AlMomin *et al.,* 2015; Bayona-Vasquez *et al.,* 2015; Santos *et al.,* 2015) have also had polymorphic microsatellite markers created using NGS techniques. Similar to this, NGS has been effectively used in several studies to identify single nucleotide polymorphisms (SNPs) (Seeb *et al.,* 2011; Shen *et al.,* 2012; Houston *et al.,* 2012; Carlsson *et al.,* 2013; Montes *et al.,* 2013; Houston *et al.,* 2014; Xiao *et al.,* 2015).

The goal of sequencing the genomes of every fish species has advanced significantly in the last few years. The NCBI database (http://www.ncbi.nlm.nih.gov) indicates that the genomes of several fish species have been assembled and sequenced (Table 1). The possibility of identifying genes that contribute to better performance qualities is one of the key benefits of having a complete genome sequence. Through the application of marker-assisted selection, these genes can be employed in selective breeding projects. Additionally, whole genome sequencing facilitates the discovery of SNPs, a type of genetic variation that is primarily responsible for behavioural variations across individuals of the same species. Researchers can now address issues related to wild animal conservation and aquaculture sustainability because of the understanding of SNPs.

Table 1. Assembling the genomes and annotating the genes of the valuable fish in aquaculture and global fisheries

Species	**Genome size**	**GenBank assembly ID**	**References**
Atlantic cod (*Gadus morhua)*	824.3Mb	CAEA01000001/ CAEA01554869	(Breadbury *et al.,* 2010; Star *et al.,* 2011)
Fugu (*Takifugu rubripes*)	390 Mb	GCA_000180615.2	-
Nile tilapia (*Oreochromis niloticus*)	1.01 Gb	GCA 000188235.1	(Conte *et al.,* 2017)
Platyfish (*Xiphophorus maculatus*)	653 Mb	GCA 000241075.1	-
Spotted gar (*Lepisosteus oculatus*)	946 Mb	GCA 000242695.1	-
Stickleback (*Gasterosteus aculeatus*)	460 Mb	GCA_000180675.1	(Glazer *et al*., 2015; Peichel *et al.,* 2017)
Tetraodon (*Tetraodon nigroviridis*)	340 Mb	GCA_000180735.1	(Jaillon *et al.,* 2004)
Zebrafish (*Danio rerio*)	1.4 Gb	GCA 000002035.3	(Howe *et al.,* 2013)
Common carp	1.7 Gb	GCA_018340385.1	(Xu *et al.,* 2014)
Grass carp	893.2 Mb	PRJNA496416	(Wu *et al.,* 2022).
Asian seabass	670 Mb	SRP069848	(Domingos *et al*., 2015 Vi *et al.,* 2016)
Tilapia	1000 Mb	GCA_001858045.3	(Brawand *et al.,* 2014)
Atlantic salmon	2970	GCA_000233375.4	(Davidson *et al.,* 2010; Lien *et al.,* 2016)

Advances in Fish Immunology Research

NGS technologies enabled the identification of genes involved in innate host defense against a pathogen and innate immune response. Using microarray and NGS techniques, for instance, Ordas *et al.* (2011) examined the gene responses of *Salmonella typhimurium* infection in zebrafish embryos, revealing genes related to innate immunity that are expressed both with and without a fully functional adaptive immune system. Zhang *et al.* (2011) used RNA-seq and whole genome sequencing data to discover the innate immunity genes in *Cyprinus carpio.* Furthermore, the effectiveness of an oral vaccine against viruses that cause fish sickness is assessed using NGS technology (Sarropoulou *et al.,* 2012). It has also been used to examine the level of significant genetic diversity in MHC (major histocompatibility complex) antigen-presenting genes and their potential interactions with the microbiota (Wegner *et al.*, 2012). The study's findings revealed limited immunogenetic diversity among marine fish populations. Another study used NGS technology to identify how pathogenic and nutritional factors control the transcription of the mucin gene (Perez-Sanchez *et al.,* 2013). These investigations lead us to the conclusion that NGS technologies are an invaluable resource for identifying and characterizing transcripts related to pathogen and vaccine-induced immunity and identifying the genes implicated in the immune response. The development of effective vaccinations for fish species that are significant to the fish industry can benefit from this knowledge.

Pathogen Identification

According to Robinson *et al.* (2014), Through the use of high-throughput sequencing, find individuals of farmed rohu (*Labeo rohita*) that are resistant to *Aeromonas hydrophila*, the agent that causes aeromoniasis. Using next-generation sequencing, Oetama *et al.* (2016) found frequent viral illnesses and possible bacterial infections that impact prawns. Significant fish and prawn pathogens can now be compared genomically because to the successful completion and publication of these two species' whole genome sequences recently (Wang *et al.,* 2009; Williams *et al.,* 2012). *Flavobacterium psychrophilum* and *Flavobacterium branchiophilum* have had their genome sequences completely sequenced and made public (Duchaud *et al.,* 2007; Touchon *et al.,* 2011). High adhesion to fish epithelial tissues, gliding movement, fast tissue disintegration, and the formation of severe muscular ulcers are the hallmarks of *F. psychrophilum* infections. The existence of completely functional genes encoding a type III secretion system (T3SS), is crucial for the virulence of the fish disease *Aeromonas salmonicida* A449. This was validated by genome sequencing of the pathogen (Burr *et al.*, 2005).

To further resolve taxonomic ambiguities among these species, the genomes of *Aeromonas hydrophila* ATCC 7966T, *Aeromonas salmonicida subsp. Salmonicida* A449, *Aeromonas veronii* strain B565, and *Aeromonas caviae* have been sequenced (Seshadri *et al.,* 2006; Reith *et al.,* 2008; Beatson *et al.,* 2011; Li *et al*., 2011). Understanding the virulence mechanisms, host adaptation, and ecological niche specialization of fish pathogenic vibrios has been made possible by genome sequencing. They have finished and published the genomes of *Vibrio anguillarum, Vibrio ordalii, Aliivibrio salmonicida*, and *Vibrio vulnificus* (Hjerde *et al.,* 2008; Gulig *et al.,* 2010; Naka *et al.,* 2011). Furthermore, the pirABvp gene, which causes acute hepatopancreatic necrosis disease (AHPND), was found in *Vibrio campbellii*. This organism was sequenced and examined (Dong *et al*., 2017).

Table 2. Illustrations of next-generation sequencing in immunological research

Host	Pathogen	Molecule	NGS Platform	References
Danio rerio	*Salmonella typhimurium*	RNA	Illumina GA II	(Ordas *et al.*, 2011)
Cyprinus carpio	*Mycobacterium marinum*	RNA	Illumina GA IIx	(Zhang *et al.*, 2011)
Dicentrarchus labrax	*Vibrio anguillarum*	RNA	Roche 454 FLX GS Titanium	(Sarropoulou *et al.*, 2012)
Sparus aurata	*Enteromyxum leei*	RNA	Illumina GA II	(Parez-Sanchez *et al.*, 2013)
Penaeus monodon	Vibrio, WSSV, IHHNV, YHV	RNA	Illumina MiSeq	(Oetama *et al.*, 2016)
Labeo rohita	*Aeromonas hydrophila*	RNA	Genome Analyser II (Illumina)	(Robinson *et al.*, 2014)
Lates calcarifer	Scale Drop Disease Virus	DNA	Illumina HiSeq	(Chokmangmeepisarn *et al.*, 2024)

Gut Bacterial Identification

Many endogenous host factors and exogenous environmental influences affect the composition and functional activities of the gut microbiota, which is necessary for host growth and immunity (Yukgehnaish *et al.,* 2020). In their 2023 study, Chew *et al.* examined the intestinal Asian seabass (Lates calcarifer) bacterial communities in both healthy and diseased individuals with bacterial infections of various kinds and persistent enteritis. Jiao *et al.* (2023) compared the digestive systems of fish with multiple feeding strategies, such as herbivorous species (*Ctenopharyngodon idellus*), carnivorous species (*Siniperca chuatsi*), omnivorous species (*Carassius auratus*), and plankton feeders (*Schizothorax graham*), to investigate digestive enzyme activity, intestinal morphology, and gut microbiota. Likewise, Vala *et al.* (2024) investigated the gut microbiome microbial diversity in *Pangasius pangasius* cultivated in cage, pond, and biofloc environments. Ofek *et al.* (2022) examined tilapia's gut and spleen microbiota in both healthy and sick conditions. Proteobacteria, Fusobacteria, Firmicutes, Bacteroidetes, Actinobacteria, Clostridium, Bacilli, and Verrucomicrobia are among the bacteria that dominate the gut microbiota of fish, according to next-generation sequencing (NGS) studies. The intestinal microbiota of different fish species is composed of up to 90% Proteobacteria, Firmicutes, and Bacteroidetes (Ringo *et al.,* 2006; Desai *et al.,* 2012; Carda-Dieguez *et al.,* 2013; Li *et al.,* 2013; Givens *et al.,* 2015; Ingerslev *et al.,* 2014). Additional research on transgenic rainbow trout (*Oncorhynchus mykiss*), gizzard shad (*Dorosoma cepedianum*), silver carp (*Hypophthalmichthys molitrix*), and common carp (*Cyprinus carpio* L.) has revealed notable differences in intestinal bacterial communities between species, even when raised under comparable living, rearing, and feeding conditions (Navarrete *et al.,* 2012; Li *et al.,* 2013; Franchini *et al.,* 2014).

Assessment and use of Genetic Resources

A vital species in aquaculture, the *Cyprinus carpio* is also one of the most well-liked decorative fish for outdoor settings because of its distinctive colouration and scale patterns (Xu *et al.,* 2014). The Songpu strain yielded a high-quality genome assembly using whole-genome sequencing. The evolutionary linkages and population structure of the species were also revealed by the genome resequencing of 33 distinct individuals, which represented important domesticated strains and populations globally. Additionally, 1.7 million tiny insertion deletions and 19 million candidate SNPs were found. It was possible to identify genetic loci related to skin color and scale patterns by combining transcriptome analysis and genomic comparison. The draft genome of the common carp provides a valuable genomic resource to enable genome-based breeding technologies in carp aquaculture and get a better understanding of the genetic basis of commercially significant traits (Xu *et al.,* 2014).

Growth and Development

In fish breeding programs, improving growth performance is a primary goal. However, growth is highly influenced by physiological, metabolic, and environmental factors, and is intricately governed by genetic factors (De-Santis and Jerry, 2007; Johnston *et al*., 2011). Morphological characteristics are intricate and frequently influenced by the interaction of environmental and genetic variables. The common carp (*Cyprinus carpio*) has effectively adapted to various environments and shows notable genetic variation. According to Xu *et al.* (2014), a reduced-scale phenotype is linked to fibroblast growth factor receptor 1a1 (fgfr1a1) gene deletions in the coding regions, discovered during the genome sequencing of common carp. Grass carp is one of the most economically important species worldwide in aquaculture. The lack of genes encoding cellulose-degrading enzymes was discovered in the grass carp genome sequence. The liver's mevalonate and steroid biosynthesis pathways, as well as the gut's circadian rhythm regulatory pathways, were significantly enriched in differential gene expression analysis during the dietary shift from carnivory to herbivory (Wang *et al.,* 2015) (Table 3). Researchers proposed that consuming nutrient-rich, high-intensity food enables grass carp to sustain its quick expansion (Wang *et al.,* 2015). The sequencing of the *C. carpio* genome has been instrumental in identifying genes linked to essential economic characteristics in herbivorous fish, aiding in the genetic enhancement of fresh inventory.

Sex Determination

There is a lot of genetic, environmental, and social effect on fish biological sex, which is a complex trait with fast evolutionary turnover (Martinez *et al.,* 2014; Mei and Gui 2015). Given its crucial role in growth performance, reproduction, product quality, the capacity to improve sex is a major component of selective breeding in fish (Budd *et al.,* 2015). The discovery of many fish sex-determining genes and processes has been made possible by advancements in molecular cytogenetics, transcriptome sequencing, and WGS (Rondeau *et al.,* 2013; Reichwald *et al.,* 2015; Li *et al.,* 2016). Fish sex-determining gene diversity results from de novo development of other genes and sexual regulatory networks (Li and Gui 2018). Asian seabasses are hermaphrodites, meaning that as they mature, their sex changes from male to female. The genomic study revealed that certain gene duplications, including nuclear factor kb2 (nfkb2) and anti-mullerian hormone (amh), were enriched for gonad development-related functions. Autosomal interferon regulating factor 9 (irf9) was duplicated to produce the sdy (salmonid master sex-determining gene). In rainbow trout, it can restrict female development by interfering with and inhibiting the activity of a crucial ovarian differentiation factor (Berthelot *et al.*, 2014). In rainbow trout, the sdy gene causes ovarian differentiation when it is inactivated, and testicular differentiation when it is overexpressed (Yano *et al.,* 2012The long-range organisation of the closely related species' ~50 Mb WZ sex-determination region on linkage group 3 (LG3) and ~9 Mb XY sex-determination region on linkage group 1 (LG1) *Oreochromis niloticus* was discovered by a novel genome assembly in Nile tilapia (*Oreochromis niloticus*) (Conte *et al.,* 2017). This work demonstrates the importance of high-quality genome assemblies for discovering genes in fish species that determine sex.

Current Status of Whole Genome Sequencing in Aquaculture

Early in the 2000s, sequencing of genome in aquaculture initiatives were undertaken in China, the United States, and several EU nations. To sequence the whole genomes of aquaculture species, many techniques were employed. An economical fosmid-pooling technique was employed, for instance, to sequence the *Crassostrea gigas*, *Ostreidae* (Zhang *et al.,* 2012). The entire genome of the rohu (*Labeo rohita*) was sequenced in order to better understand the genes that may increase aquaculture efficiency. One male rohu's blood was used to extract DNA, which was then sequenced utilizing platforms from Oxford Nanopore Technologies (ONT) and Illumina (Iqbal *et al.,* 2023).

Shotgun method of whole genome was used for *Cyprinus carpio*, incorporating information from multiple NGS platforms (Xu *et al.,* 2014). Short-read sequencing has been utilized more frequently lately (Ao *et al.,* 2015) in conjunction with extremely long sequencing (SMRT sequencing, for example) (Roberts *et al.,* 2013; Vij *et al.,* 2016). More than 24 aquaculture species, such as sea bass (*Dicentrarchus labrax*), grass carp (*Ctenopharyngodon idella*), and common carp (*Cyprinus carpio*), now have draft genome sequences available (Table 1). Nevertheless, despite sequencing efforts aimed at Both marine shrimp (*Penaeus monodon*) (Huang *et al.,* 2011) and Pacific white shrimp (*Litopenaeus vannamei*) (Yu *et al.*, 2015), no whole genome sequence has been released for any shrimp species.

The high frequency of repeated sequences in prawn genomes may cause sequencing difficulties (Huang *et al.,* 2011). The Pacific oyster (*Crassostrea gigas*), which belongs to the phylum Mollusca, is a significant food source, and the Atlantic cod (*Gadus morhua*) a huge cold-adapted teleost, is essential to fisheries is one of the 24 species that have been sequenced (Zhang *et al.*, 2012). With 10% of the world's freshwater aquaculture production, the common carp is a significant freshwater food fish species (Xu *et al.,* 2014). The Nile tilapia (*Oreochromis niloticus*, Cichlidae) is a great model for comprehending the genetic processes underlying cichlid phenotypic diversity (Brawand *et al.,* 2014). Here is a summary of several aquaculture species' key discoveries and genome sequences.

The Asian seabass's (*Lates calcarifer*) first whole-genome shotgun assembly, which included 84% of the genome, was published in 2015. Despite having an N50 length of roughly 310 kb and 22,775 contigs, the assembly included several gaps. The Asian seabass genome was assembled at the chromosomal level and published in high quality in 2016. The genome assembly resulted in an N50 contig size greater than 1 Mb and a scaffold N50 size exceeding 25 Mb. The completed genome was approximately 671 Mb, representing about

96% of the total genome, having 24 haploid chromosomes accounted for 87% of the sequences. The genome of Asian seabass is compact in comparison to the genomes of other teleosts, enabling downstream research to discover causative genes/mutations for essential features. In contrast to fish species with bigger genomes, such as Atlantic salmon (*Salmo salar*, ~3 Gb), Nile tilapia (*Oreochromis niloticus*, ~1 Gb), and zebrafish (~1 Gb), gene identification is simpler in smaller species. For upcoming genetic research, the genomic assembly of Asian seabass is a priceless resource (Yue *et al.,* 2023).

Using a fosmid-pooling technique and short reads, the oyster genome was sequenced (Zhang *et al.,* 2012). In 2012, the assembly of genomes was difficult because of the high frequency of repetitive DNA strings. The scaffold N50 and contig N50 of the finished 559 Mb assembly were 401 kb and 19.4 kb, respectively. At 1670 kb, the longest scaffolds accounted for over 90% of the entire setup. The oyster genome is highly polymorphic and contains a large percentage of repetitive nucleotides. Variation is still being actively influenced by certain transposable elements.

Das and associates used a multi-platform sequencing technique to publish the entire draft genome of the farmed rohu, revealing 1.18 Gb distributed over 1416 cM. There are 40.63% repetitions, 2516 tRNAs (estimated by tRNAScan), and 26400 protein-coding genes in the genome. Ten wild riverine Rohu populations were re-sequenced, and of the approximately 5 million SNPs found, 3193 matched 667 scaffolds at specific locations. The placement of farmed rohu (Labeo rohita) into the Otophysi clade is confirmed by a phylogenetic study in conjunction with other Otophysan species. The Rohu draft genome is an invaluable resource for planning and carrying out genome-wide association studies (Das *et al.,* 2020).

Information from cutting-edge sequencing platforms like Illumina and Roche 454, were integrated to sequence the common carp's (*Cyprinus carpio*) genome (Xu *et al*., 2014). 52,610 of the 1.69 Gb preliminary genome of the domesticated *C. carpio* genes code for proteins, making up roughly 92.3% of the entire genome. Remarkably, 2,503 big scaffolds were used to produce 90 percent of the genome assembly. *Cyprinus carpio*, the common carp, has a 37% GC content in its genome. *Carpio carpio* and its subspecies, *C. carpio haematopterus*, were discovered to have a single origin based on the genome re-sequencing of 33 individuals from various populations worldwide. For C. carpio and species that are closely related, the draft genome is an invaluable resource for improved genome-assisted breeding programs and cutting-edge molecular research. Additional information about the genome and uses of the common carp *(Cyprinus carpio)* can be found at [CarpBase] (http://

www. Carpbase.org/). There are continuous efforts to improve the reference genome's accuracy and quality.

The Illumina technology was utilized to sequence the genomes of four more cichlid species and the Nile tilapia. The assembled genome of Nile tilapia consisted of 13,517 scaffolds with an N50 value of 2.8 Mb and 77,578 contigs with an N50 value of 29.5 kb. The overall genome size was 815.7 Mb and included 21437 protein-coding genes, 22 pseudogenes, 3 retrotransposed genes, and 821 non-coding RNAs. You may find the genome sequence and annotation data at [Ensembl](http://www.ensembl.org/Oreochromis_niloticus/Info/Index) (Brawand *et al.,* 2014).

Cirrhinus mrigala is an important species in freshwater aquaculture, representing 1.08% of global production. The *C. mrigala* library was sequenced using the Illumina HiSeq X10 system (Illumina, USA). Three paired-end libraries with insert sizes of 300–350 bp, 500–650 bp, and 700–750 bp were prepared using the NEBNext® Ultra™ II DNA Library Prep Kit for Illumina® (New England Biolabs, MA, USA). The highly contiguous genome assembly, with a total size of 1.057 Gb, revealed 39,091 genes, all of which were functionally annotated. The genome sequences of *Cirrhinus mrigala* are accessible under NCBI Accession No.: GCA_036247105.143. Additionally, the complete mitochondrial genome is available under NCBI Accession No.: OP289478, while the partial COI sequence is deposited under Accession No.: OP363958 (Chowdhury *et al.,* 2024).

Applications of ESTs and Genomic Sequencing in Aquaculture

Genome sequencing makes possible the production of a broad range of DNA-based genetic markers, which helps with the management of both wild and cultivated populations. Suppressive Subtractive Hybridization (SSH) or tissue-specific sources can be used to create ESTs (expressed sequence tags) which are produced by complementary DNA (cDNA) library sequencing. Most sequences are entered into databases. (http://www.ncbi.nlm.nih.gov/Genbank/, http://www.ncbi.nlm.nih.gov/sites/entrez, http://compbio.dfci.harvard.edu.tgi/), and EST databases have been established for several significant marine species (e.g. in Atlantic salmon: http://web.uvic.ca/cbr/grasp/). EST is the initial step in extracting whole gene and cDNA sequences. Shellfish and whole fish, genome sequencing advances aquaculture, environmental genomics, and our knowledge of vertebrate and invertebrate evolution (Cossins and Crawford 2005; Crollius and Weissenbach 2005).

Some model species of fish, including the *Danio rerio* (zebrafish) (http://www.sanger.ac.uk/Projects/D_rerio/), the *Gasterosteus aculeatus* (stickleback) (http://www.genome.gov/12512292), the *Tetraodon nigroviridis* (puffer fish) (http://www.genome.gov/11008305), the *Oryzias latipes* (medaka) (http://dolphin.lab.nig.ac.jp/medaka/index.php), and the *Takifigu rubripes* (fugu) (http://www.fugu-sg.org/) have complete genome sequences. Identifying economically significant genes and polymorphisms depends on sequencing the genomes of commercially significant species. However, the only commercially important fish species to have been funded for genome sequencing is tilapia (Cichlid Genome Consortium). The goal of recent international collaborative initiatives has been to obtain Comprehensive or partial genome sequencing of fish and shellfish species that are commercially significant, including *Crassostrea gigas* (Pacific oyster) (Hedgecock *et al.,* 2005), cod (www.codgene.ca), *Dicentrarchus labrax* (sea bass) (Chini *et al.,* 2006), salmon and rainbow trout (cGRASP, www.cgrasp.org) and cod.

Full genome sequences can be obtained by shotgun methods or by sequencing and assembling BAC (bacterial artificial chromosomes) that include DNA segments among the entire target genome. BAC libraries have become accustomed to create physical maps (e.g. salmon, sea bass, channel catfish) (Ng *et al.,* 2005; Quiniou *et al.,* 2007; Volckaert *et al.,* 2006). These libraries are available for Atlantic salmon (Thorsen *et al.,* 2005), *Oncorhynchus mykiss* (Rainbow trout) (Palti *et al.,* 2004), sea bass (Whitaker *et al.,* 2006) and *C. virginica* (oysters) and *C. gigas* (Cunningham *et al.,* 2006).

Despite the publication of draft genomes for more than 24 aquaculture species and the ongoing sequencing of more than 10 species (such as Japanese flounder (*Paralichthys olivaceus*), blunt-snout bream (*Megalobrama amblycephala)*), most species of aquaculture's genome assemblies are Still in draft form. Numerous flaws, gaps, and scaffolds that have not been assembled are still present in the assemblies. The majority of aquaculture species' reference genomes lacked complete annotations. For many downstream applications, the genome assemblies and annotations need to be further improved. PacBio (Pacific Biosciences) technique for sequencing (Roberts *et al,* 2013; Vij *et al.,* 2016) and optical mapping (Gupta *et al.,* 2016) may be used to enhance these draft genomes' quality by long sequencing (e.g. SMRT sequencing). Although assembled genomes of several species used in aquaculture such as common carp, tilapia, salmon, and catfish are already available to the public and provide Research community resources, many are still inaccessible to the general public. Soon, perhaps, all aquatic species' reference genomes that have been sequenced will be publicly accessible. When starting genome sequencing for other aquaculture species, a combination sequencing strategy that combines short reads with ultra-long reads can be used to lower sequencing costs and enhance assembly quality.

Reference Genome

A reference genome, often referred to as a reference assembly, it is a database of assembled DNA sequences that contains annotations and additional data about chromosome, gene, and DNA sequence activities. A reference genome's completeness, continuity, base error rates, gaps, fragment coverage distribution, misassemblies, and low-quality sections are some metrics that can be used to assess its quality. A reference genome's quality can be evaluated using bioinformatic techniques like REAPR (Hunt *et al.,* 2013). Genetic analyses are based on high-quality reference genomes. It is usually possible to build new genomes far more quickly and affordably by using reference genomes as a guide. It is possible to address issues regarding population diversity, strain identification, population characterization, gene expression, genome structure, regulation, and the relationships between significant features and DNA variants when a reference genome is properly constructed and annotated.

A reference genome sequence of high quality can help with genomic selection (GS), positional cloning, GWAS, DNA marker discovery, basic research, resolving challenging sections, and accessing unique genetic diversity. DNA samples from resequencing individuals or different groups were combined and compared to the reference genome to find a substantial number of SNPs. For instance, the LG23 sex-determining locus was fine-mapped using the Nile tilapia reference genome, which led to the discovery of a potential gene that determines sex called Amh (anti-mullerian hormone) (Palaiokostas *et al.*, 2015). Even though the preliminary genomes of more than There are 24 known aquaculture species, however not all of these genomes are publicly accessible, which restricts their use as reference genomes for further research. Furthermore, the majority of reference genomes for animals raised for aquaculture have not yet been completely annotated. Additional annotation is needed using databases, reference genomes of model species, and RNA-seq data. Some excellent studies (Campbell and Yandell 2015) provide detailed information on genome annotation processes.

Addressing Important Issues Related to Aquaculture

Fifty percent of vertebrates on the planet are teleost fish (Nelson, 2006). The environments in which they live are distinct. Aquaculture has a lot of fascinating biological concerns and problems that need to be resolved, like speciation, environmental adaptation, pigmentation, domestication, sex determination, etc. (Bone and Moore 2008). WGS allows for a comprehensive understanding of aquaculture species' biology. It has provided answers to certain intriguing biological puzzles, creating opportunities for in-depth investigation. Stress tolerance is vital in aquaculture one of the main goals of breeding is to increase stress tolerance. In aquaculture species, sex-determining processes are complex. Environmental (such as temperature), social, and genetic factors can all influence phenotypic sexes (Baroiller and D cotta 2001).

Many marine fish species exhibit sex reversal. Growth performances and sex are frequently linked. Males develop significantly quicker than females in certain species (like tilapia), whereas females grow quicker than men in other species (like common carp). Gaining knowledge about the plasticity of sex-determination mechanisms and considering the way sex chromosomes have evolved may greatly improve growth performance. In general, defense refers to all of the ways that organisms defend themselves against outside threats, such as pollution, medications, chemicals, microbes and their byproducts, and animal behavior. As the primary endogenous defense mechanism, immunity aids in differentiating between self and non-self. Invertebrate immune responses comprise both acquired and innate immunity. The immunological defenses of fish have been the subject of remarkable advancement in recent years.

This is mostly due to aquaculture's explosive growth during this time, which makes disease control a top priority. The utilization of fishmeal and fish oil in aquaculture has been severely hampered by the anticipated shortage of these resources shortly. As a result, plant-based products are being used in aquafeeds at an accelerated rate. According to whole genome sequencing, the grass carp's switch to a herbivorous diet from a carnivorous one is probably linked to transcriptional activation of the hepatic steroid synthesis and mevalonate pathway (Wang *et al.,* 2015). This study increases the likelihood that altering a group of genes in the genome could alter how fish in captivity use plant proteins, hence decreasing the amount of fish meal used in fish feed. Selective breeding to enhance a fish species' genetic makeup and natural selection in the wild leave fingerprints in a species' genome. In both evolutionary biology and aquaculture, the identification of these signs is crucial for identifying genes that encourage the rapid advancement of genetics.

Tilapia, possess a total of 1.43 million high-quality SNPs which were acquired from the genome sequencing of 47 tilapia individuals (Xia *et al.,* 2015). Tilapia genomes had linkage disequilibrium blocks that varied in size from 10 to 100 kb. Each line of tilapia had more than 100 putative selective sweep areas. Non-coding areas of the tilapia genome contained most of the selection signatures that were found. Positive selection was applied to the integrin signaling pathways, Wnt signaling, and gonadotropin-releasing hormone receptors in genetically modified farmed tilapia strains. To accelerate genetic development and research in tilapia, it is essential to create a detailed map of the species' genetic diversity and selection footprints across the genome. This mapping is vital for understanding the genetic factors that influence tilapia's adaptability and resilience.

The reliability of the phenotypic data required to train prediction models is one of the challenges in using genomics in selection. Imaging, image processing, robotics, spectroscopy and high-performance computers have all been used to develop non-invasive methods for plant phenotyping (Cobb *et al.*, 2013). Fish have also been studied using these computer-vision techniques (Zenger *et al.*, 2017). According to Saberioon *et al.* (2017), new technologies are necessary to precisely evaluate the physical characteristics (such as freshness, texture, and color) and chemical components (such as fat, protein, and moisture) of fish. Future intelligent systems based on computer vision will accurately extract quantitative data from digital photos, greatly improving genome selection procedures (Figure 1).

Conclusion

Genetics and next-generation sequencing technology have entirely changed the sciences and methods of fisheries and aquaculture. Whole genomes have now been sequenced for many important fisheries and aquaculture species, making them accessible for comparison, research, and knowledge discovery. The genetic mechanisms underlying the use of sexual determination, germplasm resources, growth and development, disease resistance, and fisheries management have been extensively studied. However, genomic applications in fisheries management and genomic selection have not yet reached their full potential. It is anticipated that the adoption of genomic techniques, such as genome editing and genomic selection, as well as new big data and artificial intelligence systems, will significantly improve the sustainability of aquaculture and fisheries in the next decades.

References

AlMomin, S., Kumar, V., Al-Amad, S., Al-Hussaini, M., Dashti, T., Al-Enzi, K., & Akbar, A. (2015). Identification of novel microsatellite markers for P. argenteus from Kuwait waters. Conservation Genetics Resources, 7(2), 473–475. https://doi.org/10.1007/s12686-014-0399-x

Ao, J., Mu, Y., Xiang, L., Fan, D., Feng, M., Zhang, S., Shi, Q., Zhu, L., Li, T., Ding, Y., Nie, L., Li, Q., Dong, W., Jiang, L., Sun, B., Zhang, X., Li, M., Zhang, H., Xie, S., . . . Chen, X. (2015). Genome Sequencing of the Perciform Fish Larimichthys crocea Provides Insights into Molecular and Genetic Mechanisms of Stress Adaptation. PLoS Genetics, 11(4), e1005118. https://doi.org/10.1371/journal.pgen.1005118

Aparicio, S., Chapman, J., Stupka, E., Putnam, N., Chia, J. M., Dehal, P., Brenner, S. (2002). Whole-genome shotgun assembly and analysis of the genome of Fugu rubripes. Science, 297(5585), 1301–1310. https://doi.org/10.1126/science.1072104.

Astier, Y., Braha, O., & Bayley, H. (2006). Toward single molecule DNA sequencing: Direct identification of ribonucleoside and deoxyribonucleoside 5'-monophosphates by using an engineered protein nanopore equipped with a molecular adapter. Journal of the American Chemical Society, 128(5), 1705–1710. https://doi.org/10.1021/ ja057123+.

Barnes, T. C., Izzo, C., Bertozzi, T., Saint, K. M., Donnellan, S., Hammer, M. P., & Gillanders, B. M. (2014). Development of 15 microsatellite loci from mulloway, Argyrosomus japonicus (Pisces: Sciaenidae) using next generation sequencing and an assessment of their cross amplification in other sciaenids. Conservation Genetics Resources, 6(2), 345–348. https://doi.org/10.1007/s12686-013-0090-7

Baroiller, J., & D'Cotta, H. (2001). Environment and sex determination in farmed fish. Comparative Biochemistry and Physiology Part a Molecular & Integrative Physiology, 126, 10. https://doi.org/10.1016/s1095-6433(00)80018-5

Bayona-Vásquez, N. J., Díaz-Jaimes, P., & Uribe-Alcocer, M. (2015). Isolation and characterization of microsatellite loci in the common dolphinfish Coryphaena hippurus (Perciformes: Coryphaenidae) from next generation sequencing and cross amplification in pompano dolphinfish Coryphaena equiselis. Conservation Genetics Resources, 7(2), 373–375. https://doi.org/10.1007/s12686-014-0372-8

Beatson, S. A., De Luna, M. D. G., Bachmann, N. L., Alikhan, N., Hanks, K. R., Sullivan, M. J., Wee, B. A., Freitas-Almeida, A. C., Santos, P. a. D., De Melo, J. T. B., Squire, D. J. P., Cunningham, A. F., Fitzgerald, J. R., & Henderson, I. R. (2011). Genome Sequence of the Emerging Pathogen Aeromonas caviae. Journal of Bacteriology, 193(5), 1286–1287. https://doi.org/10.1128/jb.01337-10

Bernardi, G., Wiley, E. O., Mansour, H., Miller, M. R., Orti, G., Haussler, D., ... Venkatesh, B. (2012). The fishes of Genome 10K. Marine Genomics, 7, 3–6. https://doi.org/10. 1016/j. margen.2012.02.002.

Bernatchez, L., Wellenreuther, M., Araneda, C., Ashton, D. T., Barth, J. M. I., Beacham, T. D., ... Withler, R. E. (2017). Harnessing the power of genomics to secure the future of seafood. Trends in Ecology & Evolution, 32(9), 665–680. https://doi.org/10.1016/j. tree.2017.06.010.

Berthelot, C., Brunet, F., Chalopin, D., Juanchich, A., Bernard, M., Noël, B., Bento, P., Da Silva, C., Labadie, K., Alberti, A. A., Aury, J., Louis, A., Dehais, P., Bardou, P., Montfort, J., Klopp, C. C., Cabau, C., Gaspin, C., Thorgaard, G. H., Guiguen, Y. (2014). The rainbow trout genome provides novel insights into evolution after whole-genome duplication in vertebrates. Nature Communications, 5(1). https://doi.org/10.1038/ncomms4657

Bian, C., Huang, Y., Li, J., You, X. X., Yi, Y. H., Ge, W., et al. (2019). Divergence, evolution and adaptation in ray-finned fish genomes. Science China Life Sciences, 62(8),1003–1018. https://doi.org/10.1007/s11427-018-9499-5.

Bonacic, K., Campoverde, C., Sastre, M., Hachero-Cruzado, I., Ponce, M., Manchado, M., Estevez, A., Gisbert, E., & Morais, S. (2016). Mechanisms of lipid metabolism and transport underlying superior performance of Senegalese sole (Solea senegalensis, Kaup 1858) larvae fed diets containing n-3 polyunsaturated fatty acids. Aquaculture, 450, 383–396. https://doi.org/10.1016/j.aquaculture.2015.07.009

Bone, Q., & Moore, R. (2008). Biology of fishes. Taylor & Francis.https://doi. org/10.1201/9781134186310

Bradbury, I. R., Hubert, S., Higgins, B., Borza, T., Bowman, S., Paterson, I. G., Bentzen, P. (2010). Parallel adaptive evolution of Atlantic cod on both sides of the Atlantic Ocean in response to temperature. Proceedings of the Royal Society B: Biological Sciences, 277(1701), 3725–3734. https://doi.org/10.1098/rspb.2010.0985.

Brawand, D., Wagner, C. E., Li, Y. I., Malinsky, M., Keller, I., Fan, S., Simakov, O., Ng, A. Y., Lim, Z. W., Bezault, E., Turner-Maier, J., Johnson, J., Alcazar, R., Noh, H. J., Russell, P., Aken, B., Alföldi, J., Amemiya, C., Azzouzi, N., & Di Palma, F. (2014). The genomic substrate for adaptive radiation in African cichlid fish. Nature, 513(7518), 375–381. https://doi.org/10.1038/nature13726

Budd, A., Banh, Q., Domingos, J., & Jerry, D. (2015). Sex control in fish: Approaches, challenges and Opportunities for Aquaculture. Journal of Marine Science and Engineering, 3(2), 329–355. https://doi.org/10.3390/jmse3020329

Burr, S. E., Pugovkin, D., Wahli, T., Segner, H., & Frey, J. (2005). Attenuated virulence of an Aeromonas salmonicida subsp. salmonicida type III secretion mutant in a rainbow trout model. Microbiology, 151(6), 2111–2118. https://doi.org/10.1099/mic.0.27926-0

Campbell, M. S., & Yandell, M. (2015). An introduction to genome annotation. Current Protocols in Bioinformatics, 52(1). https://doi.org/10.1002/0471250953.bi0401s52

Canario, V. M. A. (2019). Editorial: Aquaculture genomics. Aquaculture and Fisheries, 4(1), 1–2. https://doi.org/10.1016/j.aaf.2018.12.005.

Cao, L., Wang, W., Yang, Y., Yang, C., Yuan, Z., Xiong, S., & Diana, J. (2007). Environmental impact of aquaculture and countermeasures to aquaculture pollution in China. Environmental Science and Pollution Research-International, 14, 452-462. https://doi.org/10.1065/espr2007.05.426

Carda-Diéguez, M., Mira, A., & Fouz, B. (2013). Pyrosequencing survey of intestinal microbiota diversity in cultured sea bass (Dicentrarchus labrax) fed functional diets. FEMS Microbiology Ecology, 87(2), 451–459. https://doi.org/10.1111/1574-6941.12236

Carlsson, J., Gauthier, D. T., Carlsson, J. E. L., Coughlan, J. P., Dillane, E., Fitzgerald, R. D., Keating, U., McGinnity, P., Mirimin, L., & Cross, T. F. (2013). Rapid, economical single-nucleotide polymorphism and microsatellite discovery based on de novo assembly of a reduced representation genome in a non-model organism: a case study of Atlantic cod Gadus morhua. Journal of Fish Biology, 82(3), 944–958. https://doi.org/10.1111/jfb.12034

Carvalho, D. C., Rodríguez-Zárate, C. J., Hammer, M. P., & Beheregaray, L. B. (2011). Development of 21 microsatellite markers for the threatened Yarra pygmy perch (Nannoperca obscura) through 454 shot-gun pyrosequencing. Conservation Genetics Resources, 3(4), 601–604. https://doi.org/10.1007/s12686-011-9413-8

Chen SongLin, C. S., Xu WenTeng, X. W., & Liu Yang, L. Y. (2019). Fish genomic research: decade review and prospect. Journal of Fisheries of China, 43, 1–14.

Chen, Y., Shi, M., Zhang, W., Cheng, Y., Wang, Y., & Xia, X. (2017). The Grass Carp Genome Database (GCGD): an online platform for genome features and annotations. Database, 2017. https://doi.org/10.1093/database/bax051

Chew, X. Z., Gibson-Kueh, S., Jerry, D. R., & Shen, X. (2023). Comparison of intestinal bacterial communities in asymptomatic and diseased Asian seabass (Lates calcarifer) with chronic enteritis and mixed bacterial infections. Aquaculture, 572, 739516. https://doi.org/10.1016/j.aquaculture.2023.739516

Chini, V., Rimoldi, S., Terova, G., Saroglia, M., Rossi, F., Bernardini, G., & Gornati, R. (2006). EST-based identification of genes expressed in the liver of adult seabass (Dicentrarchus labrax, L.). Gene, 376(1), 102–106. https://doi.org/10.1016/j.gene.2006.02.010

Chokmangmeepisarn, P., Azmai, M. N. A., Domingos, J. A., Van Aerle, R., Bass, D., Prukbenjakul, P., Senapin, S., & Rodkhum, C. (2024). Genome Characterization and Phylogenetic Analysis of Scale Drop Disease Virus Isolated from Asian Seabass (Lates calcarifer). Animals, 14(14), 2097. https://doi.org/10.3390/ani14142097

Chowdhury, L. M., Mohindra, V., Kumar, R., & Jena, J. (2024). Genome sequencing and assembly of Indian major carp, Cirrhinus mrigala (Hamilton, 1822). Scientific Data, 11(1). https://doi.org/10.1038/s41597-024-03747-6

Cobb, J. N., DeClerck, G., Greenberg, A., Clark, R., & McCouch, S. (2013). Next-generation phenotyping: requirements and strategies for enhancing our understanding of genotype–phenotype relationships and its relevance to crop improvement. Theoretical and Applied Genetics, 126, 867-887. https://doi.org/10.1007/s00122-013-2066-0

Conte, M. A., Gammerdinger, W. J., Bartie, K. L., Penman, D. J., & Kocher, T. D. (2017). A high-quality assembly of the Nile Tilapia (Oreochromis niloticus) genome reveals the structure of two sex determination regions. BMC Genomics, 18(1), 341. https://doi.org/10.1186/s12864-017-3723-510.1186/s12864-017-3723-5.

Cossins, A. R., & Crawford, D. L. (2005). Fish as models for environmental genomics. Nature Reviews Genetics, 6(4), 324-333. https://doi.org/10.1038/nrg1590

Crollius, H. R., & Weissebach, J. (2005). Fish genomic and biology. Genome Res., 15, 1675–1682. http://en.journals.sid.ir/ViewPaper.aspx?ID=346145

Cunningham, C., Hikima, J., Jenny, M. J., Chapman, R. W., Fang, G., Saski, C., Lundqvist, M. L., Wing, R. A., Cupit, P. M., Gross, P. S., Warr, G. W., & Tomkins, J. P. (2006). New Resources for Marine Genomics: Bacterial Artificial Chromosome Libraries for the Eastern and Pacific Oysters (Crassostrea virginica and C. gigas). Marine Biotechnology, 8(5), 521–533. https://doi.org/10.1007/s10126-006-6013-9

Das, P., Sahoo, L., Das, S. P., Bit, A., Joshi, C. G., Kushwaha, B., Kumar, D., Shah, T. M., Hinsu, A. T., Patel, N., Patnaik, S., Agarwal, S., Pandey, M., Srivastava, S., Meher, P. K., Jayasankar, P., Koringa, P. G., Nagpure, N. S., Kumar, R., . . . Jena, J. (2020). De novo Assembly and Genome-Wide SNP Discovery in Rohu Carp, Labeo rohita. Frontiers in Genetics, 11. https://doi.org/10.3389/fgene.2020.00386

Davidson, W. S., Koop, B. F., Jones, S. J., Iturra, P., Vidal, R., Maass, A., Jonassen, I., Lien, S., & Omholt, S. W. (2010). Sequencing the genome of the Atlantic salmon (Salmo salar). Genome Biology, 11(9). https://doi.org/10.1186/gb-2010-11-9-403

Desai, A. R., Links, M. G., Collins, S. A., Mansfield, G. S., Drew, M. D., Van Kessel, A. G., & Hill, J. E. (2012). Effects of plant-based diets on the distal gut microbiome of rainbow trout (Oncorhynchus mykiss). Aquaculture, 350–353, 134–142. https://doi.org/10.1016/j.aquaculture.2012.04.005

De-Santis, C., & Jerry, D. R. (2007). Candidate growth genes in finfish — Where should we be looking? Aquaculture, 272(1–4), 22–38. https://doi.org/10.1016/j.aquaculture.2007.08.036

Domingos, J. A., Zenger, K. R., & Jerry, D. R. (2015). Whole-genome shotgun sequence assembly enables rapid gene characterization in the tropical fish barramundi, Lates calcarifer. Animal Genetics, 46(4), 468–469. https://doi.org/10.1111/age.12312

Dong, X., Wang, H., Zou, P., Chen, J., Liu, Z., Wang, X., & Huang, J. (2017). Complete genome sequence of Vibrio campbellii strain 20130629003S01 isolated from shrimp with acute hepatopancreatic necrosis disease. Gut Pathogens, 9(1). https://doi.org/10.1186/s13099-017-0180-2

Duchaud, E., Boussaha, M., Loux, V., Bernardet, J., Michel, C., Kerouault, B., Mondot, S., Nicolas, P., Bossy, R., Caron, C., Bessières, P., Gibrat, J., Claverol, S., Dumetz, F., Hénaff, M. L., & Benmansour, A. (2007). Complete genome sequence of the fish pathogen Flavobacterium psychrophilum. Nature Biotechnology, 25(7), 763–769. https://doi.org/10.1038/nbt1313

Dunier, M., & Siwicki, A. K. (1993). Effects of pesticides and other organic pollutants in the aquatic environment on the immunity of fish: a review. Fish & Shellfish Immunology, 3(6), 423-438. https://doi.org/10.1006/fsim.1993.1042

Dupont, C., Cousin, P., & Dupont, S. (2018). IoT for Aquaculture 4.0 Smart and easy-to-deploy real-time water monitoring with IoT. In 2018 Global Internet of Things Summit (GIoTS) (Pp. 1-5)IEEE. https://doi.org/10.1109/giots.2018.8534581

Eid, J., Fehr, A., Gray, J., Luong, K., Lyle, J., Otto, G., ... Turner, S. (2009). Real-time DNA sequencing from single polymerase molecules. Science, 323(5910), 133–138. https://doi.org/10.1126/science.1162986.

Fernandez-Silva, I., Snelgrove, B. N., & Bowen, B. W. (2013). Twelve microsatellite DNA markers to resolve population structure of the Yellow-Striped Goatfish Mulloidichthys flavolineatus (family Mullidae). Conservation Genetics Resources, 5(2), 565–568. https://doi.org/10.1007/s12686-012-9853-9

Franchini, P., Fruciano, C., Frickey, T., Jones, J. C., & Meyer, A. (2014). The gut microbial community of Midas cichlid fish in repeatedly evolved Limnetic-Benthic species pairs. PLoS ONE, 9(4), e95027. https://doi.org/10.1371/journal.pone.0095027

Froese, R., & Pauly, D. (2010). FishBase. Fisheries centre. University of British Columbia. Genetics, 49(1), 119.

Ghanbari, M., Kneifel, W., & Domig, K. J. (2015). A new view of the fish gut microbiome: Advances from next-generation sequencing. Aquaculture, 448, 464–475. https://doi.org/10.1016/j.aquaculture.2015.06.033

Givens, C. E., Ransom, B., Bano, N., & Hollibaugh, J. T. (2015). Comparison of the gut microbiomes of 12 bony fish and 3 shark species. Marine Ecology Progress Series, 518, 209-223. DOI: https://doi.org/10.3354/meps11034

Gjedrem, T., & Robinson, N. (2014). Advances by selective breeding for aquatic species: a review. Agricultural Sciences, 5(12), 1152.10.4236/as.2014.512125

Glazer, A. M., Killingbeck, E. E., Mitros, T., Rokhsar, D. S., & Miller, C. T. (2015). Genome Assembly Improvement and Mapping Convergently Evolved Skeletal Traits in Sticklebacks with Genotyping-by-Sequencing. G3 Genes Genomes Genetics, 5(7), 1463–1472. https://doi.org/10.1534/g3.115.017905

Gong, G., Dan, C., Xiao, S., Guo, W., Huang, P., Xiong, Y., Mei, J. (2018). Chromosomal-level assembly of yellow catfish genome using third-generation DNA sequencing and Hi-C analysis. GigaScience, 7(11), https://doi.org/10.1093/ gigascience/giy120.

Gulig, P. A., De Crécy-Lagard, V., Wright, A. C., Walts, B., Telonis-Scott, M., & McIntyre, L. M. (2010). SOLiD sequencing of four Vibrio vulnificus genomes enables comparative genomic analysis and identification of candidate clade-specific virulence genes. BMC Genomics, 11(1), 512. https://doi.org/10.1186/1471-2164-11-512

Gupta, A., Kounovsky-Shafer, K. L., Ravindran, P., & Schwartz, D. C. (2016). Optical mapping and nanocoding approaches to whole-genome analysis. Microfluidics and Nanofluidics, 20(3). https://doi.org/10.1007/s10404-015-1685-y

Hajibabaei, M., Shokralla, S., Zhou, X., Singer, G. A. C., & Baird, D. J. (2011). Environmental Barcoding: A Next-Generation Sequencing approach for biomonitoring applications using River Benthos. PLoS ONE, 6(4), e17497. https://doi.org/10.1371/journal.pone.0017497

Harper, L., Campbell, J., Cannon, E. K. S., Jung, S., Poelchau, M., Walls, R., Andorf, C., Arnaud, E., Berardini, T. Z., Birkett, C., Cannon, S., Carson, J., Condon, B., Cooper, L., Dunn, N., Elsik, C. G., Farmer, A., Ficklin, S. P., Grant, D., Main, D. (2018). AgBioData consortium recommendations for sustainable genomics and genetics databases for agriculture. Database, 2018. https://doi.org/10.1093/database/bay088

Haussler, D., O'Brien, S. J., Ryder, O. A., Barker, F. K., Clamp, M., Crawford, A. J., Grp, A. (2009). Genome 10K: A proposal to obtain whole-genome sequence for 10 000 vertebrate species. Journal of Heredity, 100(6), 659–674. https://doi.org/10.1093/ jhered/esp086.

Heather, J. M., & Chain, B. (2016). The sequence of sequencers: The history of sequencing DNA. Genomics, 107(1), 1–8. https://doi.org/10.1016/j.ygeno.2015.11.003.

Hedgecock, D., Gaffney, P. M., Goulletquer, P., Guo, X., Reece, K., & Warr, G. W. (2005). The Case for Sequencing the Pacific Oyster Genome. Journal of Shellfish Research, 24(2), 429–441. http://doi.org/10.2983/0730-8000(2005)24

Hjerde, E., Lorentzen, M. S., Holden, M. T., Seeger, K., Paulsen, S., Bason, N., Churcher, C., Harris, D., Norbertczak, H., Quail, M. A., Sanders, S., Thurston, S., Parkhill, J., Willassen, N. P., & Thomson, N. R. (2008). The genome sequence of the fish pathogen Aliivibrio salmonicida strain LFI1238 shows extensive evidence of gene decay. BMC Genomics, 9(1). https://doi.org/10.1186/1471-2164-9-616

Houston, D. D., Elzinga, D. B., Maughan, P. J., Smith, S. M., Kauwe, J. S., Evans, R. P., & Shiozawa, D. K. (2012). Single nucleotide polymorphism discovery in cutthroat trout subspecies using genome reduction, barcoding, and 454 pyro-sequencing. BMC genomics, 13, 1-17. https://doi.org/10.1186/1471-2164-13-724

Houston, R. D., Taggart, J. B., Cézard, T., Bekaert, M., Lowe, N. R., Downing, A., & Hamilton, A. (2014). Development and validation of a high-density SNP genotyping array for Atlantic salmon (Salmo salar). BMC Genomics, 15, 1-13. https://doi.org/10.1186/1471-2164-15-90

Howe, K., Clark, M. D., Torroja, C. F., Torrance, J., Berthelot, C., Muffato, M., ... Stemple, D. L. (2013). The zebrafish reference genome sequence and its relationship to the human genome. Nature, 496(7446), 498–503. https://doi.org/10.1038/nature12111.

Huang, S. W., Lin, Y. Y., You, E. M., Liu, T. T., Shu, H. Y., Wu, K. M., Tsai, S. F., Lo, C. F., Kou, G. H., Chin, G., MA, Chen, M., Wu, D., Aoki, T., Hirono, I., & Yu, H. T. (2011). Fosmid library end sequencing reveals a rarely known genome structure of marine shrimp Penaeus monodon. BMC Genomics, 12(1). https://doi.org/10.1186/1471-2164-12-242

Hughes, L. C., Orti, G., Huang, Y., Sun, Y., Baldwin, C. C., Thompson, A. W., Shi, Q. (2018). Comprehensive phylogeny of ray-finned fishes (Actinopterygii) based on transcriptomic and genomic data. Proceedings of the National Academy of Sciences of the United States of America, 115(24), 6249–6254. https://doi.org/10.1073/pnas. 1719358115.

Hunt, M., Kikuchi, T., Sanders, M., Newbold, C., Berriman, M., & Otto, T. D. (2013). REAPR: a universal tool for genome assembly evaluation. Genome Biology, 14(5), R47. https://doi.org/10.1186/gb-2013-14-5-r47

Ingerslev, H., Von Gersdorff Jørgensen, L., Strube, M. L., Larsen, N., Dalsgaard, I., Boye, M., & Madsen, L. (2014b). The development of the gut microbiota in rainbow trout (Oncorhynchus mykiss) is affected by first feeding and diet type. Aquaculture, 424–425, 24–34. https://doi.org/10.1016/j.aquaculture.2013.12.032

Iqbal, G., Dar, S. A., & Maqbool, Y. (2023) Status of Whole Genome Sequencing in Fisheries and Aquaculture.Chronicle of Aquatic Science 1(1): 40-44

Jaillon, O., Aury, J., Brunet, F., Petit, J., Stange-Thomann, N., Mauceli, E., Bouneau, L., Fischer, C., Ozouf-Costaz, C., Bernot, A., Nicaud, S., Jaffe, D., Fisher, S., Lutfalla, G., Dossat, C., Segurens, B., Dasilva, C., Salanoubat, M., Levy, M., Crollius, H. R. (2004). Genome duplication in the teleost fish Tetraodon nigroviridis reveals the early vertebrate proto-karyotype. Nature, 431(7011), 946–957. https://doi.org/10.1038/nature03025

Jiang, Y., Gao, X., Liu, S., Zhang, Y., Liu, H., Sun, F., Liu, Z. (2013). Whole genome

Jiao, F., Zhang, L., Limbu, S. M., Yin, H., Xie, Y., Yang, Z., Shang, Z., Kong, L., & Rong, H. (2023). A comparison of digestive strategies for fishes with different feeding habits: Digestive enzyme activities, intestinal morphology, and gut microbiota. Ecology and Evolution, 13(9). https://doi.org/10.1002/ece3.10499

Johnston, I. A., Bower, N. I., & Macqueen, D. J. (2011). Growth and the regulation of myotomal muscle mass in teleost fish. Journal of Experimental Biology, 214(10), 1617-1628. https://doi.org/10.1242/jeb.038620

Kasahara, M., Naruse, K., Sasaki, S., Nakatani, Y., Qu, W., Ahsan, B., Yamada, T., Nagayasu, Y., Doi, K., Kasai, Y., Jindo, T., Kobayashi, D., Shimada, A., Toyoda, A., Kuroki, Y., Fujiyama, A., Sasaki, T., Shimizu, A., Asakawa, S., Kohara, Y. (2007). The medaka draft genome and insights into vertebrate genome evolution. Nature, 447(7145), 714–719. https://doi.org/10.1038/nature05846

Ke, R., Mignardi, M., Hauling, T., & Nilsson, M. (2016). Fourth generation of next generation sequencing technologies: Promise and consequences. Human Mutation, 37(12), 1363–1367. https://doi.org/10.1002/humu.23051.

Koepfli, K. P., Paten, B., O'Brien, S. J., & Scientists, G. K. C. (2015). The genome 10K project: A way forward. Annual Review of Animal Biosciences, 3(3), 57–111. https:// doi.org/10.1146/annurev-animal-090414-014900.

Kumar, G., & Kocour, M. (2017). Applications of next-generation sequencing in fisheries research: A review. Fisheries Research, 186, 11-22. https://doi.org/10.1016/j.fishres.2016.07.021

Lander, E. S., Linton, L. M., Birren, B., Nusbaum, C., Zody, M. C., Baldwin, J., International Human Genome Sequencing, C. (2001). International human genome sequencing initial sequencing and analysis of the human genome. Nature, 409(6822), 860–921. https://doi.org/10.1038/35057062.

Lewin, H. A., Robinson, G. E., Kress, W. J., Baker, W. J., Coddington, J., Crandall, K. A., ... Zhang, G. (2018). Earth BioGenome project: Sequencing life for the future of life. Proceedings of the National Academy of Sciences of the U S A, 115(17), 4325–4333. https://doi.org/10.1073/pnas.1720115115.

Li, X. Y., & Gui, J. F. (2018). Diverse and variable sex determination mechanisms in vertebrates. Science China Life Sciences, 61(12), 1503–1514. https://doi.org/10.1007/ s11427-018-9415-7.

Li, X., Yan, Q., Xie, S., Hu, W., Yu, Y., & Hu, Z. (2013). Gut Microbiota Contributes to the Growth of Fast-Growing Transgenic Common Carp (Cyprinus carpio L.). PLoS ONE, 8(5), e64577. https://doi.org/10.1371/journal.pone.0064577

Li, X., Zhang, Q., Zhang, J., Zhou, L., Li, Z., Zhang, X., Wang, D., & Gui, J. (2016). Extra microchromosomes play male determination role in polyploid gibel carp. Genetics, 203(3), 1415–1424. https://doi.org/10.1534/genetics.115.185843

Li, Y., Liu, Y., Zhou, Z., Huang, H., Ren, Y., Zhang, Y., Li, G., Zhou, Z., & Wang, L. (2011). Complete Genome Sequence of Aeromonas veronii Strain B565. Journal of Bacteriology, 193(13), 3389–3390. https://doi.org/10.1128/jb.00347-11

Lien, S., Koop, B. F., Sandve, NelsonS. R., Miller, J. R., Kent, M. P., Nome, T., Hvidsten, T. R., Leong, J. S., Minkley, D. R., Zimin, A., Grammes, F., Grove, H., Gjuvsland, A., Walenz, B., Hermansen, R. A., Von Schalburg, K., Rondeau, E. B., Di Genova, A., Samy, J. K. A., . . . Davidson, W. S. (2016). The Atlantic salmon genome provides insights into rediploidization. Nature, 533(7602), 200–205. https://doi.org/10.1038/nature17164

Liu, Z. J. (2010). Next generation sequencing and whole genome selection in aquaculture. In Wiley eBooks. https://doi.org/10.1002/9780470958964

Lu, G., & Luo, M. (2020). Genomes of major fishes in world fisheries and aquaculture: Status, application and perspective. Aquaculture and Fisheries, 5(4), 163–173. https://doi.org/10.1016/j.aaf.2020.05.004

Luo, W., Nie, Z., Zhan, F., Wei, J., Wang, W., & Gao, Z. (2012). Rapid Development of Microsatellite Markers for the Endangered Fish Schizothorax biddulphi (Günther) Using Next Generation Sequencing and Cross-Species Amplification. International Journal of Molecular Sciences, 13(12), 14946–14955. https://doi.org/10.3390/ijms131114946

MacKenzie, S. A., & Jentoft, S. (Eds.). (2016). future perspective. Genomics in aquaculture.

Academic Press. (pp. 275–277).

Margulies, M., Egholm, M., Altman, W. E., Attiya, S., Bader, J. S., Bemben, L. A., ... & Rothberg, J. M. (2005). Genome sequencing in microfabricated high-density picolitre reactors. Nature, 437(7057), 376-380. https://doi.org/10.1038/nature03959

Martínez, P., Viñas, A. M., Sánchez, L., Díaz, N., Ribas, L., & Piferrer, F. (2014). Genetic architecture of sex determination in fish: applications to sex ratio control in aquaculture. Frontiers in genetics, 5, 340. https://doi.org/10.3389/fgene.2014.00340

Mei, J., & Gui, J. F. (2015). Genetic basis and biotechnological manipulation of sexual dimorphism and sex determination in fish. Science China Life Sciences, 58, 124-136. https://doi.org/10.1007/s11427-014-4797-9

Metzker, M. L. (2010). Sequencing technologies—the next generation. Nature reviews genetics, 11(1), 31-46. https://doi.org/10.1038/nrg2626

Meyer, F. P., & Schnick, R. A. (1989). A review of chemicals used for the control of fish diseases. Reviews in Aquatic Sciences, 1(4), 693-710.

Mignardi, M., & Nilsson, M. (2014). Fourth-generation sequencing in the cell and the clinic. Genome Medicine, 6(4), 31. https://doi.org/10.1186/gm548

Montes, I., Conklin, D., Albaina, A., Creer, S., Carvalho, G. R., Santos, M., & Estonba, A. (2013). SNP Discovery in European Anchovy (Engraulis encrasicolus, L) by High-Throughput Transcriptome and Genome Sequencing. PLoS ONE, 8(8), e70051. https://doi.org/10.1371/journal.pone.0070051

Naka, H., Dias, G. M., Thompson, C. C., Dubay, C., Thompson, F. L., & Crosa, J. H. (2011). Complete Genome Sequence of the Marine Fish Pathogen Vibrio anguillarum Harboring the pJM1 Virulence Plasmid and Genomic Comparison with Other Virulent Strains of V. anguillarum and V. ordalii. Infection and Immunity, 79(7), 2889–2900. https://doi.org/10.1128/iai.05138-11

Navarrete, P., Magne, F., Araneda, C., Fuentes, P., Barros, L., Opazo, R., Espejo, R., & Romero, J. (2012). PCR-TTGE Analysis of 16S rRNA from Rainbow Trout (Oncorhynchus mykiss) Gut Microbiota Reveals Host-Specific Communities of Active Bacteria. PLoS ONE, 7(2), e31335. https://doi.org/10.1371/journal.pone.0031335

Nelson, J. S. (2006). Fishes of the world. http://fipak.areeo.ac.ir/site/catalogue/18344444

Ng, S. H., Artieri, C. G., Bosdet, I. E., Chiu, R., Danzmann, R. G., Davidson, W. S., Ferguson, M. M., Fjell, C. D., Hoyheim, B., Jones, S. J., De Jong, P. J., Koop, B. F., Krzywinski, M. I., Lubieniecki, K., Marra, M. A., Mitchell, L. A., Mathewson, C., Osoegawa, K., Parisotto, S. E., . . . Zhu, B. (2005). A physical map of the genome of Atlantic salmon, Salmo salar. Genomics, 86(4), 396–404. https://doi.org/10.1016/j.ygeno.2005.06.001

Oetama, V. S., Hennersdorf, P., Abdul-Aziz, M. A., Mrotzek, G., Haryanti, H., & Saluz, H. P. (2016). Microbiome analysis and detection of pathogenic bacteria of Penaeus monodon from Jakarta Bay and Bali. Marine Pollution Bulletin, 110(2), 718–725. https://doi.org/10.1016/j.marpolbul.2016.03.043

Ofek, T., Lalzar, M., Izhaki, I., & Halpern, M. (2022). Intestine and spleen microbiota composition in healthy and diseased tilapia. Animal Microbiome, 4(1). https://doi.org/10.1186/s42523-022-00201-z

Ordas, A., Hegedus, Z., Henkel, C. V., Stockhammer, O. W., Butler, D., Jansen, H. J., Racz, P., Mink, M., Spaink, H. P., & Meijer, A. H. (2011). Deep sequencing of the innate immune transcriptomic response of zebrafish embryos to Salmonella infection. Fish & Shellfish Immunology, 31(5), 716–724. https://doi.org/10.1016/j.fsi.2010.08.022org/10.1038/ncomms6770.

Palaiokostas, C., Bekaert, M., Khan, M. G., Taggart, J. B., Gharbi, K., McAndrew, B. J., & Penman, D. J. (2015). A novel sex-determining QTL in Nile tilapia (Oreochromis niloticus). BMC Genomics, 16(1). https://doi.org/10.1186/s12864-015-1383-x

Palti, Y., Gahr, S. A., Hansen, J. D., & Rexroad, C. E. (2004). Characterization of a new BAC library for rainbow trout: evidence for multi-locus duplication. Animal Genetics, 35(2), 130–133. https://doi.org/10.1111/j.1365-2052.2004.01112.x

Peichel, C. L., Sullivan, S. T., Liachko, I., & White, M. A. (2017). Improvement of the three spine stickleback genome using a Hi-C-Based Proximity-Guided assembly. Journal of Heredity, 108(6), 693–700. https://doi.org/10.1093/jhered/esx058

Pérez-Sánchez, J., Estensoro, I., Redondo, M. J., Calduch-Giner, J. A., Kaushik, S., & Sitjà-Bobadilla, A. (2013). Mucins as diagnostic and Prognostic biomarkers in a Fish-Parasite model: Transcriptional and Functional analysis. PLoS ONE, 8(6), e65457. https://doi.org/10.1371/journal.pone.0065457

Quiniou, S. M., Waldbieser, G. C., & Duke, M. V. (2007). A first generation BAC-based physical map of the channel catfish genome. BMC Genomics, 8(1). https://doi.org/10.1186/1471-2164-8-40

Reichwald, K., Petzold, A., Koch, P., Downie, B. R., Hartmann, N., Pietsch, S., Baumgart, M., Chalopin, D., Felder, M., Bens, M., Sahm, A., Szafranski, K., Taudien, S., Groth, M., Arisi, I., Weise, A., Bhatt, S. S., Sharma, V., Kraus, J. M., . . . Platzer, M. (2015). Insights into Sex Chromosome Evolution and Aging from the Genome of a Short-Lived Fish. Cell, 163(6), 1527–1538. https://doi.org/10.1016/j.cell.2015.10.071

Reid, K., Hoareau, T. B., & Bloomer, P. (2012). High-throughput microsatellite marker development in two sparid species and verification of their transferability in the family Sparidae. Molecular Ecology Resources, 12(4), 740–752. https://doi.org/10.1111/j.1755-0998.2012.03138.x

Reith, M. E., Singh, R. K., Curtis, B., Boyd, J. M., Bouevitch, A., Kimball, J., Munholland, J., Murphy, C., Sarty, D., Williams, J., Nash, J. H., Johnson, S. C., & Brown, L. L. (2008). The genome of Aeromonas salmonicida subsp. salmonicida A449: insights into the evolution of a fish pathogen. BMC Genomics, 9(1), 427. https://doi.org/10.1186/1471-2164-9-427

Ringø, E., Sperstad, S., Myklebust, R., Refstie, S., & Krogdahl, Å. (2006). Characterisation of the microbiota associated with intestine of Atlantic cod (Gadus morhua L.): the effect of fish meal, standard soybean meal and a bioprocessed soybean meal. Aquaculture, 261(3), 829-841.https://doi.org/10.1016/j.aquaculture.2006.06.030

Roberts, R. J., Carneiro, M. O., & Schatz, M. C. (2013). The advantages of SMRT sequencing. Genome Biology, 14(6). https://doi.org/10.1186/gb-2013-14-6-405

Robinson, N., Baranski, M., Mahapatra, K., Saha, J., Das, S., Mishra, J., Das, P., Kent, M., Arnyasi, M., & Sahoo, P. (2014). A linkage map of transcribed single nucleotide polymorphisms in rohu (Labeo rohita) and QTL associated with resistance to Aeromonas hydrophila. BMC Genomics, 15(1), 541. https://doi.org/10.1186/1471-2164-15-541

Roest Crollius, H., & Weissenbach, J. (2005). Fish genomics and biology. Genome Research, 15(12), 1675–1682. https://doi.org/10.1101/gr.3735805.

Rondeau, E. B., Messmer, A. M., Sanderson, D. S., Jantzen, S. G., Von Schalburg, K. R., Minkley, D. R., Leong, J. S., Macdonald, G. M., Davidsen, A. E., Parker, W. A., Mazzola, R. S., Campbell, B., & Koop, B. F. (2013). Genomics of sablefish (Anoplopoma fimbria): expressed genes, mitochondrial phylogeny, linkage map and identification of a putative sex gene. BMC Genomics, 14(1). https://doi.org/10.1186/1471-2164-14-452

Ruparel, H., Bi, L. R., Li, Z. M., Bai, X. P., Kim, D. H., Turro, N. J., et al. (2005). Design and synthesis of a 3 '-O-allyl photocleavable fluorescent nucleotide as a reversible terminator for DNA sequencing by synthesis. Proceedings of the National Academy of Sciences of the United States of America, 102(17), 5932–5937. https://doi.org/10. 1073/pnas.0501962102.

Saberioon, M., Gholizadeh, A., Cisar, P., Pautsina, A., & Urban, J. (2017). Application of machine vision systems in aquaculture with emphasis on fish: state-of-the-art and key issues. Reviews in Aquaculture, 9(4), 369-387.

Sahu, B. P., Patel, A., Sahoo, L., Das, P., Meher, P., & Jayasankar, P. (2012). Rapid and cost-effective development of SSR markers using next-generation sequencing in Indian major carp, Labeo rohita (Hamilton, 1822). Indian Journal of Fisheries, 59(3). http://epubs.icar.org.in/ejournal/index.php/IJF/article/download/15486/11414

Sahu, B., Sahoo, L., Joshi, C., Mohanty, P., Sundaray, J., Jayasankar, P., & Das, P. (2014). Isolation and characterization of polymorphic microsatellite loci in Indian major carp, Catla catla using next-generation sequencing platform. Biochemical Systematics and Ecology, 57, 357–362. https://doi.org/10.1016/j.bse.2014.09.010

Samy, J. K. A., Mulugeta, T. D., Nome, T., Sandve, S. R., Grammes, F., Kent, M. P., Lien, S., & Våge, D. I. (2017). SalmoBase: an integrated molecular data resource for Salmonid species. BMC Genomics, 18(1). https://doi.org/10.1186/s12864-017-3877-1

Santos, B. S., Romana-Eguia, M. R. R., Basiao, Z. U., & Ikeda, M. (2015). Development and characterization of nine novel microsatellite markers for the milkfish Chanos chanos. Conservation Genetics Resources, 7(2), 451–453. https://doi.org/10.1007/s12686-014-0393-3

Sarropoulou, E., Galindo-Villegas, J., García-Alcázar, A., Kasapidis, P., & Mulero, V. (2012). Characterization of European Sea Bass Transcripts by RNA SEQ After Oral Vaccine Against V. anguillarum. Marine Biotechnology, 14(5), 634–642. https://doi.org/10.1007/s10126-012-9466-z

Schultz, T. F., Fitzpatrick, C. K., Freshwater, D. W., & Morris, J. A. (2013). Characterization of 18 polymorphic microsatellite loci from invasive lionfish (Pterois volitans and P. miles). Conservation Genetics Resources, 5(3), 599–601. https://doi.org/10.1007/s12686-013-9860-5

Seeb, J. E., Carvalho, G., Hauser, L., Naish, K., Roberts, S., & Seeb, L. W. (2011). Single-nucleotide polymorphism (SNP) discovery and applications of SNP genotyping in nonmodel organisms. Molecular Ecology Resources, 11(s1), 1–8. https://doi.org/10.1111/j.1755-0998.2010.02979.x

Seo, T. S., Bai, X. P., Kim, D. H., Meng, Q. L., Shi, S. D., Ruparelt, H., ... Ju, J. Y. (2005). Four-color DNA sequencing by synthesis on a chip using photocleavable fluorescent nucleotides. Proceedings of the National Academy of Sciences of the United States of America, 102(17), 5926–5931. https://doi.org/10.1073/pnas.0501965102.

Seshadri, R., Joseph, S. W., Chopra, A. K., Sha, J., Shaw, J., Graf, J., ... & Heidelberg, J. F. (2006). Genome sequence of Aeromonas hydrophila ATCC 7966T: jack of all trades. Journal of bacteriology, 188(23), 8272-8282.DOI: https://doi.org/10.1128/jb.00621-06

Shen, Y., Catchen, J., Garcia, T., Amores, A., Beldorth, I., Wagner, J., Zhang, Z., Postlethwait, J., Warren, W., Schartl, M., & Walter, R. B. (2012). Identification of transcriptome SNPs between Xiphophorus lines and species for assessing allele-specific gene expression within F1 interspecies hybrids. Comparative Biochemistry and Physiology Part C Toxicology & Pharmacology, 155(1), 102–108. https://doi.org/10.1016/j.cbpc.2011.03.012

Shendure, J., Balasubramanian, S., Church, G. M., Gilbert, W., Rogers, J., Schloss, J. A., et al. (2017). DNA sequencing at 40: Past, present and future. Nature, 550(7676), 345–353. https://doi.org/10.1038/nature24286.

Shendure, J., Porreca, G. J., Reppas, N. B., Lin, X., McCutcheon, J. P., Rosenbaum, A. M., Wang, M. D., Zhang, K., Mitra, R. D., & Church, G. M. (2005). Accurate multiplex polony sequencing of an evolved bacterial genome. Science, 309(5741), 1728–1732. https://doi.org/10.1126/science.1117389

Slattery, P., Eschenbrenner, C., Arias-Rodriguez, L., Streit, B., Bierbach, D., Riesch, R., Tobler, M., Pfenninger, M., Feldmeyer, B., Plath, M., & Lerp, H. (2012). Twelve new microsatellite loci for the sulphur molly (Poecilia sulphuraria) and the related Atlantic molly (P. mexicana). Conservation Genetics Resources, 4(4), 935–937. https://doi.org/10.1007/s12686-012-9677-7 species. BMC Genomics, 14(1), 780.

Star, B., Nederbragt, A. J., Jentoft, S., Grimholt, U., Malmstrøm, M., Gregers, T. F., Rounge, T. B., Paulsen, J., Solbakken, M. H., Sharma, A., Wetten, O. F., Lanzén, A., Winer, R., Knight, J., Vogel, J., Aken, B., Andersen, Ø., Lagesen, K., Tooming-Klunderud, A., . . . Jakobsen, K. S. (2011). The genome sequence of Atlantic cod reveals a unique immune system. Nature, 477(7363), 207–210. https://doi.org/10.1038/nature10342

Taguchi, M., Kitamura, T., Shigenobu, Y., Ohkubo, M., Yanagimoto, T., Sugaya, T., Nakamura, Y., Saitoh, K., & Yokawa, K. (2013). Development of 15 polymorphic microsatellite markers for the shortfin mako, Isurus oxyrinchus, and cross-species amplification in lamniforme sharks. Conservation Genetics Resources, 5(3), 675–678. https://doi.org/10.1007/s12686-013-9880-1

Thorsen, J., Zhu, B., Frengen, E., Osoegawa, K., De Jong, P. J., Koop, B. F., Davidson, W. S., & Høyheim, B. (2005). A highly redundant BAC library of Atlantic salmon (Salmo salar): an important tool for salmon projects. BMC Genomics, 6(1). https://doi.org/10.1186/1471-2164-6-50

Touchon, M., Barbier, P., Bernardet, J., Loux, V., Vacherie, B., Barbe, V., Rocha, E. P. C., & Duchaud, E. (2011). Complete Genome Sequence of the Fish Pathogen Flavobacterium branchiophilum. Applied and Environmental Microbiology, 77(21), 7656–7662. https://doi.org/10.1128/aem.05625-11

Vala, F. D., Lende, S. R., Solanki, V. K., Jora, K., Desai, R., Sharma, P., Nathani, N., & Mootapally, C. (2024). Melodious tuning of microbial dynamics in biofloc, cage, and pond culturing system: a study on Pangasius pangasius fish gut microbiome. Frontiers in Microbiology, 15. https://doi.org/10.3389/fmicb.2024.1434312

Vij, S., Kuhl, H., Kuznetsova, I. S., Komissarov, A., Yurchenko, A. A., Van Heusden, P., Singh, S., Thevasagayam, N. M., Prakki, S. R. S., Purushothaman, K., Saju, J. M., Jiang, J., Mbandi, S. K., Jonas, M., Tong, A. H. Y., Mwangi, S., Lau, D., Ngoh, S. Y., Liew, W. C., . . . Orbán, L. (2016). Chromosomal-Level assembly of the Asian Seabass genome using long sequence reads and multi-layered scaffolding. PLoS Genetics, 12(4), e1005954. https://doi.org/10.1371/journal.pgen.1005954

Villanova, G. V., Vera, M., Díaz, J., Martinez, P., Calcaterra, N. B., & Arranz, S. E. (2015). Isolation and characterization of 20 polymorphic microsatellite loci in the migratory freshwater fish Leporinus obtusidens (Characiformes: Anostomidae) using 454 shotgun pyrosequencing. Journal of Fish Biology, 86(3), 1209–1217. https://doi.org/10.1111/jfb.12632

Volckaert, F., Batargias, C., Chatziplis, D., Chistiakov, D., Guo, Y., Haley, C., Hellemans, B., Kotoulas, G., Law, A., Libertini, A., McAndrew, B., Tsigenopoulos, C., & Whitaker, H. (2007). A second generation linkage map of European sea bass. Aquaculture, 272, S316. https://doi.org/10.1016/j.aquaculture.2007.07.198

Wang, J., Gaughan, S., Lamer, J. T., Deng, C., Hu, W., Wachholtz, M., ... Lu, G. (2019). Resolving the genetic paradox of invasions: Preadapted genomes and post introduction hybridization of bigheaded carp in the Mississippi River Basin. Evolutionary Applications. https://doi.org/10.1111/eva.12863.

Wang, J., Yu, X., Zhao, K., Zhang, Y., Tong, J., & Peng, Z. (2012). Microsatellite Development for an Endangered Bream Megalobrama pellegrini (Teleostei, Cyprinidae) Using 454 Sequencing. International Journal of Molecular Sciences, 13(3), 3009–3021. https://doi.org/10.3390/ijms13033009

Wang, Q., Yang, M., Xiao, J., Wu, H., Wang, X., Lv, Y., Xu, L., Zheng, H., Wang, S., Zhao, G., Liu, Q., & Zhang, Y. (2009). Genome Sequence of the Versatile Fish Pathogen Edwardsiella tarda Provides Insights into its Adaptation to Broad Host Ranges and Intracellular Niches. PLoS ONE, 4(10), e7646. https://doi.org/10.1371/journal.pone.0007646

Wang, Y. P., Lu, Y., Zhang, Y., Ning, Z. M., Li, Y., Zhao, Q., ... Zhu, Z. Y. (2015). The draft genome of the grass carp (Ctenopharyngodon idellus) provides insights into its evolution and vegetarian adaptation (vol 47, pg 625, 2015). Nature Genetics, 47(8), https://doi.org/10.1038/ng0815-962a 962-962.

Wegner, K. M., Shama, L. N., Kellnreitner, F., & Pockberger, M. (2012). Diversity of immune genes and associated gill microbes of European plaice Pleuronectes platessa. Estuarine Coastal and Shelf Science, 108, 87–96. https://doi.org/10.1016/j.ecss.2011.09.001

Whitaker, H. A., McAndrew, B. J., & Taggart, J. B. (2006). Construction and characterization of a BAC library for the European sea bass Dicentrarchus labrax. Animal Genetics, 37(5), 526. https://doi.org/10.1111/j.1365-2052.2006.01514.x

Williams, M. L., Gillaspy, A. F., Dyer, D. W., Thune, R. L., Waldbieser, G. C., Schuster, S. C., Gipson, J., Zaitshik, J., Landry, C., Banes, M. M., & Lawrence, M. L. (2012). Genome Sequence of Edwardsiella ictaluri 93-146, a Strain Associated with a Natural Channel Catfish Outbreak of Enteric Septicemia of Catfish. Journal of Bacteriology, 194(3), 740–741. https://doi.org/10.1128/jb.06522-11

Wu, C., Ma, Z., Zheng, G., Zou, S., Zhang, X., & Zhang, Y. (2022). Chromosome-level genome assembly of grass carp (Ctenopharyngodon idella) provides insights into its genome evolution. BMC Genomics, 23(1). https://doi.org/10.1186/s12864-022-08503-x

Xia, J. H., Bai, Z., Meng, Z., Zhang, Y., Wang, L., Liu, F., Jing, W., Wan, Z. Y., Li, J., Lin, H., & Yue, G. H. (2015). Signatures of selection in tilapia revealed by whole genome resequencing. Scientific Reports, 5(1). https://doi.org/10.1038/srep14168

Xia, J. H., He, X. P., Bai, Z. Y., Lin, G., & Yue, G. H. (2011). Analysis of the Asian Seabass transcriptome based on expressed sequence tags. DNA Research, 18(6), 513–522. https://doi.org/10.1093/dnares/dsr036

Xiao, S., Han, Z., Wang, P., Han, F., Liu, Y., Li, J., & Wang, Z. Y. (2015). Functional marker detection and analysis on a comprehensive transcriptome of large yellow croaker by next generation sequencing. PLoS ONE, 10(4), e0124432. https://doi.org/10.1371/journal.pone.0124432

Xu, D., Zhang, Y., & Peng, Z. (2013). High-throughput microsatellite marker development in Amur catfish (Silurus asotus) using next-generation sequencing. Conservation Genetics Resources, 5(2), 487–490. https://doi.org/10.1007/s12686-012-9834-z

Xu, P., Zhang, X., Wang, X., Li, J., Liu, G., Kuang, Y., Xu, J., Zheng, X., Ren, L., Wang, G., Zhang, Y., Huo, L., Zhao, Z., Cao, D., Lu, C., Li, C., Zhou, Y., Liu, Z., Fan, Z., . . . Sun, X. (2014). Genome sequence and genetic diversity of the common carp, Cyprinus carpio. Nature Genetics, 46(11), 1212–1219. https://doi.org/10.1038/ng.3098

Yano, A., Guyomard, R., Nicol, B., Jouanno, E., Quillet, E., Klopp, C., ... Guiguen, Y. (2012). An immune-related gene evolved into the master sex-determining gene in rainbow trout, Oncorhynchus mykiss. Current Biology, 22(15), 1423–1428. https:// doi.org/10.1016/j.cub.2012.05.045.

Yasuike, M., Fujiwara, A., Nakamura, Y., Iwasaki, Y., Nishiki, I., Sugaya, T., Shimizu, A., Sano, M., Kobayashi, T., & Ototake, M. (2016). A functional genomics tool for the Pacific bluefin tuna: Development of a 44K oligonucleotide microarray from whole-genome sequencing data for global transcriptome analysis. Gene, 576(2), 603–609.

https://doi.org/10.1016/j.gene.2015.10.023

Yin, L. N., Zhang, H., Yanagimoto, T., & Gao, T. X. (2012). Isolation and characterization of nine polymorphic microsatellite markers of the marbled rockfish Sebastiscus marmoratus (Scorpaeniformes, Scorpaenidae). Russian Journal of Genetics, 48(12), 1264–1266. https://doi.org/10.1134/s1022795412120174

Yu, J., Kim, B., Yun, Y., Kim, H., Bae, C., Kim, S., Oh, K., & Lim, C. E. (2014). Characterization of 11 novel polymorphic microsatellite loci in the threatened Korean loach, Iksookimia koreensis, isolated using a next-generation sequencing method. Biochemical Systematics and Ecology, 56, 132–137. https://doi.org/10.1016/j.bse.2014.05.016

Yu, Y., Zhang, X., Yuan, J., Li, F., Chen, X., Zhao, Y., Huang, L., Zheng, H., & Xiang, J. (2015). Genome survey and high-density genetic map construction provide genomic and genetic resources for the Pacific White Shrimp Litopenaeus vannamei. Scientific Reports, 5(1). https://doi.org/10.1038/srep15612

Yue, G. H. (2014). Recent advances of genome mapping and marker-assisted selection in aquaculture. Fish and fisheries, 15(3), 376-396.

Yue, G. H., Wang, L., Yang, Z., Sun, F., Tay, Y. X., Wong, J., & Yeo, S. (2023). Genomic resources and their applications in aquaculture of Asian seabass (Lates calcarifer). Reviews in Aquaculture, 15(2), 853–871. https://doi.org/10.1111/raq.12764

Yue, G., & Wang, L. (2017). Current status of genome sequencing and its applications in aquaculture. Aquaculture, 468, 337–347. https://doi.org/10.1016/j. aquaculture 2016.10.036

Yukgehnaish, K., Kumar, P., Sivachandran, P., Marimuthu, K., Arshad, A., Paray, B. A., & Arockiaraj, J. (2020). Gut microbiota metagenomics in aquaculture: factors influencing gut microbiome and its physiological role in fish. Reviews in Aquaculture, 12(3), 1903–1927. https://doi.org/10.1111/raq.12416

Zenger, K., Khatkar, Jerry, & Raadsma, H. (2017). The next wave in selective breeding: implementing genomic selection in aquaculture. In: Paper Presented at the Proc Assoc. Advmt. Anim. Breed. Genet. Vol. 22, (Pp. 105–112). https://researchonline.jcu.edu.au/49822

Zhang, G., Fang, X., Guo, X., Li, L., Luo, R., Xu, F., Yang, P., Zhang, L., Wang, X., Qi, H., Xiong, Z., Que, H., Xie, Y., Holland, P. W. H., Paps, J., Zhu, Y., Wu, F., Chen, Y., Wang, J., . . . Wang, J. (2012). The oyster genome reveals stress adaptation and the complexity of shell formation. Nature, 490(7418), 49–54. https://doi.org/10.1038/nature11413

Zhang, Y., Stupka, E., Henkel, C. V., Jansen, H. J., Spaink, H. P., & Verbeek, F. J. (2011). Identification of Common Carp Innate Immune Genes with Whole-Genome Sequencing and RNA-Seq Data. Berichte Aus Der Medizinischen Informatik Und Bioinformatik/Journal of Integrative Bioinformatics, 8(2), 165–175. https://doi.org/10.1515/jib-2011-169

10

Advances in Molecular Systematics Implications for Fisheries Management and Aquaculture

Rushikeshkumar J. Valmik*, Thatikonda Navya and Deepak T. Vaghela

College of Fisheries Science, Kamdhenu University, Veraval, Gujarat (362 265), India

Abstract

Molecular systematics being a vital field surpasses traditional morphological methods by providing more accurate insights into evolutionary relationships and species classification. It employs nucleic acid and amino acid-based sequences to classify organisms and utilizes molecular data to analyze the taxonomy, phylogeny, and biogeography of organisms. The important advances in this field during the last century influenced the genetic research and application of molecular methods in fishery and aquaculture, leading to great improvement in animal health, better product quality and production efficiency. Molecular systematics continues to be an influential tool in advancing fisheries and aquaculture research, contributing to sustainable resource management and the development of more efficient and environmentally responsible aquaculture practices.

Keywords: *evolutionary, phylogeny, molecular data, sequences*

Introduction

Molecular systematics is a branch of systematics that uses molecular data, particularly DNA sequences, to understand the evolutionary relationships among organisms. It utilizes nucleic acid and amino acid sequence compositions to analyse the taxonomy, phylogeny, and biogeography of biological organisms. It offers a superior alternative to morphological systematics, which relies on physical traits. This approach has significant applications in taxonomic studies, disease diagnosis, biodiversity conservation, and industrial purposes. The integration of molecular biology and bioinformatics has significantly advanced the capabilities of molecular systematics, making it a powerful tool in both microbiological research and public health.

Significant progress in molecular biology and bioinformatics has provided scientists with cutting-edge tools to tackle long-unresolved problems of biological study. Molecular systematics referred as the information derived from molecular data to determine phylogenetic relationships between organisms. A phylogeny, also known as an evolutionary tree, represents the historical relationships among different groups (lineages) of entities, such as organisms or genetic sequences (Page & Holmes, 2009). Comprehending these evolutionary relationships is crucial in comparative studies due to the statistical connections observed among entities with shared ancestral origins. Phylogenetic analyses were initially confined to the study of organismal evolution. However, they have now become indispensable tools across various research domains, including genomics, protein engineering, conservation biology, and agricultural pest control. For instance, phylogenies have been instrumental in investigating the origin and timeline of the predominant pandemic strain of the human immunodeficiency virus (HIV) (Korber *et al.*, 2000).

Molecular systematics plays a vital role in enhancing fisheries and aquaculture by providing tools for species identification, genetic diversity assessment, and management strategies. The Integration on of molecular techniques allows for better understanding of genetic relationships and population structures, which is essential for sustainable practices in aquaculture and conservation efforts. Systematics utilizes molecular markers like microsatellites and mtDNA to assess genetic variation, population structure, and species identification, aiding in conservation efforts, monitoring genetic diversity, and improving hatchery practices for sustainable fish populations.

Molecular Markers and Techniques

Overview of different molecular markers

Genetic variety (polymorphism) results from mutations that occur in all organisms due to abnormal cellular processes or interactions with the environment. A species' ability to adapt to its changing environment is improved by genetic variety, which is essential for the species' survival (Fisher, 1930). Genetic diversity between individuals occurs in conjunction with other evolutionary processes such as selection and genetic drift, resulting in differentiation at the population, species, and higher-order taxonomic group levels. Molecular genetic markers can identify genetic differences between individuals, groups, and species (Avise, 1994). These markers have enhanced the analytical ability required to examine genetic diversity. The findings from genetic diversity data have a variety of uses in studies on natural resource management, conservation, and evolution, programs for genetic enhancement, etc. There are two types of molecular markers: type I and type II. Genes with known functions are linked to type I markers, whereas anonymous genomic areas are linked to type II markers. Since the protein that allozyme markers encode has a known function, they fall under the category of type I markers. The reason RAPD markers are type II is because the polymerase chain reaction (PCR) amplifies RAPD bands from anonymous genomic areas (O'Brien, 1991).

Type II markers also include microsatellite markers, unless they are linked to genes with recognized functions. The importance of type I markers in aquaculture genetics is growing significantly. To compare and transfer genetic information from a species with a lot of maps to one with a lot less, type I markers act as a bridge. Generally, type II markers like AFLPs, microsatellites, and RAPDs are regarded as non-coding and so selectively neutral. These markers are frequently used in population genetic research to describe genetic variation both within and between populations or species (Brown & Epifanio, 2003).

Allozyme marker

Allozyme locus analysis is still one of the most widely used methods for investigating fish population genetics and stock structure issues (Suneetha, 2000). The method offers an independent assessment of the degree of variation within a population without requiring a comprehensive morphological and quantitative survey and it is quick and reasonably priced (Menezes *et al.*,1993). Isohyets are molecular versions of an enzyme system that differ structurally but have the same catalytic function and are encoded by one or more loci. "Allozymes" or "alloenzymes" are isohyets that are encoded by distinct alleles of the same gene locus (Starck,1998). The polypeptide chains of the various allelic variants of an enzyme differ in amino acids, which correspond to variations in the underlying DNA sequence. The resultant protein products may migrate through a gel exposed to an electrical field at varying rates (due to charge and size differences) depending on the type of amino acid modifications. Genetic variation is measured and genetic units are distinguished at the population, species, and higher taxonomic levels using differences in the relative frequency of alleles. The occasional heterozygote deficits brought on by null (enzymatically inactive) alleles and sensitivity to the quantity and quality of tissue samples are drawbacks of allozymes. Additionally, certain DNA sequence variations are obscured at the protein level, which lowers the amount of variation that may be detected. Certain polypeptide alterations do not affect the protein's mobility in an electrophoretic gel (synonymous substitutions), while certain nucleotide sequence changes do not affect the encoded polypeptide (silent substitutions). Currently, 75 isozyme systems that correspond to hundreds of genetic loci are recognized (Murphy *et al.*, 1996). Because of its codominant marker strength, affordability, and convenience of application, allozyme markers are widely used in population Phylogenetic and structural analyses, but has limited application in aquaculture genetics.

Mitochondrial DNA Markers

The use of mitochondrial DNA (mtDNA) analysis has grown in recent phylogenetic and population surveys of organisms. Research on vertebrates has typically demonstrated that sequence divergence builds up in mitochondrial DNA more quickly than in nuclear DNA (Brown, 1985). This has been linked to a smaller effective population size because of the strict maternal inheritance of the haploid mitochondrial genome (Birky *et al.*, 1989). and a greater mutation rate in mtDNA that could be caused by a lack of repair mechanisms during replication. Because of its quick evolution, mtDNA analysis has helped elucidate the relationships of closely related species. It is known that distinct regions of the mitochondrial genome evolve at varying speeds (Meyer, 1993). The majority of the mtDNA molecule is transcribed, except the 1-kb regulatory region (D-loop), which initiates replication and transcription. Non-coding segments, such as the D-loop, have higher levels of diversity than coding regions like the cytochrome b gene (Brown *et al.*, 1993). This is likely due to fewer functional limitations and relaxed selection pressure. The 16S rRNA gene in the mitochondrial genome evolves slowly, while regulatory areas evolve rapidly. The mtDNA is considered a single locus because of its Non-Mendelian method of inheritance. Furthermore, because mtDNA is inherited, the phylogenies and population structures produced from it may not provide a complete picture of the nuclear genome due to gender-biased migration, selection, or introgression. mtDNA markers have been extensively employed to study stock structure in vertebrates including as fish, birds, mammals, and reptiles (Avise *et al.*, 1998).

Random amplified polymorphic DNA (RAPD) markers

The amplified products of less functional regions of the genome that do not exhibit a robust phenotypic response to selection are known as RAPD markers. These DNA segments might develop more nucleotide mutations, which could be used to measure genetic differences between populations. High amounts of DNA polymorphisms can be found using RAPD, which amplifies genomic DNA by applying PCR using arbitrary nucleotide sequence primers (Williams *et al.*, 1990). Both coding and non-coding DNA sequences can be detected by the technique, and many of the most instructive polymorphic sequences come from repetitive (non-coding) DNA sequences found in the genome. The majority of the expanded loci are expected to be selection neutral as 90% of the nuclear genome (Haymer, 1994) of vertebrates is non-coding. Present/absent scores are assigned to RAPD loci, which are dominantly inherited as Mendelian markers. With the extra benefit of commercially accessible primers, RAPDs offer all the benefits of a PCR-based marker. You don't need to know the target DNA sequence or genome structure beforehand. The simplicity with which many loci and individuals can be tested at once is one of the additional benefits of RAPDs. The difficulty in differentiating between homozygotes and heterozygotes and the challenge of proving Mendelian inheritance of the loci are two drawbacks of the marker. The analysis is predicated on the idea that the populations being studied adhere to Hardy-Weinberg expectations. Furthermore, the use of this marker in fisheries science has been restricted by the existence of paralogous PCR products, which are distinct DNA sections with the same lengths that appear to be a single locus, as well as low repeatability brought on by the low annealing temperature utilized in the PCR amplification (Wirgin & Waldman, 1994).

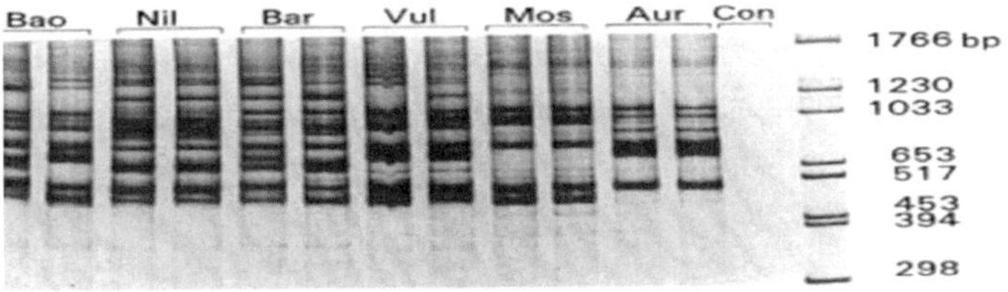

Fig. 1: RAPD patterns from each of six populations of tilapia fish using primer OPA 04 (Bardakci & Skibinski, 1994)

Single nucleotide polymorphism (SNP)

Polymorphisms resulting from point mutations that produce distinct alleles with alternative bases at a specific nucleotide position within a locus are referred to as single nucleotide polymorphisms (SNPs). The most prevalent polymorphism in the genome of any organism (including coding and non-coding sections) is represented by SNPs, which are increasingly becoming a focal point in the development of molecular markers. capable of being automated and exposing latent polymorphism that other markers and techniques are unable to identify (Morin *et al.*, 2004). A single nucleotide polymorphism (SNP) inside a locus has the potential to generate up to two alleles, each of which contains one of two potential base pairs at the SNP location. SNPs are inherited as co-dominant markers and they are categorized as bi-allelic. SNP discovery methods include SSCP analysis, heteroduplex analysis, and direct DNA sequencing. SNPs have limits and may only give minor utility in certain applications, such as relatedness (Morin *et al.*, 2004).

Microsatellite markers

Microsatellites are made up of several copies of simple sequence repeats (SSRs) that are organized in tandem and vary in size from one to six base pairs [ACA or GATA] (Tautz, 1989); Litt & Luty (1989). Microsatellite motifs are widespread in all species that have been investigated to date, and in fish, they have been predicted to appear as frequently as once every 10 kb (Wright, 1993). Microsatellites are uniformly dispersed across the genome, covering all chromosomes and regions. This statement is contradicted by data from whole genome sequencing. They have been discovered in non-gene sequences, introns, (Liu *et al.*, 2001) and gene coding regions. With a few to several hundred repeats, the majority of microsatellite loci are quite small. Changes in the number of repeat units can produce a huge number of alleles at each microsatellite locus in a population, regardless of the precise processes. As codominant markers, microsatellites have been inherited in a Mendelian manner. In several species that displayed nearly no variation at other markers, microsatellites were found to be useful (Taylor *et al.*, 1994). Nevertheless, using microsatellite markers requires a significant upfront financial outlay and effort. To create PCR primers, each microsatellite locus must be located and its flanking region sequenced. Small size variations between alleles of a particular microsatellite locus (as few as 2 bp in a locus composed of di-nucleotide repeats) may occur because of polymerase slippage during replication. Recently, microsatellites have gained a lot of popularity as a marker type in a range of genomic studies.

Expressed Sequence Tags (EST)

Knowing the genomes of fish species that are important for aquaculture is essential for learning more about the performance and producing characteristics. Scientists still view ESTs as the best technique for identifying target genes in aquaculture, despite the rise of techniques like serial analysis of gene expression (SAGE). Several innovative sequencing systems have recently been created that enable the de novo sequencing of the entire genomic transcriptome to produce expressed sequence tags. One of the main methods used for transcriptomics analysis in a variety of species, including fish, is EST sequencing (Liu, 2006). ESTs are thought to be significant in gene sequencing and full-length cDNA synthesis. It is one of the fastest molecular approaches for quickly identifying a lot of genes. EST analysis has gained extraordinary acceptance due to its unparalleled rate of gene discovery. Single-pass sequences produced by random sequencing of cDNA clones are known as expressed sequence tags (ESTs). Genes are identified using the EST, and expression profiling is used to analyze their expression. A rapid and useful study of genes expressed in particular tissue types, physiological circumstances, or developmental stages is facilitated by it. ESTs provide the creation of cDNA microarrays, which enable systematic investigation of genes with differential expression, in addition to their significant utility in genome mapping. ESTs are most suited for linkage mapping and physical mapping in animal genomics, such as those of pigs and cattle, where non-polymorphic DNA markers can be mapped using radiation hybrid panels. For species like oysters (*Crassostrea gigas*) and Atlantic salmon (*Salmo salar*), EST databases have already been created. similarly, the public now has access to databases of cDNAs and ESTs that provide the creation of essentially free markers for rainbow trout (Venkatesh *et al.*, 2005). The EST database has been among NCBI's fastest-mounting repositories in comparison. ESTs have been developed from species such as Atlantic salmon, rainbow trout, catfish, oysters, and shrimp.

Amplified Fragment Length Polymorphism (AFLP)

According to Vos *et al.* (1995), AFLP is a PCR-based method that can produce trustworthy, multi-locus genomic fingerprints. The fundamental idea of AFLP polymorphism, like that of RFLPs, is insertion and deletion (indels) and nucleotide replacement between and at restriction sites; like RAPD, it also involves base substitutions at primer binding sites during PCR. Because it enables the binding of adaptors with known sequences to DNA fragments produced by thorough digestion of genomic DNA, the technique is quite unusual. Despite having the same primary goal as RFLP, namely polymorphism, it permits the simultaneous investigation of multiple loci as an alternative to examining one locus at a time. The sections that follow provide a summary of the many benefits of the AFLP technique. There is no requirement to know any sequence information beforehand. Each experiment produces a number of bands. Across the genome, these bands are generated. The method can be repeated (Blears *et al.*, 1998). possess a strong capacity for discrimination. For comparison, the data can be kept in a database such as AmpliBASE MT (Majeed *et al.*, 2004). When determining an individual's origin from a known population, Campbell *et al.* (2003) investigated the efficacy of AFLP and discovered that it was more effective than microsatellite markers. These can be viewed as markers in genetic and physical mapping since the majority of AFLP segments match up to unmatched genomic sites. The method can also be useful for mapping genes and distinguishing closely related individuals at the subspecies level (Althoff *et al.*, 2007). The majority of genetic research has been committed to assessing and choosing superior features through interspecific hybridization and cross-breeding (Bondari, 1984; Dunham & Smitherman 1983; Hallerman *et al.*, 1986; Wolters & Johnson

1994; Padi, 1995). Researchers must accelerate the development of superior markers in conjunction with genetic loci controlling economically viable traits of superior breed stocks to increase selective competence, particularly for traits like disease resistance, feed conversion efficiency, and processing yields (Waldbieser *et al.*, 1998).

Quantitative Trait Loci (QTL) Mapping

A QTL is an area of the genome that contains genes linked to a specific attribute, such as disease resistance, growth, yield, or environmental stress tolerance. The availability of linkage maps is crucial to help select the required features to locate QTL for critical traits throughout the complete genome of the species of interest. Yue (2013) discussed that QTL for characteristics like cold and salinity tolerance, sex determination, growth, and disease resistance have been mapped in over 20 aquaculture fish species. These investigations may offer tools for marker-assisted selection in addition to the essential data regarding patterns and the rate of evolutionary changes (Groenen *et al.*, 2000). The progeny of hybrids between the divergent crosses usually exhibit segregation patterns and differing crosses as a result of mapping. The position of particular phenotypic features, which may be useful in breeding programs, is the applied purpose of mapping as it is now understood. There is fundamental and practical interest in mapping genes and QTLs in aquaculture species. Yue (2013) stated that to create a linkage map of any species, polymorphic markers, genotyping platforms, reference families, and software applications are necessary. By identifying the impact and number of genes influencing a trait, QTL mapping aims to support selective breeding (Naish & Hard, 2008). For over 20 aquaculture species, including finfish, mussels, and crustaceans, QTL analyses have been performed for critical traits such as thermal tolerance, spawning time, embryonic development, growth traits, stress, salinity tolerance, infectious hematopoietic necrosis, and bacterial cold-water disease. According to Sakamoto *et al.* (2000), QTL mapping is the actual use of marker-assisted selection in aquaculture. Of the 20 species that have been examined to date, rainbow trout and tilapia are among the most significant freshwater food fish species, whereas salmons are the most common marine fish species that are cultivated (Gjedrem & Baranski, 2009).

Array-Based Systems

Molecular markers have evolved during the last few decades in response to demands for increased throughput, lower costs per data point, and improved map resolution (Kumar, 1999; Gupta & Rustgi, 2004). A microarray is essentially a carefully designed collection of probes arranged on a hard surface where cellular cDNA copies of RNA are hybridized. Microarray technology offers a careful understanding of the patterns of gene expression in space and time in response to various stimuli that an organism encounters. It is an effective method for measuring the pattern of expression of many genes in an organism at once, which eventually aids in understanding the transcriptome and proteome. In addition to finding new genes linked to phenotypic traits, this method may reveal global patterns of expressed genes. According to Zhang *et al.* (2009), microarray technology is a potent technique in fisheries research for quickly screening genes or transcriptional profiles or for describing economically significant features, such as genes connected to growth, feed conversation, illness control, etc. There are two main platforms for DNA microarray in fisheries and/or aquaculture: Gene fragments are identified on High-Density Arrays or Expressed Sequence Tags (EST), which offer a "ready-to-use chip" for examining the simultaneous expression of a sizable number of genes. Synthetic oligonucleotides or cDNA fragments based on ESTs spotted on glass slides or nylon membranes were amplified using PCR in customized cDNA microarrays (Ju *et al.*, 2007). A cDNA microarray containing 4512 clones in zebra fish (Ton *et al.*, 2002), 11060 clones spotted in duplicate in European flounder (*Platichthys flesus*) (Cohen *et al.*, 2008), and a chip with roughly 4500 features for the African cichlid fish, *Astatotilapia burtoni* (Renn *et al.*, 2004) are just a few examples of the cDNA microarrays derived from many fish species. Similarly, the transcriptomics response to fast-twitching muscle atrophy in gravid trout versus sterile fishes has been studied using salmonid microarrays comprising roughly 160y (Ewart *et al.*, 2005; Salem *et al.*, 2006). Each of these microarrays is unique. Additionally, a DNA microarray assay was created by Pancker *et al.* (2004) for the real-time identification of five marine fish pathogens: *Aeromonas salmonicida*, *Vibrio vulnificus*, *Vibrio anguillarum*, *Vibrio parahaemolyticus*, and *Photobacterium damselae*. Additionally, González *et al.* (2004) developed a gene-specific DNA microarray to detect three shellfish pathogens: *Vibro vulnificus, V. cholerae*, and *V. parahaemolyticus*. Studies by Lindenstrom *et al.* (2006), Rise *et al.* (2004), Sigh *et al.* (2004), Fast *et al.* (2006), and Martin *et al.* (2006) describe how target gene expression patterns based on microarrays are used to examine immunological response, disease processes, and resistance in salmonids. The response of the rainbow trout transcriptome to chemical pollutants (Koskinen *et al.*, 2004), toxicogenomic profiling of hepatic tumour promoters (Tilton *et al.*, 2006), and gene expression in atrophying muscle (Salem *et al.*, 2006) have also been investigated using microarray technology. Through the identification of new genes, their expression patterns, and their response to different environmental stimuli at the genome level, the use of DNA microarray in fish biology and aquaculture could be extremely significant.

Molecular Techniques

DNA based Methods

In the 20th century, DNA-based methods emerged as a superior alternative to protein analysis for species identification. DNA's inherent stability and heat resistance surpass those of proteins. Furthermore, the polymerase chain reaction (PCR) enables the amplification of small DNA fragments, yielding sufficient information for accurate species identification (Lenstra, 2003). DNA's ubiquitous presence within an organism facilitates its extraction from various sources, including muscles, fins, and blood.

DNA-based methods play a crucial role in fish breeding programs. They enable family tracking, pedigree analysis of individual fish, identification of lineages, and the discovery of genetic markers associated with economically valuable traits. These traits include growth rate, survival rates, sexual maturity, disease resistance, the occurrence of deformities, and the efficient utilization of nutrients (Davis & Hetzel, 2000).

There are a few molecular techniques that have been mostly used during the past decade for fish identification as RFLP (restriction fragment length polymorphism), PCR sequencing (known also as FINS- forensically informative nucleotide sequencing) and PCR-specific primers. More recently, two methods have been developed, real-time PCR and microarray, both used in gene expression research as well as in species identification (Teletchea, 2009).

DNA Extraction and Purification

Methods

Phenol-chloroform extraction is the traditional method that separates nucleic acids from proteins, While Alcohol Precipitation, utilizes alcohol to precipitate DNA, allowing for its collection (Joshi *et al.*, 2024). While commercial kits are effective, they can be expensive. Alternative in-house protocols have been created to lower costs while ensuring high yield and purity (Fatima *et al.*, 2024).

Purification of DNA

DNA purification is a critical process in molecular biology, essential for isolating DNA from cellular components for various applications, including cloning, sequencing, and diagnostics. The evolution of purification techniques has significantly improved the yield and

purity of DNA, making it suitable for sensitive downstream applications. DNA purification is based on cell disruption, Nuclease Inactivation (Laughner, 2023) and Protein denaturation (Li & Liu, 2024).

Recent Advancements

- **Cost-Effective Protocols**: New methods have been optimized for high-yield DNA purification from PCR products, demonstrating comparable results to commercial kits (Fatima *et al.*, 2024).
- **Industrial Methods**: Innovations in industrial purification, such as using sodium caprylate, enhance recovery rates and reduce environmental impact (Tujing *et al.*, 2021).

PCR (Polymerase Chain Reaction)

The polymerase chain reaction (PCR) was invented in 1985 which amplifies specific DNA sequences. Because of the sensitivity, specificity, speed, and simplicity of the reaction, PCR has been used in a variety of applications in the past years; such applications include characterising the structure and expression of genes; identifying disease-causing genes and pathogens; diagnosing inherited disease prenatally; and DNA fingerprinting in forensics, agriculture, and archaeology.

DNA can be analysed using techniques such as sequencing, DNA-DNA hybridization and the polymerase chain reaction (PCR), which are based on the detection of species-specific DNA sequences in food products. In particular, PCR-based techniques have a high potential because of their rapidity, increased sensitivity and specificity (Lockley & Bardsley, 2000).

The amplification products or amplicons can then be analysed further by a variety of methods such as PCR-sequencing, multiplex-PCR, PCR restriction fragment length polymorphism (PCR-RFLP), PCR-single strand conformation polymorphism (PCR SSCP), Random Amplified Polymorphic DNA (RAPD), Real-Time PCR and PCR lab-on-a-chip (Gil, 2007).

Real-time PCR

Many processed fish products, like cakes, soups, and fishmeal, often contain a mix of different fish species. This can lead to fraudulent practices where cheaper fish are added to products labelled as more expensive ones. To ensure accurate labelling, reliable methods for identifying and quantifying the specific fish species in these products are crucial. Since heat processing can damage proteins, DNA-based methods like real-time PCR are preferred for analysis. Real-time PCR has proven effective in food authentication and quantification, making it a valuable tool for accurately identifying and quantifying the different fish species present in processed food products (Brodmann & Moor, 2003; Lo´pez-Calleja *et al.*, 2007).

Real-time PCR often employs TaqMan probes, which are short DNA strands labelled with a reporter dye and a quencher dye. These probes attach to specific target DNA sequences within the region amplified by PCR primers. During PCR, the Taq DNA polymerase, while synthesizing new DNA, also degrades the bound probe. This degradation separates the reporter dye from the quencher, allowing the reporter to fluoresce. The increase in fluorescence signal, measured in real-time, directly correlates with the amount of target DNA present (Holland *et al.*, 1991).

PCR lab-on-a-chip

Miniaturizing PCR reactions offer significant advantages by accelerating the heating and cooling cycles. This is due to improved heat and mass transfer within smaller volumes. As a result, the PCR process can be performed much faster, leading to quicker results and increased throughput. Additionally, the rapid temperature changes enhance the accuracy and efficiency of crucial steps like primer binding and DNA synthesis (De Mello, 2001).

This technology that uses microfluidic devices, also called PCR lab-on-a-chips, has been recently used for fish species authentication. Thus, Dooley *et al.* (2005) used this technology to discriminate admixtures of salmon/trout. These authors achieved the detection of 5% salmon DNA in trout DNA. This technology permitted an improvement in PCR-RFLP fragment resolution and detection in comparison to conventional gel-based methods.

PCR is utilized in fisheries and aquaculture for family tracking, pedigree determination, and identifying genetic lineages. It enhances understanding of economically important traits such as growth, disease resistance, and nutrient utilization, contributing to improved production efficiency and animal health (Zorka *et al.*, 2019).

DNA Sequencing

DNA sequencing is revolutionizing numerous scientific fields, from archaeology and anthropology to genetics, biotechnology, and forensics. This technology is driving groundbreaking discoveries that are reshaping our understanding of various disciplines. However, alongside these advancements, critical ethical, public health, and safety concerns are emerging.

DNA Sequencing Strategies

Dideoxy Sequencing/ sanger sequencing

Dideoxy sequencing, also known as Sanger sequencing, proceeds by primer-initiated, polymerase-driven synthesis of DNA strands complementary to the template whose sequence is to be determined complementary to the template whose sequence is to be determined (Shendure *et al.*, 2008).

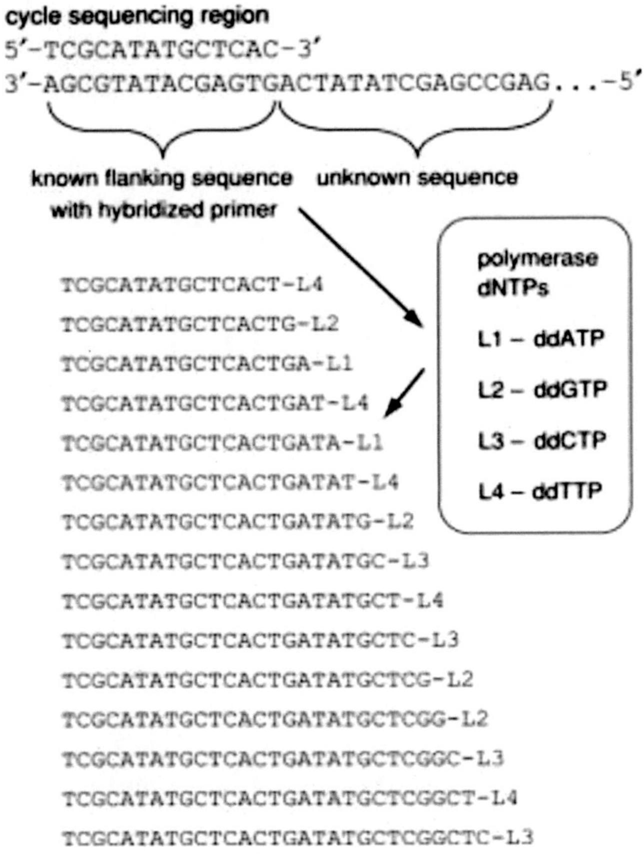

Fig. 2: Schematic representation of the basic principle involved in dideoxy sequencing. The sequencing template consists of an unknown region whose sequence is to be determined, flanked by known sequence to which a sequencing primer can be hybridized. Cycle sequencing (multiple cycles of primer annealing, primer extension, and denaturation) are performed with polymerase, dNTPs, and fluorescently labelled ddNTPs (where a different label is present on each species of ddNTP) (Shendure *et al*, 2008).

Sanger sequencing relies on generating many DNA fragments of varying lengths, each ending with a specific type of chain-terminating dideoxynucleotide (ddNTP).

- Template Preparation: The DNA to be sequenced (the template) is typically amplified using PCR or isolated from a plasmid.
- Sequencing Reaction:
  - The sequencing reaction includes the template DNA, a primer, DNA polymerase, normal nucleotides (dNTPs), and a small amount of one of the four types of ddNTPs (ddATP, ddGTP, ddCTP, or ddTTP).

- When a ddNTP is incorporated into the growing DNA strand, it prevents further elongation, creating a fragment of a specific length.
- Four Reactions: Four separate reactions are performed, each containing a different ddNTP. This results in four sets of DNA fragments, each ending with a specific type of ddNTP.
- Gel Electrophoresis: The fragments from each reaction are separated by size using gel electrophoresis.
- Sequence Determination: By analysing the pattern of bands on the gel, the order of nucleotides in the original DNA template can be determined (Shendure *et al.*, 2008).

Cyclic Array Sequencing

Cyclic array sequencing technologies, exemplified by platforms from companies like Illumina and Roche, revolutionize DNA sequencing by simultaneously processing millions of DNA fragments on a single surface. This "massively parallel" approach significantly reduces sequencing costs.

- These platforms immobilize numerous DNA fragments, each potentially unique, on a surface, creating a dense array.
- Parallel Processing: Reagents are applied to the entire array at once, enabling simultaneous processing of all DNA fragments
- Cyclic Process: The sequencing process involves repeated cycles of enzymatic reactions and imaging, where the identity of one nucleotide at a time is determined for each fragment.
- Data Acquisition: Images of the array are captured after each cycle, providing information about the nucleotide sequence of each fragment (Shendure *et al.*, 2008).

RFLPs (Restriction Fragment Length Polymorphisms)

RFLPs are genetic markers with several advantages compared to conventional markers. They describe directly the genotype instead of the phenotype and, therefore, are not influenced by the environment. The number of RFLP markers which can be mapped is only limited by the molecular differences existing between available genotypes; it is also proportional to the effort applied to detect them (Shendure *et al.*, 2008).

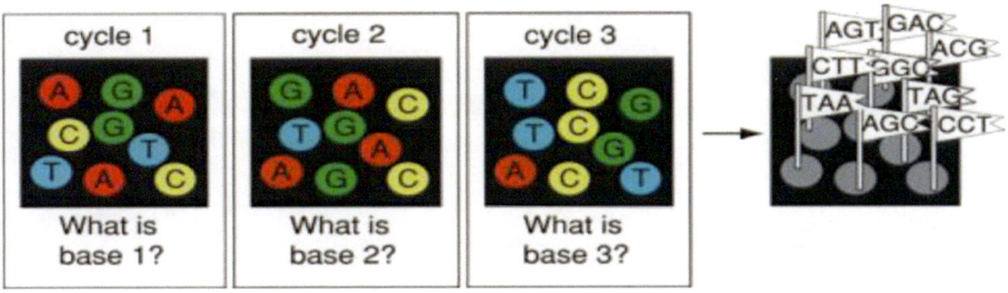

Fig. 3: The concept of cyclic array sequencing platforms involves an array of DNA features to be sequenced, immobilized to constant locations on a solid substrate. At each cycle, the identity of a single base position is interrogated at each feature. Data are collected at each cycle by imaging of the array (Shendure *et al.*, 2008).

Application of Restriction Fragment Length Polymorphism (RFLP) in Aquaculture and Fishery

RFLP (Restriction Fragment Length Polymorphism) is a straightforward, reliable, and relatively cost-effective technique (Aranishi, 2005). It exploits variations in the lengths of DNA fragments generated by specific restriction enzymes (Rasmunsen *et al.*, 2008). The target genes are amplified using PCR and subsequently cleaved by these enzymes into smaller fragments of varying sizes (Liu & Cordes, 2004). These size differences are then used to distinguish between different fish lineages or species. The fragments are separated based on size using agarose gel electrophoresis. The gel is then visualized under UV light and photographed using a gel documentation system. The sizes of the DNA fragments are determined by comparing them to a known DNA size standard.

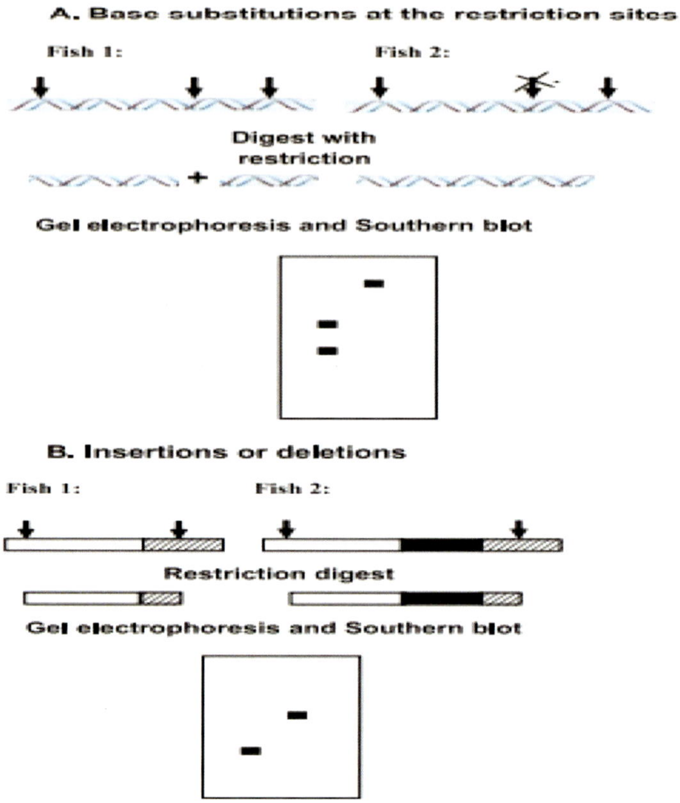

Fig. 4: Molecular basis of restriction fragment length polymorphism (RFLP). Genomic DNA is digested with a particular restriction enzyme (arrows) (A) Base substitution within the restriction site can knock out the site (conversely, new sites can arise by base substitutions as well). Loss of a restriction within the locus of interest leads to loss of one fragment and increase of fragment size that can be resolved (by gel electrophoresis. (B) Insertion of a piece (black bar) between two restriction sites within the locus leads to an increase in the fragment size that can be resolved by gel electrophoresis (reversely a deletion should reduce the fragment size). In both cases, classical means of detection relies on Southern blot using specific probes (Liu *et al.*, 2004).

This method is particularly valuable for determining the genetic differences between distinct populations of the same species that inhabit geographically diverse regions. The fish populations must be studied exhibit high levels of genetic diversity, indicating that they have not undergone extensive interbreeding.

The integration of RFLP in genetic management allows for the estimation of effective population sizes and inbreeding coefficients, crucial for maintaining genetic diversity in aquaculture (Taniguchi *et al.*, 2003).

RFLP (restriction fragment length polymorphism) is utilized in fisheries for determining genetic lineages in polymorphic fish species, enabling family tracking, pedigree determination, and identification of markers linked to economically important traits such as growth, disease resistance, and nutrient utilization (Zokra *et al.*, 2019).

Application of Molecular Systematics in Aquaculture and Fisher

Applications in Fisheries

Intraspecific Variation

Molecular markers have been widely used to study genetic differences both within and between species. For instance, RAPD markers have been employed (Bardakci & Skibinski, 1994) to differentiate between tilapia species and subspecies, while isozymes have been used for species identification in Sparidae (Alarcón & Alvarez, 1999). Ideally, multiple diagnostic markers should be used to minimize the risk of misidentification due to potential variations within the same species.

Phylogeography and Phylogenetic

The primary goal of quantitative genetic stock identification is to objectively classify individuals or samples into distinct groups. These analyses also aim to understand the evolutionary history and geographic distribution of these groups. Research in this field began in the mid-1970s with the advent of mtDNA analysis in population genetics. Interpreting the distribution of genetic lineages requires a multidisciplinary approach, drawing upon knowledge from fields such as molecular genetics, population genetics, phylogenetics, demography, animal behaviour, and historical geography (Avise, 1994).

Population Structure: Between and Within Population Variations

Understanding genetic variation within and between populations is crucial for both fisheries management and conservation efforts. A key goal is to identify distinct groups within a species that have limited interbreeding. Scientists need to identify populations that do not interbreed, assess the level of gene exchange between different groups, and monitor how the genetic makeup of these populations' changes over time (Carvalho & Hauser, 1995).

Mixed Fishery and Genetic Stock Identification

Species are often composed of distinct subpopulations or "stocks," which are the units that fisheries target or manage. In some cases, these distinct stocks can mix together. Managing fisheries where multiple species, life stages, and/or individuals from different stocks are present (mixed fisheries) presents significant challenges for fisheries managers. Examples of mixed fisheries include salmonids in feeding grounds or during their migration in rivers (Cross *et al.*, 2004). The primary goal in these mixed fisheries is often to determine the origin of individual fish. Over the past 10-15 years, genetic data has become increasingly valuable in studying stock structure and understanding the contributions of different stocks to these mixed fisheries (Shaklee & Currens, 2003). Genetic stock identification (GSI) is a system specifically designed to identify the species or population of origin for individual fish.

Applications in Aquaculture

Molecular markers have a broad range of applications in aquaculture. Some are described below.

Manage and conserve stocks

Molecular markers can be used to pinpoint the location of genes influencing traits like growth rate and disease resistance (QTLs). This information is valuable for marker-assisted selection in breeding programs. Gene expression analysis can identify genes that play a role in specific traits, providing targets for further investigation. Genetic markers can also be used to identify the origin of fish, which is crucial for tracking hatchery stocks, protecting intellectual property, and monitoring the impact of fish releases into natural environments (Hallerman, 2006).

Diagnose Diseases

DNA-based techniques, particularly PCR, have proven effective in diagnosing various viral diseases in cultured fish and shrimp. These methods have been developed for a wide range of pathogens, including CCV, IHNV, IPNV, VHSV, VNNV, and bacterial infections like those caused by *Renibacterium salmoninarum*. For example, in Japan, PCR screening of striped jack broodstock for VNNV has been successful in preventing the transmission of this virus to offspring. By selecting virus-free broodstock (PCR-negative), the vertical transmission of the disease can be effectively controlled (Muroga, 1997).

Improve Disease Resistance

Identify genes associated with disease resistance (Genes that are resistant to pathogens like Major Histo-Compatibility genes (MHC) and breed for these traits) can be identified and used for selection programmes to produce disease-resistant strains of fin or shellfishes (Sukumaran & Gopalakrishnan, 2019).

Conclusion

Molecular systematics has emerged as a cornerstone of modern evolutionary biology. Molecular systematics has revolutionized our understanding of evolutionary relationships by employing a diverse array of molecular methods. These methods, encompassing DNA sequencing, PCR, and comparative genomics, provide a wealth of data for constructing robust phylogenetic trees. By analyzing genetic material, researchers can overcome the limitations of morphological data, such as homoplasy and convergent evolution, and gain a deeper understanding of evolutionary processes. The ongoing development and refinement of molecular techniques, coupled with the ever-growing availability of genomic data, promise to further enhance our understanding of the tree of life and drive continued advancements in evolutionary biology.

References

Alarcón, J. A., & Alvarez, M. C. (1999). Genetic identification of sparid species by isozyme markers: application to interspecific hybrids. Aquaculture, 173(1-4), 95-103.

Aranishi, F. (2005). Rapid PCR-RFLP method for discrimination of imported and domestic mackerel. Marine Biotechnology, 7(6), 571–575. DOI: https:// doi.org/10.1016/j. aquaculture.2004.05.027

Avise, J. C. (1994). Molecular Markers, Natural History and Evolution. Chapman and Hall, New York.

Avise, J.C., Helfman, G.S., Saunders, N.C. and Hales, L.S. (1986). Mitochondrial DNA differentiation in North Atlantic eels: Population genetic consequences of an unusual life history pattern. Proceeding of the National Academy Science (USA), 83(12), 4350-4354.

Avise, J.C., Walker, D. and Johns, G.C. (1998). Speciation durations and Pleistocene effects on vertebrate phylogeography. Proceedings of the Royal Society of London Series B, 265(1407), 1707-1712.

Baker, A.J. and Marshall, H.D. (1997). Molecular evolution of the mitochondrial genome. In: Mindell, D.P. Ed., Avian Molecular Evolution and Systematics, Academic Press, San Diego, 51-82.

Bardakci, F. and Skibinski, D.O.F. (994). Application of the RAPD technique in tilapia fish: species and subspecies identification. Heredity, 73: 117–123.

Birky, C.W., Fuerst, P. and Maruyama, T. (1989). Organelle gene diversity under migration, mutation, and drift: equilibrium expectations, approach to equilibrium, effect of heteroplasmic cells, and comparison to nuclear genes. Genetics, 121(3), 613-627.

Brodmann, P. D., & Moor, D. (2003). Sensitive and semi-quantitative TaqMan™ real-time polymerase chain reaction systems for the detection of beef (Bos taurus) and the detection of the family Mammalia in food and feed. Meat Science, 65(1), 599-607.

Brown, B. and Epifanio, J. (2003). Nuclear DNA. In: Hallermann, E.M. Ed., Population Genetics: Principles and Applications for Fisheries Scientists, American Fisheries Society, Bethesda, 458-472.

Brown, J.R., Bechenbach, A.T. and Smith, M.J. (1993). Intraspecific DNA sequence variation of the mitochondrial control region of white sturgeon (Acipenser transmontanus). Molecular Biology Evolution, 10(2), 326-341.

Brown, W.M. (1985). The mitochondrial genome of animals. In: MacIntyre, R.J. Ed., Molecular Evolutionary Genetics, Plenum, New York, 95-130.

Carvalho, G.R. and Hauser, L. (1995). Molecular genetics and the stock concept in fisheries. G.R. Carvalho and T.J. Pitcher (Eds.), Molecular Genetics in Fisheries. Chapman & Hall, London: 55-80.

Chow, S. and Kishino, H. (1995). Phylogenetic relationships between tuna species of the genus Thunnus (Scombriidae: Teleosrei): Inconsistent implications from morphology, nuclear and mitochondrial genomes. Journal of Molecular Evolution, 41, 741-748.

Chow, S., Clarke, M.E. and Walsh, P.J. (1993). PCR-RFLP analysis of thirteen western Atlantic snappers (subfamily Lutjaninae): A simple method for species and stock identification. Fisheries Bulletin (US), 91, 619-627.

Chow, S., Okamoto, H., Uozumi, Y., Takeuchi, Y. and Takeyama, H. (1997). Genetic stock structure of the swordfish (Xiphias gladius) inferred by PCR-RFLP analysis of the mitochondrial DNA control region. Marine Biology, 127(3), 359-367.

Davis, G. P., & Hetzel, D. S. (2000). Integrating molecular genetic technology with traditional approaches for genetic improvement in aquaculture species. Aquaculture research, 31(1), 3-10.

De Mello, A. J. (2001). DNA amplification: does 'small' really mean 'efficient'? Lab on a Chip, 1, 24Ne29N

Dooley, J. J., Sage, H. D., Brown, H. M., & Garrett, S. D. (2005). Improved fish species identification by use of lab-on-a-chip technology. Food Control, 16, 601e607

Ferguson, A., Taggart, J.B., Prodohl, P.A., McMeel, O., Thompson, C., Stone, C., McGinnity, P. and Hynes, R.A. (1995). The application of molecular markers to the study and conservation of fish populations with special reference to Salmo. Journal of Fish Biology, 47(A), 103-126.

Fisher, R.A. (1930). The Genetical Theory of Natural Selection. Oxford University Press, UK.

Gil, L. A. (2007). PCR-based methods for fish and fishery products authentication. Trends in Food Science & Technology, 18(11), 558-566.

Gold, J.R. Sun, F. and Richardson, L.R. (1997). Population structure of red snapper from the Gulf of Mexico as inferred from analysis of mitochondrial DNA. Transaction of American Fisheries Society, 126(3), 386-396.

Gold, J.R., Richardson, L.R., Furman, C. and King, T.L. (1993). Mitochondrial DNA differentiation and population structure in red drum (Sciaenops ocellatus) from the Gulf of Mexico and the Atlantic Ocean. Marine Biology, 116(2), 175-185.

Graves, J.E., McDowell, J.R. and Jones, M.L. (1992). A genetic analysis of weakfish Cynoscion regalis stock structure along the mid-Atlantic coast. Fisheries Bulletin, 90, 469-475.

11

Genetic Linkage Analysis for Trait Improvement in Aquaculture Species

Binal Rajeshbhai Khalasi*, Annappaswamy T.S, Sheethal K.U, Prabhutva Chaturvedi, Bhavy A. Dalsaniya and Mayurkumar U. Tandel

College of Fisheries, Mangaluru, Karnataka Veterinary, Animal and Fisheries Sciences University, Bidar, Karnataka (575 002), India

Abstract

Aquaculture, the practice of cultivating aquatic organisms for human consumption, has become a vital aspect of global food supply. This growth highlights the need for improved breeding strategies that focus on increasing yield, enhancing disease resistance, and promoting environmental sustainability. In aquaculture genetics, linkage mapping has emerged as a crucial method for identifying quantitative trait loci (QTL) linked to important characteristics in aquatic organisms. This technique has helped analyse genetic markers associated with vital traits such as rate of growth, resistance against diseases, and tolerance to abnormal conditions called stress. This approach advances our knowledge of the genetic structures underlying these characteristics and aids in the creation of selective breeding initiatives to increase aquaculture productivity. Nevertheless, linkage mapping encounters several challenges, such as the complications of polyploidy, a lack of high-density genetic markers and complete reference genomes, extended generation times, and significant environmental effects on trait manifestation. Furthermore, low marker density and high resource requirements can diminish the efficacy of conventional mapping methods. To tackle these issues, there is a noticeable trend towards adopting more advanced genomic approaches, including Genomic Selection, Genome-Wide Association Studies (GWAS) and Whole Genome Sequencing. While linkage mapping remains a valuable resource, especially for species with limited genomic information, its combination with these cutting-edge techniques is crucial for speeding up genetic advancements and promoting sustainable aquaculture practices.

Keyword: *Aquaculture, Genetics, Linkage mapping, Genetic marker, Quantitative trait loci*

Introduction

Aquaculture is rapidly outpacing all other sectors in agricultural development. The study of fish genetics gained traction in the 20th century, driven by advancements in the understanding of Inheritance and reproduction. During the 1960s, selective breeding programmes aimed at enhancing genetic qualities became more common (Gjedrem and Baranski, 2009). Currently, more than 60 fish and shellfish species are going through selective breeding (Gjedrem and Baranski, 2009). Despite this, many critical traits like resistance to disease, efficiency of feed conversion, fatty acid composition and quality of meat are challenging to assess in breeding prospects but significantly impact the yield and profitability of several aquaculture species. Genetic mapping refers to the process of creating a genetic map (Van Ooijen and Jansen, 2013). Making genetic linkage maps is still essential for carrying out genetic and genomic research and many aquaculture species have had such maps developed. Developments in genotyping and sequencing technologies have made it the creation of higher-density, high-resolution (HR) genetic linkage maps. However, there are now important tasks in linkage analysis due to the usage of a large number of molecular markers. High-density linkage maps are valuable tools for studying genomic biology and evolutionary genetics because they serve as the foundation for QTL (quantitative trait loci) mapping. They facilitate the quantification of recombination rates, linkage disequilibrium levels, and chromosomal rearrangements across chromosomes, between sexes, and among different populations. The development of marker-Assisted Selection (MAS) has gained significant consideration and has significantly accelerated genetic breeding efforts in numerous cultured aquatic organism (Li, 2019; Yu, 2019). Quantitative trait loci (QTL) linked to economically important variables are frequently mapped using genetic linkage maps, which are an invaluable tool for examining genomic regions. (Huang, 2020; Wan 2017; Zhou, 2021). Growth rate, flesh quality, and disease resistance are only a few of the important features that are regulated by several genes or loci. The use of genetic maps in QTL (quantitative trait loci) mapping is essential for determining the locations of genes linked to these traits. (Huang, 2020; Li, 2019; Wan, 2017; Wang, 2018). The mapping of trait-related quantitative trait loci has been greatly improved by recent developments in high-throughput next-generation sequencing techniques (You, 2020). Research in this field has greatly advanced through the use of a variety of techniques, including restriction site-associated DNA sequencing (RAD-seq), genotyping-by-sequencing, specific-locus amplified fragment sequencing, and genome-wide association studies (GWAS). Of these, RAD-seq is thought to be an appropriate, robust, and economical approach that has demonstrated efficacy in detecting genotypes and single nucleotide polymorphisms (SNPs) (Andrews, 2016). Linkage mapping in aquaculture is a valuable tool that has revolutionized selective breeding. By revealing the genetic foundations of key traits, it allows for more efficient and sustainable aquaculture practices. The integration of linkage mapping with other genomic technologies is expected to enhance productivity, improve disease management, and ensure the long-term sustainable growth of aquaculture industries worldwide.

Genetics: Basic Concepts

Cells are the basic functional components of all alive organisms. They follow instructions that control their activities and are found within the deoxyribonucleic acid (DNA) sequence. DNA, which is composed of similar chemical components in all organisms, is abbreviated as A, T, C, and G. Although qualities can be impacted by more than one gene, a gene is a section of DNA that codes for a particular trait. The location of a gene on a chromosome is referred to as its locus. Each chromosome contains numerous genes, which are essential physical and functional elements of heredity. Geneticists use maps to indicate the specific locations of genes on chromosomes. One type of map, known as a cytogenetic map, identifies a gene's position by analyzing the unique banding patterns that appear when chromosomes are dyed with certain chemicals. Additionally, by sequencing the DNA nucleotides that compose the chromosome, molecular maps can precisely determine the chromosomal position of a gene.

What is Linkage or Linkage Mapping?

Linkage is the phenomenon in which two or more genes are situated on the same chromosome, resulting in their tendency to be transmitted together. A linkage map is a Graphical illustration of genes on a chromosome created through linkage analysis which represents their relative positions rather than physical distances. These positions are determined by how frequently two gene loci are inherited together. In linkage mapping, the hereditary units, or genes, that influence an individual's traits are carried on chromosomes, with each individual having numerous genes responsible for different characteristics. Finding genes linked to quantitative traits like growth, sex determination, and illness resistance requires the use of linkage mapping. When homologous chromosomes cross across, the frequency of recombination events between genetic markers is analysed to create linkage maps. The key principle is that the more distant two markers are on the same chromosome, the higher the likelihood of recombination between them, while markers that are closer together have a lower chance of recombination.

Types of Linkage

Linkage is categorized into two types: complete linkage and incomplete linkage.

Complete Linkage: Linkage occurs when genes on the same chromosome are inherited together due to the limited occurrence of crossing over. Complete linkage allows for the inheritance of parental traits in their original combinations. Although it is rare, complete linkage has been observed in some heterogametic animals.

Incomplete Linkage: Genes located on similar chromosomes often separate due to crossing over, resulting in the production of recombinant offspring alongside parental types. Generally, the number of recombinant individuals is smaller than what would be anticipated with independent assortment. Each of the four distinct gamete types—two parental and two recombinants—would arise with an equal frequency of 25% under the independent assortment principle. However, with linkage, the frequency of each parental type exceeds 25%, while the frequency of each recombinant type falls below 25%.

Historical Background

With a history of fewer than 160 years, genetics is a relatively new science that came into being in the middle of the 1800s thanks to Gregor Johann Mendel's pioneering work. Mendel carried out painstaking research on pea plants, closely examining how different features were passed down over the generations. His perceptive research resulted in the development of basic ideas that are now referred to as Mendel's Laws. "First Law" of segregation, "Second Law" of independent assortment, and "Third Law" of dominance are some of these laws. The law of segregation states that during gamete creation, the alleles of each gene separate, leaving just one allele per gene in each gamete. The law of independent assortment states that genes corresponding to different traits segregate independently during gamete formation. There are recessive and dominant alleles; the law of dominance states that dominant alleles outnumber recessive alleles (Bateson & Mendel 190; Miko; 2008). The field of genetics was founded on these ideas. Given our knowledge of genetic linkage, it may be concluded that Mendel's observations were only partially accurate because genes that are adjacent to one another on a chromosome frequently do not assort separately because of their connection. Unlike Mendel's first theory in his law of independent assortment (Morgan, 1911), the concept of genetic linkage originated from the finding that two genes located near to one another on the same chromosome do not assort independently. With this insight, Alfred Sturtevant, a student of Morgan, created the first linkage map. The history of genetic linkage analysis parallels the advancements in technologies for detecting polymorphisms. Initially, polymorphisms were identified at the phenotypic level, which limited the ability to detect linkage to simple traits due to the absence of molecular markers. The elucidation of DNA's structure by Watson and Crick in 1953 marked the beginning of genetic analysis at the molecular level. However, for more than 20 years following this discovery, progress in linkage analysis was slow because high molecular weight genomic DNA was difficult to analyze. The breakthrough came in 1973 with the discovery of restriction endonucleases, which allowed high-molecular-weight DNA to be assimilated into smaller fragments, facilitating the examination of polymorphisms.

Single-nucleotide polymorphism detection was greatly enhanced in 1977 with the development of DNA sequencing technology (Maxam & Gilbert, 1977; Sanger, Nicklen & Coulson, 1977). In the following ten years, linkage mapping using DNA markers advanced quickly, especially with restriction fragment length polymorphisms (RFLPs), which by the middle of the 1980s were the most widely used marker system. When polymerase chain reaction (PCR) technology was introduced in 1986, it significantly accelerated the development of DNA marker technologies, especially when microsatellites were used in linkage mapping.

Using a range of polymorphic DNA markers, including amplified fragment length polymorphism (AFLP), RFLPs, microsatellites, and rapid amplification of polymorphic DNA (RAPD), and others, linkage mapping continued to be a major focus of genetics research for almost another 15 years. But in contrast to modern marker technologies, these were comparatively costly, labour-intensive, and slow.

Previously, the scarcity of polymorphic markers and high-throughput technologies hindered linkage mapping advancements. Although SNPs, or single-nucleotide polymorphisms were detected in the 1970s, productive and cost-efficient techniques for analyzing SNPs did not become available until the early 21st century. Similarly, the techniques needed to identify greater numbers of SNPs were lacking. It was only with current developments in next-generation sequencing (Ren, 2000) that genome-wide SNP identification became feasible and cost-effective (Davey, 2011). The creation of genetic linkage maps with high density and high resolution has been made easier by the availability of extremely effective genotyping array platforms.

It is reasonably easy to manage a linkage mapping project with a few hundred markers, but it becomes much more difficult when dealing with hundreds of thousands or even a large number of markers. The use of numerous markers complicates bioinformatics analyses due to computational challenges. More importantly, it becomes more challenging to discern genuine linkage from pseudo-linkage when there are many markers.

Genetic Organization

Genes are systematically arranged on chromosomes, with each gene occupying a particular "address" or position that can be identified through genetic mapping. When genes are situated on different chromosomes, their inheritance follows the principles related to independent assortment. Conversely, linked genes found on the same chromosome are more likely to cosegregate, meaning they tend to be inherited together by offspring more frequently than would be forecasted based on independent assortment. The biological mechanism that separates linked genes is called crossing over, which occurs during meiosis and results in genetic recombination.

Crossing-Over

During meiosis in diploid organisms, the chromatids of homologous chromosomes align and form bivalents. In Meiosis I, these homologous chromatids align to physically exchange segments of chromosomes. The point where this reciprocal exchange occurs is referred to as a chiasma. The process of crossing over involves complex mechanisms that are not entirely understood, including the identification of homologous sites on chromatids, the breakage of chromosomes, and their subsequent rejoining. Crossing-over events occur somewhat randomly during meiosis, with most species experiencing one to a few crossing-over events per chromosome per meiosis. Therefore, genes on the same chromosome that are physically closer to one another are less likely to separate, which lowers the possibility of creating genetically recombinant gametes. This principle is fundamental to genetic mapping, where the genetic distance between genes reflects the likelihood of crossing over occurring between linked genes.

The Role of Recombination in Gene Mapping

Physical crossingover is a common event during Meiosis I that results in the rearrangement of heterozygous homologous chromosomes into new combinations. This process is referred to as recombination. Any two genes on a chromosome are capable of recombination, and the frequency of crossing over is determined by the proximity of the genes. When two genes are positioned far apart, such as at opposite ends of the chromosome, the incidence of crossover and non-crossover events is equal. Conversely, genes that are located closer together experience a lower frequency of crossing over, resulting in a higher number of non-crossover gametes compared to crossover gametes. This concept is shown in the figure below.

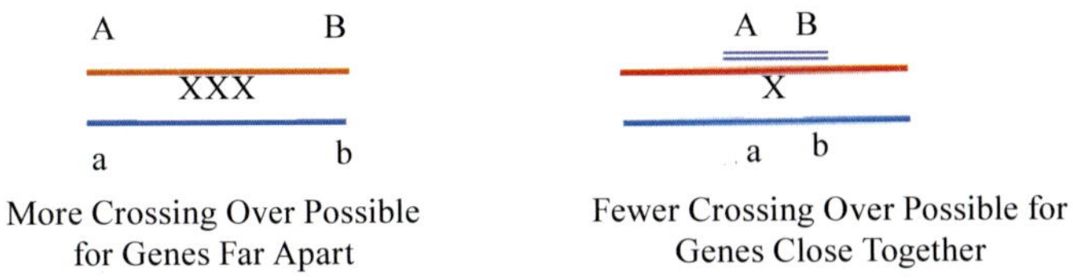

Fig. 1: Overview of crossing over
Source: https://www.ndsu.edu/pubweb/~mcclean/plsc431/linkage/linkage2.htm

Linkage Phase

Genes or Alleles that are Located on the Same Shromosome Produce two Different Kinds of Gametes

a) The absence of crossing over leads to the production of parental gametes.

b) The occurrence of crossing over leads to the production of recombinant gametes.

On homologous chromosomes, there are two potential allele configurations known as "linkage phases." Both parental and recombinant gametes' allelic composition is strongly influenced by the early allele arrangement (coupling or repulsion phase). (Semagn, 2006). Two stages of a linkage relationship-the coupling phase and the repulsion phase—are associated with the arrangement of genes on the chromosome.

The term **coupling** refers to the arrangement where two dominant alleles from one parent and two recessive alleles are found on the same homologous chromosome, represented as AB/ab. In this situation, both dominant alleles are transmitted together in one gamete from that parent, while the two recessive alleles are transmitted together in the other parent. Conversely, **repulsion** refers to a situation where the parents are heterozygous (Aa/Bb). Here, the two dominant alleles originate from different parents. Essentially, the genes are positioned on opposite homologous chromosomes, which is why it is called the repulsion phase. It should be possible to classify gametes as either parental or recombinant based on the observed distribution of gamete types.

Generally, parental gametes are usually the most prevalent, while recombinant gametes are less frequent. The reason for this is that the proportion of recombinant gametes cannot surpass 50% of the total gamete population. By analyzing the frequencies of different gamete types, accurately assessing linkage distances requires knowing whether genes involved in the coupling or repulsion phase were present in the original cross.

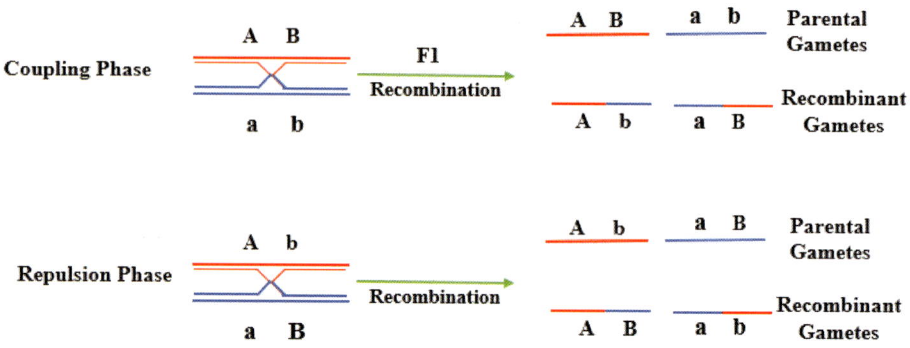

Fig. 2: The concept of linkage can be illustrated by examining gamete composition in coupling and repulsion crosses (Refer to the colour plate section for a coloured version of this figure, *Source:* (Yun Li 2017)

Linkage Mapping Includes the Given Steps

Determination of Linkage Groups (Mapping Populations): A mapping population refers to a collection of individuals utilized for gene mapping purposes. Any gene mapping study must choose the right mapping population in order to be successful. Before starting the mapping of a species' chromosomes, it is important to know the precise number of chromosomes. Additionally, the total number of genes within that species must be determined through hybridization experiments involving wild and mutant strains. This same hybridization technique can also identify which phenotypic traits consistently remain linked during inheritance, allowing for the determination of different linkage groups within a species.

Determination of Map Distance: Given that conventional units of measurement used in light microscopy are not applicable for measuring distances between genes on a chromosome, geneticists have adopted an arbitrary unit known as a map unit. This measure is the frequency of recombination events happening at a rate of 1%, represents the distance between genes.

Determination of Gene Order: Once the relative distances between genes within a linkage group are known, their linear order can be more easily determined

Combining Map Segments: Ultimately, the individual map segments derived from a complete chromosome are integrated to generate a comprehensive genetic map.

Interference & Coincidence: In higher organisms, the formation of one chiasma decreases the likelihood of another chiasma occurring in a nearby region. Interference is the phenomenon that occurs when one crossover event influences the likelihood of another crossover occurrence. The centromere exhibits a similar interference effect. Interference is quantified using the coefficient of coincidence.

Linkage Groups

A linkage group is a linearly organized set of related genes that are typically inherited together, except for crossing over. This is analogous to a chromosome that has a linear gene sequence that is connected and passed down as a whole. Both homologous chromosomes belong to the same linkage group because they have genes at the same loci that are either comparable or allelic. Consequently, the number of linkage groups in an individual's genome is equal to the number of chromosomes (all chromosomes in haploid animals or homologous pairs in diploid species). This idea is referred to as the linkage group limitation principle. The size of a linkage group is determined by the chromosome's size; shorter chromosomes tend to have smaller linkage groups, while longer chromosomes have larger linkage groups.

The LOD Score

Linkage maps are constructed by estimating genetic distances between markers based on their recombination frequencies. It is essential to establish statistical significance when evaluating these probabilities to ensure confidence in the map's reliability. The LOD score, the main statistical method for establishing whether two loci are linked, is a basic idea in genetic linkage research. The technique, initially introduced by Newton E. Morton in 1955, derives several LOD scores from various suggested link distances through an iterative approach. The LOD score compares the likelihood of witnessing the same data by chance alone to the probability of acquiring the observed data assuming that the two loci are linked. The calculation of the LOD score follows a specific formula:

$$\text{LOD} = \text{Log}_{10}\frac{\text{Likelihood if the loci are linked}}{\text{likelihood if unlinked}} = \log_{10}\frac{(1-\theta)^{\text{NR}}\text{X}\theta^{\text{R}}}{0.5^{(\text{NR}+\text{R})}}$$

In this context, **R** represents the number of offspring exhibiting recombinant genotypes, **NR** indicates the number of offspring exhibiting non-recombinant genotypes and θ Represents the recombinant fraction. In linkage analysis, a LOD score of more than (> 3) is considered strong evidence for linkage, indicating that the likelihood of linkage is 1000 times higher than the likelihood of no linkage. Conversely, a LOD score below -2 is necessary to exclude the possibility of linkage. However, this threshold should be adjusted based on factors such as the species type, trait complexity, and the number of genetic markers involved.

Mapping Function

In linkage mapping, the recombination fraction of two genes is used to determine their distance from one another. This mapping process involves both the distance and the arrangement of genetic markers. However, using only recombination fractions as a linkage metric can be problematic since these fractions are not additive. To overcome this limitation, mathematical transformations called mapping functions have been established, with the Haldane (1919) and Kosambi (1943) functions being the most recognized. The Haldane function assumes that crossovers occur independently, with no interference between them.

The mathematical expression for the Haldane mapping function is

$D = -1/2 \ln (1 - 2\theta)$

The Kosambi map function, on the other hand, allows for the presence of interference

$D = 1/4 \ln [(1 + 2\theta) (1 - 2\theta)]$

In this context, D denotes the inter-locus genetic distance, while θ signifies the recombination fraction. These recombination frequencies are transformed into map distances in centimetres (cM) using a mapping function. Conceptually, a 1% recombination rate corresponds to around 1 cM. These two mapping functions operate within the series of $0<\theta<0.5$. Practically, the Kosambi function is preferred due to its effectiveness in accounting for interference compared to the Haldane function. Additionally, research has shown that the distances measured using the Kosambi function tend to be more precise than those calculated with the Haldane function, especially in multi-point analyses (Huehn, 2010; Vinod, 2011).

Construction of Linkage Maps

These are visual representations that display the positions of genetic markers within a genome, helping researchers understand the relationships and connections between genes. These maps show the distance between genes, usually expressed in centimorgans (cM) or recombination frequency. Genes located closely on the linkage map are more likely to be inherited together. In genetic research, constructing a genetic linkage map is an essential step that provides important information on how different traits are genetically controlled. These maps can be developed for any genetically studied species and are especially useful in breeding programs and human genetic studies.

Through experimental crosses, genetic markers are grouped, ordered, arranged, and spaced apart to create a genetic linkage map. Understanding the chromosomal structure, genome evolution, and recombination, among other genetic processes, depends on these maps. According to Fierst (2015) and Song *et al.* (2016), genetic maps are essential for verifying physical maps and improving new genome assemblies, even if physical maps can offer information regarding marker grouping and layout. In addition to providing intermarker distances, it incorporates markers that might not be seen on actual maps. Additionally, genetic maps are crucial for more effective QTL mapping methods, especially in plant and animal breeding populations (Paterson *et al.*, 1988).

Factors Required for Genetic Mapping

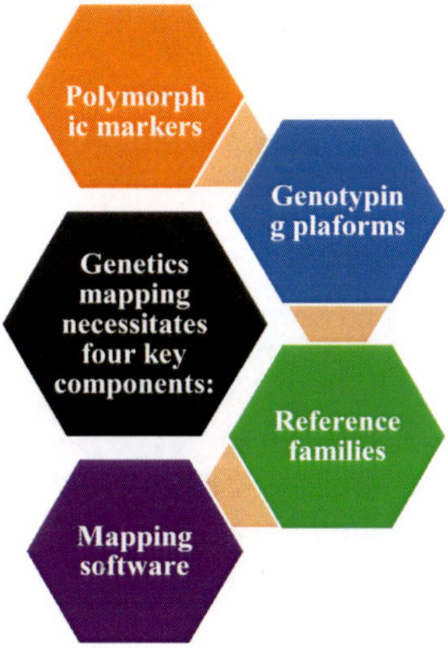

Fig. 3: Genetic map construction

Constructing a genetic linkage map requires multiple steps, starting with the identification of genetic or polymorphic markers. This is followed by genotyping individuals using appropriate genotyping platforms. Reference families are also needed to establish the inheritance patterns. Finally, statistical analysis is conducted using specialized mapping software to analyze the data and create the map.

Step: 1 Identification of Genetic Markers/ Polymorphic Markers

Identifying genetic markers is the foundational step in the creation of a genetic linkage map. These markers are individual-specific DNA sequences linked to particular attributes or characteristics, allowing researchers to trace the inheritance of genes.

To build a genetic linkage map, a variety of genetic markers can be utilized, such as:

Simple Sequence Repeats (SSRs)

Microsatellite DNA, commonly known as Simple Sequence Repeats (SSRs), are eukaryotic genomes that include repeating sequences of one to six nucleotide motifs (Abdelkrim, 2009). The majority of them are found in non-coding areas, significantly outnumbering those found in coding regions (Tautz 1984). SSRs are characterized by short DNA sequences that vary in length between individuals and are repeated multiple times. hey can be identified through the process of PCR amplification. SSRs are considered excellent molecular markers due to a range of advantageous features, including extensive robustness, high reproducibility, co-dominant inheritance, high levels of polymorphism, conservation across closely related species, and extensive genomic coverage across strongly related species (Abdelkrim, 2009). SSRs are a very flexible and economical option for genotyping because of these characteristics, especially in domains like as molecular the field of ecology, biology of conservation, and molecular breeding (Li, 2002).

Single Nucleotide Polymorphisms (SNPs)

Variations arising from point mutation at certain nucleotide locations within a locus are referred to as single nucleotide polymorphisms (Sachidanandam *et al.* 2001). SNPs are presently the preferred markers for extensive, high-throughput genomic analysis due to their frequency, common occurrence throughout the genome, and automation compatibility. Exploring large databases of Expressed Sequence Tags (ESTs) from different animals was a common method used in the past to identify SNPs, however this methodology was typically limited by issues with sequence coverage and depth (Liu, 2011). SNP discovery has changed dramatically as a result of next-generation sequencing technology, which improve transcriptomic and genomic analysis. SNPs are categorized as bi-allelic because, in theory, a single SNP can produce two alleles, each of which would represent one of two base pair pairings at that position. Numerous techniques have been used to elucidate SNPs, including direct DNA sequencing, heteroduplex analysis, and Single Strand Conformation Polymorphism (SSCP) study (Hecker, 1999). DNA sequencing is the most dependable and frequently used technique. Notwithstanding their benefits, SNPs have drawbacks and might only be marginally useful in some applications.

Restriction Fragment Length Polymorphisms (RFLPs)

Restriction Fragment Length Polymorphism investigation is a technique employed to circuitously collect data on genetic alterations occurring within specific short DNA sequences, known as restriction sites (Botstein, 1980). This method enables the identification of DNA sequence variations and is considered a premier genetic marker, facilitating the creation of highly precise linkage maps. In RFLP, organisms are distinguished based on the patterns that arise from the cleavage of their DNA by restriction enzymes. When two organisms have different distances between their cleavage sites for a specific restriction's endonuclease, the resulting lengths of the DNA fragment determination vary following digestion with the enzyme. The commonalities observed in the resulting patterns can help distinguish between different species and even specific strains. Once RFLP markers are mapped, they can be linked to phenotypic traits like accelerated growth rates, improved feed efficiency or disease resistance. These markers are particularly useful in MAS (Marker-Assisted Selection), enabling breeders to choose individuals with favourable genetic traits for breeding programs.

Rapid Amplification of Polymorphic DNA (RAPD)

RAPD-based markers result from amplifying genomic regions that are less functionally significant and have limited genetic variation for traits under selection. These areas may experience increased rates of nucleotide mutation, which can be useful for evaluating genetic differentiation between populations. By employing arbitrary nucleotide sequence primers for PCR amplification, RAPD allows for the detection of significant levels of DNA polymorphism. This technique is capable of detecting both coding and non-coding DNA sequences, characterized by a high frequency of informative polymorphic sequences originating from repetitive DNA (non-coding regions). RAPDs offer several advantages typical of PCR-based markers, including the availability of commercial primers and the absence of a need for pre-existing knowledge of the target DNA's sequence and the organization of the genome. Furthermore, RAPDs enable the concurrent screening of many loci across many individuals. However, there are limitations to using RAPDs in fisheries science, including the occurrence of PCR products derived from paralogous genes, where alternative DNA section produce fragments of the similar length and are mistakenly identified as a single locus and Low annealing temperatures during PCR amplification contribute to poor reproducibility.

Amplified Fragment Length Polymorphisms (AFLPs)

AFLPs are DNA fragments amplified via PCR utilizing primers that hybridize with specific DNA sequences. These fragments are subsequently resolved by gel electrophoresis. Once genetic markers are identified within these AFLPs, genotyping can be performed to ascertain their inheritance patterns in individuals. Linkage maps, which display the relative locations of genetic markers, are formed using AFLP. According to Erikson *et al.* (2004), AFLP is one of the most widely utilized markers for creating linkage maps for novel species. AFLP offers the advantage of quickly developing a large number of polymorphic markers and is frequently more dependable than other methods, such as randomly amplified polymorphic DNA (RAPD; Bensch and Åkesson, 2005). The AFLP technique's great efficiency per reaction and dependability make it an effective tool for genomic mapping. Compared to SSRs and RAPDs, the assay efficiency index of AFLP is more than ten times higher (Pejic 1998). Nonetheless, because of their dominant and biallelic characteristics, AFLPs typically have a low polymorphism information content (PIC) (Bensch & Åkesson 2005). Since AFLPs generate a huge number of markers per detection and don't require any prior knowledge of DNA sequence, they are perfect for linkage mapping of species for which microsatellites are unavailable.

Step: 2 Genotyping of Individuals/ Genotyping Platform

Genotyping individuals to determine their genetic makeup is the next stage in creating a genetic linkage map. Primers that target particular genetic markers are typically used in PCR to amplify DNA samples in this technique. The amplified PCR product is then analysed, either through gel electrophoresis, to identify the genotypes at apiece marker for each discrete. Genotyping is frequently carried out on a large number of individuals to increase the statistical strength of the analysis. The linkage associations between genetic markers are ascertained using the genotyping data that is produced.

With the abundance of SNP and their high-resolution analysis using dense SNP arrays platforms have been established to enable technologies for rapid and effective genotyping. Alternatively, genotyping by sequencing (GBS), which is based on next-generation sequencing (Baird 2008; Davey 2011; Miller 2007), offers an equally efficient method. Both SNP arrays and Genotyping-by-Sequencing (GBS) have proven valuable for large-scale SNP genotyping in genetic mapping and association studies.

Types of Genetic Markers and Genotypes Used in Aquaculture

Microsatellites (SSR)

Microsatellite DNA (SSR) is frequently used as a molecular marker in fields like as molecular ecology, conservation biology, and molecular breeding. Programs like RepeatFinderMicrosatellite DNA (SSR) is frequently used as a molecular marker in fields like as molecular ecology, conservation biology, and molecular breeding. Programs like RepeatFinder (Volfovsky *et al.* 2001) and SciRoko (Kofler *et al.* 2007) can easily identify microsatellites in genome sequences (Sharma *et al.*, 2007). Numerous methods are available for the determination of microsatellite genotypes (Guichoux *et al.* 2011), DNA sequencers such as the ABI3130 and ABI3730xl, which provide the most efficient platform for SSR genotyping, are currently in use in many labs. In aquaculture, SSR markers play a key role in constructing genetic linkage maps, advancing paternity testing, clarifying phylogenetic relationships, preserving genetic diversity, mapping quantitative characters and enhancing marker-assisted selection (MAS) in breeding programs (Chistiakov, 2006). Large-scale screening is supported by these co-dominant markers' high polymorphism and variability, which enable effective identification with a small collection of markers (Glaubitz, 2003). However, challenges arise in a group of taxa, like mollusks, where null alleles occur frequently complicate the use of SSR markers (Hedgecock, 2004). Microsatellites have been shown to be extensively used in Atlantic salmon to construct solid linkage maps. For example, Moen *et al.* (2004) Constructed genetic linkage maps utilizing microsatellite markers data. to identify QTLs linked to characteristics, such as tolerance to infectious pancreatic necrosis virus and growth. In Nile tilapia, SSR markers have been instrumental in genetic linkage mapping for breeding programs, such as the mapping of QTLs influencing body weight and growth (Lee *et al.* 2005), which has been used to improve selective breeding for faster-growing strains. Despite these successes, genotyping microsatellites in species like prawns, shrimp, crabs, clams, oysters and mussels pose challenges. These species' microsatellites are challenging to genotype because of significant slippage effects and non-amplification of one of the alleles at a locus. According to Scarbrough *et al.* (2002), these problems are probably caused by the long microsatellite repeats and high rates of mutation in the flanking areas' primer binding sites.

Single Nucleotide Polymorphisms (SNPs)

Although whole-genome sequence data are still unavailable for most aquaculture organisms (Liu, 2011), RAD-seq and other next-generation sequencing techniques can produce a significant number of SNPs (Miller *et al.* 2007, 2012; Davey *et al.*2011). RAD markers, which are primarily SNPs, are particularly useful for constructing linkage maps (Le Bras *et al.* 2011). To apply RAD markers to linkage mapping, RAD tags must first be isolated. These tags consist of genomic DNA sequences that flank each occurrence of a target restriction site across the genome (Miller *et al.* 2007). Bioinformatics tools are employed to genotype DNA sequence polymorphisms in isolated RAD tags (Houston *et al.*, 2012).

The release of draft genome sequences for Pacific oysters (*Crassostrea gigas*) and Atlantic cod (*Gadus morhua*) marks a breakthrough in aquaculture research

(Star *et al.*, 2011; Zhang *et al.*, 2012). It is noteworthy that at least 11 other aquaculture species are presently undergoing genome sequencing. Catfish (*Ictalurus punctatus*, Ictaluridae), salmon (*Salmo salar*, Salmonidae), rainbow trout (*Oncorhynchus mykiss*, Salmonidae), Nile tilapia (*Oreochromis niloticus*, Cichlidae), grass carp (*Ctenopharyngodon idella*, Cyprinidae), common carp (*Cyprinus carpio*, Cyprinidae), orange-spotted grouper (*Epinephelus coioides*, Serranidae), large yellow croaker (*Pseudosciaena crocea*, Sciaenidae), Japanese flounder (*Paralichthys olivaceus*, Paralichthyidae), and half-smooth tongue sole (*Cynoglossus semilaevis*, Cynoglossidae) are among the species on this list and Asian seabass (*Lates calcarifer*, Latidae). Numerous SNPs are anticipated to be produced by these sequencing efforts. Currently, six technologies and more than 30 distinct methodologies are available for SNP genotyping (Perkel, 2008).

SNPs have played a crucial role in Developing high-resolution genetic linkage maps for Atlantic salmon (*Salmo salar*). For example, Lien *et al.* (2011) created a linkage map that incorporated more than 7,000 SNPs, which significantly improved the detection of QTL correlated with traits like disease resistance and growth. Nile tilapia (*Oreochromis niloticus*), SNPs have also been incorporated into the construction of high-resolution linkage maps, facilitating the identification of QTL correlated to growth, reproduction and environmental stress tolerance. Notably, Joshi *et al.,* (2018) discovered SNP markers linked to salt tolerance in tilapia, a trait that is essential for expanding tilapia farming into brackish water environments. Various platforms have been employed for SNP genotyping in aquaculture species. (Liu, 2011).

One of the most cost-effective and high-throughput methods for analyzing a large number of SNPs in a large number of samples is SNP genotyping with next-generation sequencing technologies. However, genotyping a significant number of SNPs (say 10,000) in a large number of individuals (say, 200) still comes at a hefty price, often exceeding $75,000. Soon, next-generation sequencing is anticipated to gradually replace other platforms for large-scale SNP genotyping as sequencing costs fall further.

Restriction Fragment Length Polymorphisms (RFLPs)

Since more efficient marker systems have been developed, the use of RFLP markers, which have long been a common tool in genetic research, has declined. Their use in aquaculture genetics has been very limited, despite their widespread use in the past. To differentiate between strains, populations, or species, RFLP markers have often been employed (Papakostas *et al.* 2006; Lehoczky *et al.* 2005; Apte *et al.* 2003; Chow *et al.* 2006; Klinbunga *et al.* 2005). Many of these studies focused on mitochondrial DNA or 16S rDNA, which are not very useful for genomic study (de los Angeles *et al.* 2005). Numerous of these investigations concentrated on 16S rDNA or mitochondrial DNA, which are not very helpful for genomic research (de los Angeles *et al.* 2005). Given the availability of several more efficient marker systems and the challenges associated with developing RFLP markers from nuclear genes, the future usage and significance of RFLP in aquaculture genomics are expected to be limited. Nonetheless, RFLP markers have been extensively utilized in mapping the genome of Atlantic salmon species, contributing to the determination of QTL related to growth and disease resistance. For instance, Lien *et al.* 1999 Created a genetic linkage map for Atlantic salmon with the aid of RFLP markers. Similar to this, Agresti *et al.* (2000) created a genetic linkage map for Nile tilapia using RFLP markers, which was crucial in locating genes linked to important characteristics including body size and reproduction. All things considered, the application of RFLP markers has made it easier to identify the gene areas that affect desired aquaculture features.

Amplified Fragment Length Polymorphisms (AFLPs)

A popular molecular marker method for linkage mapping among different aquaculture species is AFLP. Because they enable the simultaneous evaluation of numerous genetic loci, these markers are extremely useful for assessing genetic diversity. The creation of genetic maps and the subsequent discovery of QTLs underpinning significant traits like growth, disease resistance, and environmental tolerance have been made possible in large part by AFLP markers. AFLP-based linkage maps have been developed for several organisms, including tilapia, rainbow trout, and medaka (Kocher *et al.* 1998; Naruse *et al.* 2000; Robison *et al.* 2001). The clustering of AFLP markers, which have been seen in a variety of species such as rainbow trout (Young, 1998), tilapia (Agresti, 2000), and channel catfish (Liu 2003), is a noteworthy feature of AFLP maps. The following distortion rates for aquaculture species were noted in the mapping family that was the subject of the study: Eight per cent of AFLP and microsatellite markers in tilapia were found to be associated with desirable features (Kocher, 1998). This proportion was higher in rainbow trout (13.3%, Young, 1998), channel catfish (16%, Liu, 2003), and eastern oysters (Yu and Guo, 2003) with AFLPs.

Rapid Amplification of Polymorphic DNA (RAPD)

Successful generation of species-specific diagnostic RAPD fingerprints has been achieved in numerous fish species., facilitating the analysis of their taxonomic relationships. Despite their similar morphologies, three endemic Spanish species of Barbus—*Barbus bocagei*, *B. graellsii*, and *B. sclateri*— were distinguished using the RAPD-PCR approach (Callejas and Ochando

1998). Five species from the Cyprinidae family were also identified using RAPD markers: *Barbus bocagei, Barbus comizo, Leuciscus pyrenaicus, Chondrostoma lemmingii*, and the introduced species *Alburnus alburnus*. All of these species are native to the Iberian Peninsula (Callejas and Ochando, 2002). The method has also been used to investigate the genetic diversity and relationships between the four principal carp species found in India: mrigal (*Cirrhinus mrigala*), kalbasu (*L. calbasu*), catla (*catla*), and rohu (*Labeo rohita*) (Barman, 2003). In addition, three eel species: *A. japonica, A. australis* and *A. bicolor*, were identified using RAPD markers (Takagi and Taniguchi, 1995). Additionally, eight species of the genus Barbus were examined for population structure and evolutionary relationships using RAPD markers (Callejas and Ochando, 2002). McConnell *et al.,* (2000) applied RAPD markers to investigate genetic diversity and conduct linkage mapping in tilapia, establishing a foundation for future genomic research utilizing more advanced markers like a microsatellites and SNPs. Additionally, Sun *et al.* 2002 developed an initial linkage map for common carp by RAPD markers, which was later supplemented with microsatellite and SNP markers. Tassanakajon *et al.,* 1998 also used RAPD markers to assess genetic diversity and map disease resistance loci in shrimp populations.

Copy Number Variation (CNV)

CNV represents a type of structural dissimilarity in the genome (Sebat *et al.* 2004). CNVs involve relatively large genomic regions that have undergone deletions or duplications on specific chromosomes. While CNVs in humans have been linked to various diseases, there is potential for these SNPs to be associated with significant traits in aquaculture species (Shirak *et al.* 2008; *Bai et al.* 2011). The use of next-generation sequencing technology has sped up the identification of CNVs (Mills *et al.* 2011). Next-generation sequencing (Mills *et al.*, 2011), microarray techniques (Komura *et al.* 2006; Wang and Bucan, 2010), and Taqman technology (Graubert *et al.* 2007) are some of the approaches used to genotype CNVs. In the near future, next-generation sequencing is anticipated to be the most attractive method for genotyping a large number of CNVs.

Step: 3 Establishing Reference Families

The formation of reference families is essential for any genetic mapping project, as the characteristics of these families can significantly impact both the quality and associated costs of linkage map construction produced (Da, 1998). In aquaculture species, the most commonly utilized mapping populations are F1 families, produced by mating the F1 hybrids with one of the parental lines, backcross populations produced by mating the F1 hybrids with genetically diverse individuals, and F2 populations resulting from these F1 hybrids (Yue, 2014). While some research has examined the ideal population size and composition for precise map creation (Ferreira, Silva & Cruz, 2006; van der Beek, Groen & van Arendonk, 1993), there are no common guidelines for creating reference families. Nevertheless, based on the preceding findings, a few general principles applicable to any mapping population type can be suggested:

- The quantity of useful information for mapping is contingent upon the number of informational meioses. Generally, informational meioses refer to the ability to determine whether a gamete is recombinant or not.
- It is advantageous to incorporate several mapping families because a only parental pair may exhibit homozygosity at certain loci, leading to a lack of informative meiosis.
- A larger number of individuals contributes to improved map resolution. While any population size can technically be used, achieving high-resolution mapping typically requires a substantial population size (van der Beek 1993).

Step: 4 Statistical Analysis/ Mapping Software

Statistical analysis of genotyping data is crucial for the successful completion of a genetic linkage map. This step entails calculating the recombination frequency between genetic markers, which is crucial for constructing the linkage map. Recombination frequency is determined through a comparison of individual genotypes. The closer the genetic linkage between two markers, the greater the tendency for their co-inheritance; in contrast, markers that are farther apart are more likely to undergo recombination during meiosis due to recombination. Several software packages, such as JoinMap, Mapchart and QTL Cartographer, can be used for the statistical analysis of genotyping data. These tools employ different statistical procedures to compute the recombination frequencies between genetic markers and to aid in the development of the genetic linkage map.

JoinMap

JoinMap is a user-friendly program specifically developed for constructing genetic linkage maps in diploid species, operating on the Microsoft Windows platform. It allows users to conduct various diagnostic tests during the mapping process, enabling the easy removal of loci and individuals identified as having erroneous data based on the calculations. Users can employ two main approaches for determining marker order: the algorithms for regression mapping and maximum likelihood mapping. However, earlier versions of the software were not suitable for analyzing full-sibling mating from outbred population families (CP), which are commonly applied in aquaculture research investigations. JoinMap version 4.1's ability to apply the multipoint maximum likelihood mapping technique, especially for CP populations is a major enhancement.

OneMap

OneMap is a newly created, open-source software program that manages a variety of experimental crosses, like recombinant inbred lines (RILs), F2 populations, backcrosses, and outcrossing. Users should be familiar with the fundamentals of the R programming language because OneMap is based on it. The study of OneMap employs a revolutionary multipoint-based methodology that uses hidden Markov models to increase the accuracy of genetic mapping.

MergeMap

The main goal of traditional software tools is to create genetic maps for specific mapping populations. Nonetheless, having more than one map for the same organism is becoming increasingly common. According to Wu, Close, and Lonardi (2008), combining these maps can result in a consensus map that offers better genome coverage and higher marker density. MergeMap was created especially to facilitate the creation of consensus genetic maps from a set of individual genetic maps (Wu & Close, 2008). Within the MergeMap framework, individual genetic maps are modelled as directed acyclic graphs (DAGs). Based on shared vertices these are then aggregated into a consensus graph, where any cycles indicate ordering conflicts between the separate mappings. The parsimonious approach used by MergeMap to resolve these conflicts considers both local reshuffles and global displacements, two kinds of errors that exist in separate maps. Local reshuffles are errors in the arrangement of markers that are close together, but global displacements happen when certain markers are placed too far away from their actual places. MergeMap eliminates these problems by removing the minimum number of required marker occurrences. For combining individual genomic maps, MergeMap is now acknowledged as the fastest and most precise software on the market today.

MapChart

Once genetic linkage maps are constructed, it's essential to generate high-quality outputs suitable for publication and demonstration. MapChart, a Microsoft Windows-based program made especially for making genetic linkage and QTL maps, is a common tool for this purpose. The visual output produced by MapChart consists of vertical bars that represent different linkage groups. To use MapChart, users need to prepare text files containing linkage information, including the loci and their respective positions, which can be generated using genetic mapping software like JoinMap.

QTL Cartographer

QTL Cartographer is a specialized software suite for identifying and mapping QTLs genomic regions linked to specific quantitative traits, such as an organism's productivity, growth rate, or disease resistance. It analyses genetic markers to find statistical correlations between genetic variations and phenotypic traits. The software supports various interval mapping techniques, such as SIM (simple interval mapping) and CIM (composite interval mapping), which help enhance the accuracy of QTL detection. QTL Cartographer also provides graphical outputs like linkage maps, QTL plots, and likelihood curves to visually represent the findings and facilitate easier interpretation of the results.

Statistical Tools and Mapping Software for Map Construction Used in Aquaculture

Linkage maps can be made using a variety of software programs; the choice is primarily determined by the experimental design, which takes into account factors like population structure and the number of genetic markers. Programs like JoinMap (Stam, 1993), CRI-MAP (Green, Falls & Crooks, 1990), MapMaker (Lincoln, 1992), Map Manager QTX (Manly, Cudmore & Meer, 2001), and LINKMFEX (Danzmann, 2001) are frequently used in aquaculture genetics. MapChart is a popular tool for making visual linkage maps (Voorrips 2002). In order to handle the complexity of higher-density linkage maps, more modern software has been developed, such as OneMap (Margarido, Souza & Garcia, 2007), MSTmap (Wu and Bhat, 2008), and HighMap (Liu, 2014). This is because higher marker densities result in larger processing demands.

Table 1. Examples of how different software packages have been applied to linkage mapping studies in aquaculture species. *Source*: (Yun Li, 2017)

Species	Software	Experimental Crosses	Genetic Markers	References
Common carp	JoinMap 4.0	Outbreeding, full-sib family	Microsatellites, SNPs	(Zheng *et al.*, 2011)
Common carp	JoinMap 4.0	Outbreeding, full-sib family	Microsatellites, SNPs	(Zhao *et al.*, 2013)
Atlantic salmon	JoinMap 4.0	Outbreeding, full-sib family	SNPs	(Hubert *et al.*, 2010)
Atlantic salmon	CRIMAP 2.4	Outbreeding, full-sib family	SNPs	(Lien *et al.*, 2011)
Atlantic salmon	CRIMAP 2.4 (modified version)	Outbreeding, full-sib family,	SNPs	(Gonen *et al.*, 2014)
Channel catfish	JoinMap 4.0	Backcross, family	Microsatellites, SNPs	(Ninwichian *et al.*, 2012)
Channel catfish	OneMap, JoinMap 4.0	Outbreeding, full-sib family	SNPs	(Li *et al.*, 2015)
Rainbow trout	MULTIMAP. CRIMAP	Outbreeding. full-sib family	Microsatellites, SNPs	(Palti *et al.*, 2011)
Rainbow trout	CARTHAGENE	Doubled-haploid lines.	Microsatellites SNPs	(Guyomard *et al.*, 2012)
Red drum	LINKMFEX 2.1	Outbreeding' full-sib family	Microsatellites	(Portnoy *et al.*, 2010)
Red drum	JoinMap 4.1	Outbreeding. full-sib family	Microsatellites	(Hollenbeck, Portnoy & Gold, 2015)
Asian sea bass	CRIMAP 2.4	Outbreeding, full-sib family	Microsatellites, SNPs	(Wang *et al.*, 2011)
Zhikong, scallop	JoinMap 4.1	Outbreeding, full-sib family	SNPs	(Jiao *et al.*, 2013)
Pearl oyster	JoinMap 4.0	Outbreeding, full-sib family	SNPs	(Shi *et al.*, 2014)
Pearl oyster	CRIMAP 2.4 (modified version)	Full-sib family, Half-sib family	SNPs	(Jones *et al.*, 2013)
White shrimp	CRIMAP	F2 family	SNPs	(Du *et al.*, 2010)

Overviews of Linkage Maps in Aquaculture Species

Aquaculture species have traditionally undergone slower genome mapping than terrestrial animals (Tong and Chu, 2002). However, significant advancements have been achieved in recent years, with the gap closing rapidly. The salmonids, catfish, tilapia, shrimp and oysters are among the five important aquaculture species that were recognized at the very first Aquaculture Species

Genome Mapping Workshops (Alcivar-Warren *et al.* 1997.) and for them moderate to high-density linkage maps have been created. Mapping efforts have subsequently broadened to include additional species such as Japanese flounders, European sea bass and scallops. The presentations at recent International Symposiums on Genetics in Aquaculture suggest that the Primary linkage map for numerous additional commercially significant species, namely flat oyster, Atlantic halibut and sea bream, will soon be accessible. Simple sequence repeats and amplified fragment length polymorphisms were initially the most often used markers for map building.

However, the increasing use of single nucleotide polymorphisms (SNPs) suggests that SNPs will be used more and more in mapping efforts in the future to improve accuracy and resolution. Large-scale selective breeding operations in aquaculture began in the 1960s with the development of Atlantic salmon breeding projects in Norway. Selective breeding is currently used by around 60 species of fish and shellfish (Gjedrem & Baranski, 2009; Abdelrahman, 2017). Traditional breeding methods, based on phenotypic or genetic selection using estimated breeding values (EBVs), have significantly improved aquaculture production. Through improving selection accuracy and efficiency, the incorporation of genetic markers into Marker-Assisted Selection techniques was anticipated to greatly increase genetic variation (Laghari, 2014; Yue, 2014; Abdelrahman, 2017). However, the utility of MAS in breeding programs is primarily limited to the detection of genes with large effects that are closely linked to specific loci. Typically, QTL mapping and association studies are used to identify these gene effects. Finding genes and polymorphisms responsible for the variation in a trait is mostly dependent on QTL investigation. By uncovering these genetic contributors, MAS helps breeders focus on specific genetic regions, improving breeding outcomes.

The primary goal of selective breeding strategies in aquaculture is to increase growth and body size because of their moderate-to-high heredity and ease of evaluation (Gjedrem, 2000; Gjedrem & Baranski, 2009). QTLs have been found in a variety of species, including rainbow trout (Wringe *et al.* 2010), Atlantic salmon (Tsai *et al.* 2015), common carp (Laghari *et al.* 2015), Asian and European seabass (Wang *et al.* 2015), scallop (Li *et al.*, 2012), and Pacific oyster (Guo *et al.* 2012). Many studies have focused on QTL mapping for traits linked to growth. Few of these QTLs have been confirmed across several populations, and the majority only contribute a small percentage of the total additive genetic variance. Delaying sexual maturity is a major focus on aquaculture breeding activities because premature maturation can have a detrimental impact on productivity and quality of meat. Several species, such as rainbow trout (Easton *et al.* 2011), Atlantic salmon (Pedersen *et al.* 2013), and Atlantic char (Küttner *et al.* 2011), have successfully undergone QTL analysis for sexual maturation. Interestingly, a significant QTL for Atlantic salmon sexual maturity was discovered in two different studies (Barson *et al.*, 2015).

Disease outbreaks pose a significant challenge in aquaculture, and identifying disease-resistant stocks has become a key objective for breeding activities. Disease-challenging research is frequently used to assess the response of aquatic species to specific diseases by tracking the survival rates of animals from recognized pedigrees. According to this research, heritability range from moderate to high for disease resistance, suggesting significant potential for improvement through selective breeding (Jorgen, 2011). However, despite valuable insights from QTL studies in species like *Litopenaeus vannamei* and *Penaeus monodon*, the most recent study by Guppy *et al.* (2018), decided that these studies have not yet provided the critical information needed to implement effective commercial marker-assisted selection (MAS) programs. Although QTLs have provided information about the genetic makeup of factors related to disease tolerance and survival characteristics, translating these findings into practical breeding strategies remains a challenge.

Table 2. Current Status of Linkage maps in Aquaculture Species, Source: (Danzmann & Gharbi 2007)

Species,	**Mapping panel,**	**Marker types,**	**Reference,**
Rainbow trout (*Oncorhynchus mykiss*),	Double haploids,	AFLP, SSR, genes,	(Young *et al.*, 1998; Nichols *et al.*, 2003).
	Inter-strainbackcross,	SSR, AFLP, genes,	(Sakamoto *et al.*, 2000; Danzmann *et al.*, 2005).
Rainbow trout,		microsatellites and SNPs,	(Guyomard *et al.*, 2012).
		SNPs.	(Gonzalez-Pena *et al.*, 2016).
Atlantic salmon, (*Salmo salar*).	Outcross,	AFLP, SSR.	(Moen *et al*, 2004).
	Inter-strain, backcross.	SSR, genes"	(Gibley *et al.*, 2004).
		SNPs"	(Lien *et al.*, 2011).
Brown trout, (*Salmo trutta*)"	Inter-strain, backcross"	SSR, genes"	(Gharbi *et al.*, 2006).
Arctic char, (*Salvelinus alpinus*)"	Inter-strain" backcross"	SSR, AFLP, genes"	(Woram *et al.*, 2004).

Nile tilapia”	Haploids”	AFLP, SSR”	(Kocher *et al.*, 1998).
Tilapia”		Microsatellites,”genes”	(Lee *et al.*, 2005).
Oreochromis” *mossambicus* x *Oreochromis* spp”		Microsatellites”	(Liu *et al.*, 2013).
Channel catfish” (*Ictalurus punctatus*)”	Outcross”	SSR, genes”	(Waldbieser *et al.*, 2001).
		SNPs”	(Li *et al.*, 2015).
		SNPs”	(Zeng *et al.*, 2017).
	Interspecific” backcross”	SNPs”	(Liu *et al.*, 2016).
Walking catfish” (*Clarias macrocephalus*)”	Haploids”	AFLP”	(Poompuang and Nakorn, 2004).
European sea bass” (*Dicentrarchus labrax*)”	Outcross”	SSR, genes”	(Chistiakov *et al.*, 2005).
European seabass”		Microsatellites,” AFLP,” SNP”	(Chistiakov *et al.*, 2008).
Common carp”		SNPs”	(Peng *et al.*, 2016).
Grass carp”		Microsatellites and SNPs”	(Xia *et al.*, 2010).
Ayu:(*Plecoglossus altivelis*):	Inter-population: Hybrids:	AFLP, SSR:	(Watanabe *et al.*, 2004).
Japanese flounder:	Offspring:	Microsatellites, “SNPs, genes:	(Castano-Sanchez *et al.*, 2010).
Scallop”		SNPs:’	(Jiao *et al.*, 2013).
Sea bream:		Microsatellites, “ESTs and SNPs:	(Tsigenopoulos *et al.*, 2014).
Giant tiger prawn:	“Offspring	SNPs;	(Baranski *et al.*, 2014).
Black tiger (*Penaeus monodon*)”	“outcross;	AFLP’	(Wilson *et al.*, 2003).
White shrimp: (*Penaeus vannamei*)”	Outcross>	AFLP’	(Pérez *et al.*, 2004).
Pacific white shrimp>		AFLP, microsatellites”	(Andriantahina *et al.*, 2013).
		SNPs>	(Yu *et al.*, 2015).
Chinese shrimp> (*Penaeus chinensis*)>	Inter-population< Hybrids>	AFLP<	(Li *et al.*, 2006).
Yellowtail<		Microsatellites>	(Ohara *et al*, 2005).

		microsatellites and SNPs>	(Aoki *et al,* 2015).
		SNPs>	(Ozaki *et al,* 2016).
Pacific oyster< (*Crassostrea gigas*)>	Inter-line double hybrids>	SSR	(Huber and Hedgecock, 2004).
	Inter-strain> Backcross<	AFLP>	(Li and Guo, 2004).
		SNPs and microsatellites	(Hedgecock *et al.,* 2015).
Sea urchin>	Interspecific> hybrid>	AFLP>	(Zhou *et al.,* 2006).

Limitation of Linkage Mapping in Aquaculture

Linkage mapping serves as an effective technique for unravelling the genetic structure of characters in different species, including aquaculture species like fish and shellfish. Nonetheless, it faces several limitations and challenges.

Complex Genetic Architecture

- **Polyploidy**: Many aquaculture species, including certain fish and shellfish, are polyploid, meaning they have multiple sets of chromosomes. This complexity makes it difficult to establish linkage maps and identify QTL, as most standard methods are tailored for diploid organisms.
- **High Genetic Diversity**: High genetic diversity within populations of many aquaculture species can make it challenging to develop clear connections between markers and traits.

Limited Genetic Resources

- **Marker Availability**: In numerous aquaculture species, there is a limited availability of high-density genetic markers, such as SNPs or microsatellites, compared to well-researched species like crops or model organisms. This limitation reduces the resolution and effectiveness of linkage maps.
- **Reference Genomes**: Many aquaculture species do not have high-quality reference genomes, which are crucial for precise linkage mapping and QTL detection. Even when these genomes exist, they may be incomplete or fragmented.

Reproductive and Breeding Challenges

- **Long Generation Time**: In some aquaculture species, extended generation times pose challenges in rapidly forming large mapping populations, which in turn delays the multigenerational studies required for producing high-quality linkage maps.

- **Large Offspring Number**: Although numerous aquaculture species yield a high volume of offspring, which aids in QTL mapping, effectively managing these large populations within breeding programs can be difficult.
- **Inbreeding and Genetic Drift**: In small or isolated aquaculture populations, inbreeding and genetic drift can alter the genetic variation accessible for mapping, resulting in decreased accuracy when identifying genuine QTLs.

Environmental and Phenotypic Variation

- **Environmental Effects on Traits**: Many traits of interest in aquaculture, including growth rate, resistance to sickness and feed effectiveness, are significantly affected by environmental factors. This complicates the differentiation between genetic influences and environmental effects, which can hinder accurate QTL detection.

Low Marker Density and Resolution

- **Limitations of Marker-Assisted Selection (MAS):** The limited resolution of linkage maps diminishes their effectiveness in marker-assisted selection programs. To improve selection accuracy, fine-mapping or employing more advanced techniques, like agenome-wide association studies (GWAS), may be required.

High Costs and Technical Requirements

- **Expensive and Labor-Intensive**: Linkage mapping demands considerable resources, both in terms of finances and labor. Creating mapping populations, genotyping individuals, and performing statistical analyses are all resource-heavy tasks, which can pose challenges for small-scale aquaculture operations.
- **Computational Challenges**: Evaluating complex linkage maps, particularly for species with large genomes or high heterozygosity, can be resource-intensive and demands expertise in advanced bioinformatics.

Conclusion

Linkage mapping has emerged as an essential tool in aquaculture genetics, enabling researchers to identify QTL related to economically crucial characteristics include resilience to illness, growth, and stress. It has established a foundation for understanding the genetic basis for these characteristics and has facilitated the development of selective breeding programs aimed at boosting aquaculture productivity. By identifying QTL linked to crucial traits, linkage mapping allows researchers to trace inheritance patterns, thereby enhancing breeding strategies. However, there are significant limitations to linkage mapping in aquaculture. These include complex genetic architectures, such as polyploidy, restricted availability of genetic markers and reference genomes, extended generation times and the significant impact that environmental influences have on characteristics. Additionally, issues like low marker density, challenges in phenotyping, and high costs can compromise the effectiveness and resolution of traditional linkage mapping. To overcome these challenges, more advanced genomic methods, including GWAS (genome-wide association studies), genomic selection, and whole-genome sequencing are increasingly being utilized in aquaculture breeding programs. While linkage mapping will continue to play a role, particularly in species with limited genomic resources, its integration with these newer technologies will be vital for accelerating genetic improvements and enhancing the sustainability of aquaculture practices.

References

Abdelkrim, J., Robertson, B. C., Stanton, J. A. L., & Gemmell, N. J. (2009). Fast, cost-effective development of species-specific microsatellite markers by genomic sequencing. Biotechniques, 46(3), 185–192.

Agresti, J. J., Seki, S., Cnaani, A., & Hulata, G. (2000). Linkage map of the tilapia (Oreochromis niloticus) genome based on AFLP and microsatellite markers. Aquaculture, 180(3), 329-338.

Agresti, J. J., Seki, S., Cnaani, A., Poompuang, S., Hallerman, E. M., Umiel, N., Hulata, G., Gall, G. A. E., & May, B. (2000). Breeding new strains of tilapia: development of an artificial center of origin and linkage map based on AFLP and micro*satellite loci. Aquacu*lture, 185, 43-56.

Alcivar-Warren, A., Dunham, R., Gaffney, P., Kocher, T., & Thorgaard, G. (1997). First Aquaculture Species Genome Mapping Workshop. Animal Genetics, 28, 451–452.

Andrews, K. R., Good, J. M., Miller, M. R., Luikart, G., & Hohenlohe, P. A. (2016). Harnessing the power of RADseq for ecological and evolutionary genomics. Nature Reviews Genetics, 17(2), 81–92. https://doi.org/10.1038/nrg.2015.28

Andriantahina, F., Liu, X., & Huang, H. (2013). Genetic map construction and quantitative trait locus (QTL) detection of growth-related traits in Litopenaeus vannamei for selective breeding applications. PLoS One, 8(9), e75206.

Aoki, J. Y., Kai, W., Kawabata, Y., Ozaki, A., Yoshida, K., Koyama, T., ... & Araki, K. (2015). Second generation physical and linkage maps of yellowtail (Seriola quinqueradiata) and comparison of synteny with four model fish. Bmc Genomics, 16, 1-11.

Apte, S., Star, B., & Gardne*r, J. P.* A. (2003). A comparison of genetic diversity between cultured and wild populations, and a test for genetic introgression in the New Zealand greenshell mussel Perna canaliculus (Gmelin 1791). Aquaculture, 219, 193-220.

Aquaculture Genomics, Genetics and Breeding Workshop, Abdelrahman, H., ElHady, M., Alcivar-Warren, A., Allen, S., Al-Tobasei, R., ... & Zhou, T. (2017). Aquaculture genomics, genetics and breeding in the United States: current status, challenges, and priorities for future research. BMC genomics*, 18, 1-23.* https://doi.org/10.1186/s12864-017-3557-1

Baird, N. A., Etter, P. D., Atwood, T. S., Currey, M. C., Shiver, A. L., Lewis, Z. A., ... & Johnson, E. A. (2008). Rapid SNP discovery and genetic mapping using sequenced RAD markers. PloS one, 3(10), e3376.

Baird, N. A., Etter, P. D., Atwood, T. S., et al. (2008). Rapid SNP discovery and genetic mapping using sequenced RAD markers. PloS One, 3(10), e3376. https://doi.org/10.1371/journal.pone.0003376

Baranski, M., Gopikrishna, G., Robinson, N. A., Katneni, V. K., Shekhar, M. S., Shanmugakarthik, J., ... & Ponniah, A. G. (2014). The development of a high-density linkage map for black tiger shrimp (Penaeus monodon) based on cSNPs. PLoS One, 9(1), e85413.

Barman, H. K., Barat, A., Yadav, B. M., Banerjee, S., Meher, P. K., Reddy, P. V. G. K., & Jana, R. K. (2003). Genetic variation between four species of Indian major carps as revealed by random amplified polymorphic DNA assay. Aquaculture, 217(1-4), 115-123.

Barson, N. J., Aykanat, T., Hindar, K., Baranski, M., Bolstad, G. H., Fiske, P., ... & Primmer, C. R. (2015). Sex-dependent dominance at a single locus maintains variation in age at maturity in salmon. Nature, 528(7582), 405-408. https://doi.org/10.1038/nature16062

Bateson, W., & Mendel, G. (1909). Mendel's principles of heredity. Putnam's Sons.

Bensch, S., & Åkesson, M. (2005). Ten years of AFLP in ecology and evolution: Why so few animals? Molecular Ecology, 14(10), 2899–2914. https://doi.org/10.1111/j.1365-294X.2005.02655.x

Botstein, D., White, R. L., Skolnick, M., & Davis, R. W. (1980). Construction of a genetic linkage map in man using restriction fragment length polymorphisms. American Journal of Human Genetics, 32(3), 314–331.

Callejas, C., & Ochando, M. D. (1998). Identification of Spanish barbel species using the RAPD technique. Journal of Fish Biology, 53(1), 208-215.

Callejas, C., & Ochando, M. D. (2002). Phylogenetic relationships among Spanish Barbus species (Pisces, Cyprinidae) shown by RAPD markers. Heredity, 89(1), 36-43.

Castaño-Sánchez, C., Fuji, K., Ozaki, A., Hasegawa, O., Sakamoto, T., Morishima, K., ... & Okamoto, N. (2010). A second generation genetic linkage map of Japanese flounder (Paralichthys *olivaceus*). BMC genomics, 11, 1-11.

Chistiakov, D. A., Hellemans, B., & Volckaert, F. A. M. (2006). Microsatellites and their genomic distribution, evolution, function and applications: A review with special reference to fish genetics. Aquaculture, 255(1-4), 1–29. https://doi.org/10.1016/j.aquaculture.2005.11.031

Chistiakov, D. A., Hellemans, B., Haley, C. S., Law, A. S., Tsigenopoulos, C., Kotoulas, G., Bertotto, D., Libertini, A., & Volckaert, F. A. (2005). A microsatellite linkage map of the European sea bass Dicentrarchus *labrax* L. Genetics, 170(4), 1821–1826.

Chistiakov, D. A., Tsigenopoulos, C. S., Lagnel, J., Guo, Y. M., Hellemans, B., Haley, C. S., ... & Kotoulas, G. (2008). A combined AFLP and microsatellite linkage map and pilot comparative genomic analysis of European sea bass Dicentrarchus *labrax* L. Animal Genetics, 39(6), 623-634.

Chow, S., Suzuki, N., Imai, H., & Yoshimura, T. (2006). Molecular species identification of spiny lobster phyllosoma larvae of the genus Panulirus from the northwestern Pacific. Marine Biotechnology. https://doi.org/10.1007/s10126-006-5085-1

Da, Y., VanRaden, P. M., Li, N., Beattie, C. W., Wu, C., & Schook, L. B. (1998). Designs of reference families for the construction of genetic linkage maps. Animal Biotechnology, 9(3), 205-228.

Danzmann, R. (2001). LINKMFEX: Linkage analysis package for outcrossed mapping families with male or female exchange of the mapping parent. Retrieved from www.uoguelph.ca/rdanzman/software.

Danzmann, R. G., & Gharbi, K. (2001). Gene mapping in fishes: a means to an end. Genetica, 111, 3–23.

Danzmann, R. G., & Gharbi, K. (2007). Linkage mapping in aquaculture species. Aquaculture genome technologies, 139-168.

Davey, J. W., Hohenlohe, P. A., Etter, P. D., Boone, J. Q., Catchen, J. M., & Blaxter, M. L. (2011). Genome-wide genetic marker discovery and genotyping using next-generation sequencing. Nature Reviews Genetics, 12(7), 499–510. https://doi.org/10.1038/nrg3012

Du, Z. Q., Ciobanu, D. C., Onteru, S. K., Gorbach, D., Mileham, A. J., Jaramillo, G., & Rothschild, M. F. (2010). A gene-based SNP linkage map for pacific white shrimp, Litopenaeus vannamei. Animal genetics, 41(3), 286-294.

Easton, A. A., Moghadam, H. K., Danzmann, R. G., & Ferguson, M. M. (2011). The genetic architecture of embryonic developmental rate and genetic covariation with age at maturation in rainbow trout (*Oncorhynchus mykiss*). Journal of Fish Biology, 78, 602–623. https://doi.org/10.1111/j.1095-8649.2010.02881.

Erickson, D. L., Fenster, C. B., Stenoien, H. K., & Price, D. (2004). Quantitative trait locus analyses and the study of evolutionary process. Molecular Ecology, 13(9), 2505–2522. https://doi.org/10.1111/j.1365-294X.2004.02257.x

Ferreira, A., Silva, M. F. D., & Cruz, C. D. (2006). Estimating the effects of population size and type on the accuracy of genetic maps. Genetics and Molecular Biology, 29, 187–192.

Fierst, J. L. (2015). Using linkage maps to correct and scaffold de novo genome assemblies: methods, challenges, and computational tools. Frontiers in Genetics, 6, 220–227.

Gharbi, K., Gautier, A., Danzmann, R. G., Gharbi, S., Sakamoto, T., Høyheim, B., ... & Guyomard, R. (2006). A linkage map for brown trout (Salmo trutta): chromosome homeologies and comparative genome organization with other salmonid fish. Genetics, 172(4), 2405-2419.

Gilbey J, E Verspoor, A McLay, and D Houlihan. (2004). A microsatellite linkage map for Atlantic salmon (*Salmo salar*). Anim. Genet, 35, pp. 98–105.

Gjedrem, T. (2000). Genetic improvement of cold-water fish species. Aquaculture Research, 31, 25–33. https://doi.org/10.1046/j.1365-2109.2000.00389.x

Gjedrem, T., & Baranski, M. (2009). Selective Breeding in Aquaculture: An Introduction. Amsterdam: Elsevier Science B.V.

Gjedrem, T., Robinson, N., & Rye, M. (2012). The importance of selective breeding in aquaculture to meet future demands for animal protein: A review. Aquaculture, 350–353, 117–129. https://doi.org/10.1016/j.aquaculture.2012.04.008

Glaubitz, J. C., Rhodes, O. E., & DeWoody, J. A. (2003). Prospects for inferring pairwise relationships with single nucleotide polymorphisms. Molecular Ecology, 12(4), 1039–1047. https://doi.org/10.1046/j.1365-294X.2003.01790.x

Gonen, S., Lowe, N. R., Cezard, T., Gharbi, K., Bishop, S. C., & Houston, R. D. (2014). Linkage maps of the Atlantic salmon (Salmo salar) genome derived from RAD sequencing. BMC genomics, 15, 1-17.

Gonzalez-Pena, D., Gao, G., Baranski, M., Moen, T., Cleveland, B. M., Kenney, P. B., ... & Leeds, T. D. (2016). Genome-wide association study for identifying loci that affect fillet yield, carcass, and body weight traits in rainbow trout (*Oncorhynchus mykiss*). Frontiers in genetics, 7, 203.

Graubert, T. A., Cahan, P., Edwin, D., Selzer, R. R., Richmond, T. A., Eis, P. S., ... & Trask, B. J. (2007). A high-resolution map of segmental DNA copy number variation in the mouse genome. PLoS Genetics, 3(1), e3.

Green, P., Falls, K., & Crooks, S. (1990). Documentation for CRI-MAP, version 2.4. Washington University School of Medicine, St Louis, MO.

Guichoux, E., Lagache, L., Wagner, S., Chaumeil, P., Léger, P., Lepais, O., ... & Petit, R. J. (2011). Current trends in microsatellite genotyping. Molecular ecology resources, 11(4), 591-611.

Guo, X., Li, Q., Wang, Q., & Kong, L. (2012). Genetic mapping and QTL analysis of growth-related traits in the Pacific oyster. Marine Biotechnology, 14, 218–226. https://doi.org/10.1007/s10126-011-9405-4

Guppy, J. L., Jones, D. B., Jerry, D. R., Wade, N. M., Raadsma, H. W., Huerlimann, R., & Zenger, K. R. (2018). The state of "Omics" research for farmed penaeids: advances in research and impediments to industry utilization. Frontiers in Genetics, 9, 282. https://doi.org/10.3389/fgene.2018.00282

Guyomard, R., Boussaha, M., Krieg, F., Hervet, C., & Quillet, E. (2012). A synthetic rainbow trout linkage map provides new insights into the salmonid whole genome duplication and the conservation of synteny among teleosts. BMC genetics, 13, 1-12.

Haldane, J. (1919). The combination of linkage values, and the calculation of distances between the loci of linked factors. Journal of Genetics, 8, 299–309.

Hecker, K. H., Taylor, P. D., & Gjerde, D. T. (1999). Mutation detection by denaturing DNA chromatography using fluorescently labeled polymerase chain reaction products. Analytical Biochemistry, 272(2), 156–164.

Hedgecock, D., Li, G., Hubert, S., Bucklin, K., & Ribes, V. (2004). Widespread null alleles and poor cross-species amplification of microsatellite DNA loci cloned from the Pacific oyster, Crassostrea *gigas*. Journal of Shellfish Research, 23(2), 379–385.

Hollenbeck, C. M., Portnoy, D. S., & Gold, J. R. (2015). A genetic linkage map of red drum (Sciaenops ocellatus) and comparison of chromosomal syntenies with four other fish species. Aquaculture, 435, 265–274.

Houston, R. D., Davey, J. W., Bishop, S. C., Lowe, N. R., Mota-Velasco, J. C., Hamilton, A., ... & Taggart, J. B. (2012). Characterisation of QTL-linked and genome-wide restriction site-associated DNA (RAD) markers in farmed Atlantic salmon. BMC genomics, 13, 1-15.

Huang, W., Cheng, C., Liu, J., Zhang, X., Ren, C., Jiang, X., Chen, T., Cheng, K., Li, H., & Hu, C. (2020). Fine mapping of the high-pH tolerance and growth trait-related quantitative trait loci (QTLs) and identification of the candidate genes in Pacific white shrimp (*Litopenaeus vannamei*). Marine Biotechnology, 22(1), 1–18. https://doi.org/10.1007/s10126-019-09932-8

Hubert, S., & Hedgecock, D. (2004). Linkage maps of microsatellite DNA markers for the Pacific oyster Crassostrea gigas. Genetics, 168, 351–362.

Hubert, S., Higgins, B., Borza, T., & Bowman, S. (2010). Development of a SNP resource and a genetic linkage map for Atlantic cod (*Gadus morhua*). BMC Genomics, 11, 1.

Huehn, M. (2010). Random variability of map distances based on Kosambi's and Haldane's mapping functions. Journal of Applied Genetics, 51, 27–31.

International SNP Map Working Group Cold Spring Harbor Laboratories: Sachidanandam Ravi 1 Weissman David 1 Schmidt Steven C. 1 Kakol Jerzy M. 1 Stein Lincoln D. 1, National Center for Biotechnology Information: Marth Gabor 2 Sherry Steve 2, Sanger Centre: Mullikin James C. 3 Mortimore Beverley J. 3 Willey David L. 3 Hunt Sarah E. 3 Cole Charlotte G. 3 Coggill Penny C. 3 Rice Catherine M. 3 Ning Zemin 3 Rogers Jane 3 Bentley David R. drb@ sanger. ac. uk 3 m, & Washington University in St. Louis: Kwok Pui-Yan 4 Mardis Elaine R. 4 Yeh Raymond T. 4 Schultz Brian 4 Cook Lisa 4 Davenport Ruth 4 Dante Michael 4 Fulton Lucinda 4 Hillier LaDeana 4 Waterston Robert H. 4 McPherson John D. 4. (2001). A map of human genome sequence variation containing 1.42 million single nucleotide polymorphisms. Nature, 409(6822), 928-933. https://doi.org/10.1038/35057149

Jiao, W., Fu, X., Dou, J., Li, H., Su, H., Mao, J., & Bao, Z. (2014). High-resolution linkage and quantitative trait locus mapping aided by genome survey sequencing: building up an integrative genomic framework for a bivalve mollusc. DNA research, 21(1), 85-101.

Jiao, W.; Fu, X.; Dou, J.; Li, H.; Su, H.; Mao, J.; Yu, Q.; Zhang, L.; Hu, X.; Huang, X.; Wang, Y.; Wang, S.; Bao, Z. (2013). Hig*h-Resolution Linkage* and Quantitative Trait Locus Mapping Aided by Genome Survey Sequencing: Building Up An Integrative Genomic Framework for a Bivalve Mollusc. DNA Research, 21(1), 85–101. doi:10.1093/dnares/dst043

Jones, *D. B., Jerry,* D. R., Khatkar, M. S., Raadsma, H. W., & Zenger, K. R. (2013). A high-density SNP genetic linkage map for the silver-lipped pearl oyster, Pinctada maxima: a valuable resource for gene localisation and marker-assisted selection. BM*C genomics, 14, 1-18.*

Joshi, R., Bastiaansen, J. W., Jong, M. C., & Crooijmans, R. P. (2018). QTL for salt tolerance in Nile tilapia (Oreochromis niloticus) using SNP markers. Aquaculture, 484, 85-93.

Jørgen, Ø., Matthew, B., Bjarne, G., & Trygve, G. (2011). Methodology for genetic evaluation of disease resistance in aquaculture species: challenges and future prospects. Aquaculture Research, 42, 103–114. https://doi.org/10.1111/j.1365-2109.2010.02669.x

Klinbunga, S., Siludjai, D., Wudthijinda, W., Tassanakajon, A., Jarayabhand, P., & Menasveta, P. (2001). Genetic h*eterogeneity of th*e giant tiger shrimp (Penaeus monodon) in Thailand revealed by RAPD and mitochondrial DNA RFLP analyses. Marine Biotechnology, 3, 428-438.

Kocher, T. D., Lee, W., Sobolewska, H., Penman, D., & McAndrew, B. (1998). A genetic linkage map of a cichlid fish, the tilapia (Oreochromis niloticus). Genetics, 148(3), 1225-1232.

Kofler, R., Schlotterer, C., & Lelley, T. (2007). SciRoKo: A new tool for whole genome microsatellite search and investigation. Bioinformatics, 23(13), 1683–1685. https://doi.org/10.1093/bioinformatics/btm157

Kosambi, D. (1943). The estimation of map distances from recombination values. Annals of Eugenics, 12, 172–175.

Küttner, E., Moghadam, H. K., Skúlason, S., Danzmann, R. G., & Ferguson, M. M. (2011). Genetic architecture of body weight, condition factor and age of sexual maturation in Icelandic Arctic charr (Salvelinus alpinus). Molecular Genetics and Genomics, 286, 67–79. https://doi.org/10.1007/s00438-011-0628-x

Laghari, M. Y., Lashari, P.*, Zhang, X.*, Xu, P., Narejo, N. T., Xin, B., ... & Sun, X. (2015). QTL mapping for economically important traits of common carp (Cyprinus carpio L.). Journal of applied genetics, 56, 65-75. https://doi.org/10.1007/s13353-014-023*2-y*

Laghari, M. Y., Lashari, P., Zhang, Y., & Sun, X. (2014). Identification of Quantitative Trait Loci (QTLs) in aquaculture species. Reviews in Fisheries Science & Aquaculture, 22, 221–238. https://doi.org/10.1080/23308249.2014.931172

Le Bras, Y., Dechamp, N., Krieg, F., Filangi, O., Guyomard, R., Boussaha, M., ... & Quillet, E. (2011). Detection of QTL with effects on osmoregulation capacities in the rainbow trout (Oncorhynchus mykiss). BMC genetics, 12, 1-14.

Lee, B. Y., Lee, W. J., Streelman, J. T.*, Carleton, K. L., H*owe, A. E., Hulata, G., & Kocher, T. D. (2005). A second-generation genetic linkage map of tilapia (Oreochromis spp.). Genetics, 170(1), 237–244. https://doi.org/10.1534/genetics.104.035527

Lehoczky, I., Jeney, Z.*, Mag*yary, I., Hancz, C., & Kohlmann, K. (2005). Preliminary data on genetic variability and purity of common carp (Cyprinus carpio L.) strains kept at the live gene bank at Research Institute for Fi*sheries*, Aquaculture and Irrigation (HAKI) Szarvas, Hungary. Aquaculture, 247, 45-49.

Li, B. J., Zhu, Z. X., Gu, X. H., Lin, H. R., & Xia, J. H. (2019). QTL mapping for red blotches in Malaysia red tilapia (Oreochromis spp.). Marine Biotechnology, 21(3), 384–395. https://doi.org/10.1007/s10126-019-09888-9

Li, H., Liu, X., & Zhang, G. (2012). A consensus microsatellite-based linkage map for the hermaphroditic bay scallop (Argopecten irradians) and its application in size-related QTL analysis. PLoS One, 7, e46926. https://doi.org/10.1371/journal.pone.0046926

Li, L., & Guo, X. M. (2004). AFLP-based genetic linkage maps of the Pacific oyster Crassostrea gigas Thunberg. Marine Biotechnology, 6, 26–36.

Li, L., Xiang, J. H., Liu, X., Zhang, Y., Dong, B., & Zhang, X. J. (2005). Construction of AFLP-based genetic linkage map for Zhikong scallop, Chlamys farreri Jones et Preston and mapping of sex-linked markers. Aquaculture, 245, 63–73.

Li, Y., Byrne, K., Miggiano, E., Whan, V., Moore, S., Keys, S., ... & Lehnert, S. (2003). Genetic ma*pping of* the kuruma prawn Penaeus japonicus using AFLP markers. Aquaculture, 219(1-4), 143-156.

Li, Y., Liu, S., Qin, Z., Waldbieser, G., Wang, R., Sun, L., ... & Liu, Z. (2015). Construc*tion of a h*igh-density, high-resolution genetic map and its integration with BAC-based physical map in channel catfish. DNA Research, 22(1), 39-52.

Li, Y., Liu, S., Wang, R., Qin, Z., & Liu, Z. (2017). Bioinformatics considerations and approaches for high-density link*age mapping* in aquaculture. In Bioinformatics in Aquaculture (pp. 356–379).

Li, Y. C., Korol, A. B., Fahima, T., Beiles, A., & Nevo, E. (2002). Microsatellites: genomic distribution, putative functions and mutational mechanisms: A review. Molecular Ecology, 11(12), 2453–2465.

Li, Z., Li, J., Wang, Q., He, Y., & Liu, P. (2006). AFLP-based genetic linkage map of marine shrimp Penaeus (Fenneropenaeus) chinensis. Aquaculture, 261, 463–472.

Lien, S., Gidskehaug, L., Moen, T., Hayes, B., Berg, P., Davidson, W. S., & Omholt, S. W. (1999). A dense linkage map of Atlantic salm*on (Salmo s*alar) based on AFLP and microsatellite markers. Genetics, 153(4), 1989–1998.

Lien, S., Gidskehaug, L., Moen, T., Hayes, B., Berg, P., Davidson, W. S., & Omholt, S. W. (2011). A dense SNP-based linkage map for Atlantic salmon (Salmo salar) reveals extended chromosome segments with conserved synteny. BMC Genomics, 12(1), 615. https://doi.org/10.1186/1471.

Lincoln, S. (1992). Mapping genes controlling quantitative traits with MapMaker/QTL 1.1. Whitehead Institute Technical Report.

Liu, D., Ma, C., Hong, W., Huang, L., Liu, M., Liu, H., ... & Zheng, H. (2014). Construction and analysis of high-density linkage map using high-throughput sequencing data. Plos one, 9(6), e98855.

Liu, F., Sun, F., Li, J., Xia, J. H., Lin, G., Tu, R. J., & Yue, G. H. (2013). A microsatellite-based linkage map of salt tolerant tilapia (Oreochromis mossambicus x Oreochromis spp.) and m*apping* of sex-determining loci. BMC genomics, 14, 1-14.

Liu, P., Wang, L., Wan, Z. Y., Ye, B. Q., Huang, S., Wong, S. M., & Yue, G. H. (2016). Mapping QTL for resistance against viral nervous necrosis disease in Asian seabass. Marine biotechnology, 18, 107-116.

Li*u, S., Li, Y., Qin,* Z., Geng, X., Bao, L., Kaltenboeck, L., ... & Liu, Z. (2016). High density interspecific genetic linkage mapping provides insights into genomic incompatibility between channel catfish and blue catfish. Animal genetics, 47(1), 81-90.

Liu, S., Zhou, Z., Lu, J., Sun, F., Wang, S., Liu, H., ... & Liu, Z. (2011). Generation of genome-scale gene-associated SNPs in catfish for the construction of a high-density SNP array. BMC genomics, 12, 1-13.

Liu, X., Guo, X., Gao, Q., Zhao, H., & Zhang, G. (2006). A preliminary genetic linkage map of the Pacific abalone Haliotis discus hannae Ino. Marine Biotechnology, *8, 386–397.*

Liu, Z., Karsi, A., Li, P., Cao, D., & Dunham, R. (2003). An AFLP-based genetic linkage map of channel catfish (Ictalurus punctatus) constructed by using an interspecific hybrid resource family. Genetics, 165(2), 687–694.

Liu, Z. J. (2011). Next-generation sequencing and whole genome selection in aquaculture. Wiley-Blackwell.

Manly, K. F., Cudmore, Jr., R. H., & Meer, J. M. (2001). Map Manager QTX, cross-platform software for genetic mapping. Mammalian Genome, 12, 930–932.

Margarido, G., Souza, A. P., & Garcia, A. (2007). OneMap: software for genetic mapping in outcrossing species. Hereditas, 144, 78–79.

Maxam, A. M., & Gilbert, W. (1977). A new method for sequencing DNA. Proceedings of the National Academy of Sciences, 74, 560–564.

McConnell, S. K., Woolliams, J. A., & Taggart, J. B. (2000). Genetic linkage map of the Nile tilapia (Oreochromis niloticus) using RAPD markers. Aquaculture, 180(3-4), 271-279.

Miko, I. (2008). Gregor Mendel and the principles of inheri*tance. Natu*re Education, 1, 134.

Miller, M. R., Brunelli, J. P., Wheeler, P. A., Liu, S., Rexroad III, C. E., Palti, Y., ... & Thorgaard, G. H. (2012). A conserved haplotype controls parallel adaptation in geographically distant salmonid populations. Molecular ecology, 21(2), 237-249.

Miller, M. R., Dunham, J. P., Amores, A., Cresko, W. A., & *Johnso*n, E. A. (2007). Rapid and cost-effective polymorphism identification and genotyping using restriction site associated DNA (RAD) markers. Genome Research, 17(2), 240–248. https://doi.org/10.1101/gr.5681207

Mills, R. E., Walter, K., Stewart, C.*, Hands*aker, R. E., Chen, K., Alkan, C., ... & Eichler, E. E. (2011). Mapping copy number variation by population-scale genome sequencing. Nature, 470(7332), 59-65.

Moen, T., Hoyheim, B., Munck, H., & Meuwissen, T. H. E. (2004). A linkage map of Atlantic salmon (Salmo salar) reveals an uncommonly large difference in recombination rate between the sexes. Animal Genetics, 35(1), 81–92. https://doi.org/10.1046/j.1365-2052.2003.01089.x

Moore, S. S., Whan, V., Davis, G. P., Byrne, K., Hetzel, D. J. S., & Preston, N. (1999). The development and application of genetic markers for the Kuruma prawn Penaeus japonicus. Aquaculture, 173, 19–32.

Morgan, T. H. (1911). Random segregation versus coupling in Mendelian inheri*tance.* Science, 34, 384.

Naruse, K., Fukamachi, S., Mitani, H., Kondo, M., & Matsuoka, T. (2000). A detailed linkage map of medaka, Oryzias latipes: Comparative genomics and genome evolution. Genetics, *154(4), 1773–1*784.

Nichols, K. M., Young, W. P., Danzmann, R. G., Robison, B. D., Rexroad, C., Noakes, M., ... & Thorgaard, G. H. (2003). A consolidated linkage map for rainbow trout (Oncorhynchus mykiss). Animal genetics, 34(2), 102-115.

Ninwichian, P., Peatman, E., Liu, H., Kucuktas, H., Somridhivej, B., Liu, S., ... & Liu, Z. (2012). Second-generation genetic linkage map of catfish and its integration with the BAC-based physical map. G3: Genes| Genomes| Genetics, 2(10), 1233-1241.

Ohara, E., Nishimura, T., Nagakura, Y., Sakamoto, T., Mushiake, K., & Okamoto, N. (2005). Genetic linkage maps of two yellowtails (Seriola quinqueradiata and Seriola lalandi). Aquaculture, 244(1-4), 41-48.

Ozaki A, Aoki J-y, Usuki H, Yoshida K, Noda T, Hotta T, Mizuochi H, Chujo T, Shima Y, Araki K. (2016). A high density genetic linkage map for yellowtail (Seriola quinqueradiata) containing EST-based SNPs. In: Plant and Animal Genome XXIV Conference: Plant and Animal Genome. San Diego

Palti, Y., Genet, C., Luo, M. C., Charlet, A., Gao, G., Hu, Y., ... & Rexroad, C. E. (2011). A first generation integrated map of the rainbow trout genome. BMC genomics, 12, 1-9.

Papakostas, S., Dooms, S., Triantafyllidis, A., Deloof, D., Kappas, I., Dierckens, K., De Wolf, T., Bossier, P., Vadstein, O., Kui, S., & others. (2006). Evaluation of DNA methodologies in identifying Brachionus species used in European hatcheries. Aquaculture, In Press.

Paterson, A. H., Lander, E. S., Hewitt, J. D., Peterson, S., Lincoln, S. E., & Tanksley, S. D. (1988). Resolution of quantitative traits into Mendelian factors by using a complete linkage map of restriction fragment length pol*ymorph*isms. Nature, 335(6192), 721-726.

Pedersen, S., Berg, P. R., Culling, M., Danzmann, R. G., Glebe, B., Leadbeater, S., ... & Boulding, E. G. (2013). Quantitative trait loci for pre*cocious* parr maturation, early smoltification, and adult maturation in double-backcrossed trans-Atlantic salmon (Salmo salar). Aquaculture, 410, 164-171. https://doi.org/10.1016/j.aquaculture.2013.06.039

Pejic, I., Ajmone-Marsan, P., Morgante, M., Kozumplick, V., & Motto, M. (1998). Comparative an*alysis of gen*etic similarity among maize inbred lines detected by RFLP, RAPD, SSR, and AFLPs. Theoretical and Applied Genetics, 97(7), 1248–1255. https://doi.org/10.1*007/s0012*20051033

Peng, W., Xu, J., Zhang, Y., Feng, J., Dong, C., Jiang, L., ... & Xu, P. (2016). An ultra-high density linkage map and QTL mapping for sex and growth-related traits of common carp (Cyprinus carpio). Scientific reports, 6(1), 26693.

Perez, F., Erazo, C., Zhinaula, M., Volckaert, F., & Calderon, J. (2004). A sex-specific linkage map of the white shrimp Penaeus (Litopenaeus) vannamei based on AFLP markers. Aquaculture, 242, 105–118.

Perkel, J. (2008). SNP genotyping: Six technologies that keyed a revolution. Nature Methods, 5(5), 447–454. https://doi.org/10.1038/nmeth0508-447

Poompuang, S., & Na-Nakorn, U. (2004). A preliminary genetic map of walking catfish (Clarias macrocephalus). Aquaculture, 232(1-4), 195–203.

Portnoy, D. S., Renshaw, M. A., Hollenbeck, C. M., & Gold, J. R. (2010). A genetic linkage map of red drum, Sciaenops ocellatus. Animal Genetics, 41, 630–641.

Ren, B., Robert, F., Wyrick, J. J., Aparicio, O., Jennings, E. G., Simon, I., ... & Young, R. A. (2000). Genome-wide location and function of DNA binding proteins. Science, 290(5500), 2306-2309.

Robison, B. D., Wheeler, P. A., Sundin, K., Sikka, P., & Thorgaard, G. H. (2001). Composite interval mapping reveals a major locus influencing the embryonic development rate in rainbow trout (Oncorhynchus mykiss). Journal of Heredity, 92(1), 16–22.

Sakamoto, T., Danzmann, R. G., Gharbi, K., Howard, P., Ozaki, A., Khoo, S. K., ... & Hoyheim, B. (2000). A microsatellite linkage map of rainbow trout (Oncorhynchus mykiss) characterized by large sex-specific differences in recombination rates. Genetics, 155(3), 1331-1345.

Sanger, F., Nicklen, S., & Coulson, A. R. (1977). DNA sequencing with chain-terminating inhibitors. Proceedings of the National Academy of Sciences, 74, 5463–5467.

Scarbrough, J. R., Cowles, D. L., & Carter, R. L. (2002). Microsatellites in shrimp species. In Modern Approaches to the Study of Crustacea (pp. 291–299). Kluwer Academic Publishers.

Sebat, J., Lakshmi, B., Troge, J., Alexander, J., Young, J., Lundin, P., ... & Wigler, M. (2004). Large-scale copy number polymorphism in the human genome. Science, 305(5683), 525-528.

Semagn, K., Bjørnstad, Å., & Ndjiondjop, M. (2006). Principles, requirements and prospects of genetic mapping in plants. African Journal of Biotechnology, 5, 2569–2587.

Sharma, P. C., Grover, A., & Kahl, G. (2007). Mining microsatellites in eukaryotic genomes. Trends in Biotechnology, 25(11), 490–498. https://doi.org/10.1016/j.tibtech.2007.08.017

Shi, Y., Wang, S., Gu, Z., et al. (2014). High-density single nucleotide polymorphisms linkage and quantitative trait locus mapping of the pearl oyster, Pinctada *fucata* martensii Dunker. Aquaculture, 434, 376–384.

Shirak, A., Golik, M., Lee, B. Y., Howe, A. E., Kocher, T. D., Hulata, G., & Ron, M. (2008). Copy number variation of lipocalin family genes for male-specific proteins in tilapia and its association with gender. Heredity, 101(5), 405-415.

Song, Q., Jenkins, J., Jia, G., Hyten, D. L., Pantalone, V., Jackson, S. A., ... & Cregan, P. B. (2016). Construction of high resolution genetic linkage maps to improve the soybean genome sequence assembly Glyma1. 01. Bmc Genomics, 17, 1-11.

Stam, P. (1993). Construction of integrated genetic linkage maps by means of a new computer package: JoinMap. The Plant Journal, 3, 739–744.

Star, B., Nederbragt, A. J., Jentoft, S., Grimholt, U., Malmstrøm, M., Gregers, T. F., ... & Jakobsen, K. S. (2011). The genome sequence of Atlantic cod reveals a unique immune system. Nature, 477(7363), 207-210.

Sun, X., Liang, L., & Hu, W. (2002). Genetic linkage map of the common carp (Cyprinus carpio) based on RAPD and AFLP markers. Aquaculture, 204(3-4), 379-392.

Takagi, M., & Taniguchi, N. (1995). Random amplified polymorphic DNA (RAPD) for identification of three species of Anguilla, A. japonica, A. australis, and A. *bicor*. Fisheries Science, 61(5), 884-885.

Tassanakajon, A., Jarayabhand, P., & Klinbunga, S. (1998). Genetic diversity and linkage mapping in shrimp (*Penaeus monodon*) using RAPD markers. Molecular Marine Biology and Biotechnology, 7(2), 199-206.

Tautz, D., & Renz, M. (1984). Simple sequences are ubiquitous repetitive components of eukaryotic genomes. Nucleic Acids Research, 12(10), 4127–4138.

Tong, J., & Chu, K. H. (2002). Genome mapping in aquatic animals: Progress and future perspectives. Russian Journal of Genetics, 38, 612–621.

Tsai, H. Y., Hamilton, A., Tinch, A. E., Guy, D. R., Gharbi, K., Stear, M. J., ... & Houston, R. D. (2015). Genome wide association and genomic prediction for growth traits in juvenile farmed Atlantic salmon using a high density SNP array. BMC genomics, 16, 1-9. https://doi.org/10.1186/s12864-015-2117-9

Tsigenopoulos, C. S., Louro, B., Chatziplis, D., Lagnel, J., Vogiatzi, E., Loukovitis, D., ... & Kotoulas, G. (2014). Second generation genetic linkage map for the gilthead sea bream Sparus aurata L. Marine genomics, 18, 77-82.

Van der Beek, S., Groen, A. F., & Van Arendonk, J. A. (1993). Evaluation of designs for reference families for livestock linkage mapping experiments. Animal Biotechnology, 4, 163–182.

Van Ooijen, J. W., & Jansen, J. (2013). Genetic mapping in experimental populations. Cambridge University Press.

Van Os, H., Stam, P., Visser, R. G., & Van Eck, H. J. (2005). SMOOTH: a statistical method for successful removal of genotyping errors from high-density genetic linkage data. Theoretical and Applied Genetics (TAG), Theoretische und angewandte Genetik, 112, 187–194.

Vinod, K. (2011). Kosambi and the genetic mapping function. Resonance, 16, 540–550.

Volfovsky, N., Haas, B. J., & Salzberg, S. L. (2001). A clustering method for repeat analysis in DNA sequences. Genome Biology, 2(8), 0027.1–0027.11. https://doi.org/10.1186/gb-2001-2-8-research0027

Voorrips, R. (2002). MapChart: software for the graphical presentation of linkage maps and QTLs. Journal of Heredity, 93, 77–78.

Waldbieser, G. C., Bosworth, B. G., Nonneman, D. J., & Wolters, W. R. (2001). A microsatellite-based genetic linkage map for channel catfish, Ictalurus *punctatus*. Genetics, 158(2), 727–734.

Wan, S. M., Liu, H., Zhao, B. W., Nie, C. H., Wang, W. M., & Gao, Z. X. (2017). Construction of a high-density linkage map and fine mapping of QTLs for growth and gonad-related traits in blunt snout bream. Scientific Reports, 7(1), 1–10. https://doi.org/10.1038/srep46509

Wang, C. M., Bai, Z. Y., He, X. P., Lin, G., Xia, J. H., Sun, F., ... & Yue, G. H. (2011). A high-resolution linkage map for comparative genome analysis and QTL fine mapping in Asian seabass, Lates calcarifer. Bmc Genomics, 12, 1-18.

Wang, K., & Bucan, M. (2010). Copy number variation detection via high-density SNP genotyping. Cold Spring Harbor Protocols, pdb.top46.

Wang, L., Wan, Z. Y., Bai, B., Huang, S. Q., Chua, E., Lee, M., ... & Yue, G. H. (2015). Construction of a high-density linkage map and fine mapping of QTL for growth in Asian seabass. Scientific reports, 5(1), 16358. https://doi.org/10.1038/srep16358

Wang, L. L., Song, L. S., Chang, Y. Q., Xu, W., Ni, D. J., & Guo, X. M. (2005). A preliminary genetic map of Zhikong scallop (Chlamys farreri Jones et Preston 1904). Aquaculture Research, 36, 643–653.

Wang, X., Fu, B., Yu, X., Qu, C., Zhang, Q., & Tong, J. (2018). Fine mapping of growth-related quantitative trait loci in Yellow River carp (Cyprinus carpio *haematopterus*). Aquaculture, 484, 277–285. https://doi.org/10.1016/j.aquaculture.2017.11.016

Watanabe, T., Fujita, H., Yamasaki, K., Seki, S., & Taniguchi, N. (2004). Preliminary study on linkage mapping based on microsatellite DNA and AFLP markers using homozygous clonal fish in ayu (Plecoglossus *altivelis*). Marine Biotechnology, 6(4), 327–334.

Watson, J. D., & Crick, F. H. (1953). Molecular structure of nucleic acids. Nature, 171, 737–738.

Wilson, K., Li, Y. T., Whan, V., Lehnert, S., Byrne, K., Moore, S., Pongsomboon, S., Tassanakajon, A., Rosenberg, G., Ballment, E., Fayazi, Z., Swan, J., Kenway, M., & Benzie, J. (2003). Genetic mapping of the black tiger shrimp Penaeus monodon with amplified fragment length polymorphism. Aquaculture, 204(3-4), 297–309.

Woram, R. A., McGowan, C., Stout, J. A., Gharbi, K., Ferguson, M. M., Hoyheim, B., ... & Danzmann, R. G. (2004). A genetic linkage map for Arctic char (Salvelinus alpinus): evidence for higher recombination rates and segregation distortion in hybrid versus pure strain mapping parents. Genome, 47(2), 304-315.

Wringe, B. F., Devlin, R. H., Ferguson, M. M., Moghadam, H. K., Sakhrani, D., & Danzmann, R. G. (2010). Growth-related quantitative trait loci in domestic and wild rainbow trout (Oncorhynchus *mykiss*). BMC Genetics, 11, 63. https://doi.org/10.1186/1471-2156-11-63

Wu, Y., Bhat, P. R., Close, T. J., & Lonardi, S. (2008). Efficient and accurate construction of genetic linkage maps from the minimum spanning tree of a graph. PLoS Genetics, 4, e1000212.

Wu, Y., Close, T. J., & Lonardi, S. (2008). On the accurate construction of consensus genetic maps. Computational Systems Bioinformatics Conference, 7, 285–296.

Xia, J. H., Liu, F., Zhu, Z. Y., Fu, J., Feng, J., Li, J., & Yue, G. H. (2010). A consensus linkage map of the grass carp (Ctenopharyngodon *idella*) based on microsatellites and SNPs. BMC genomics, 11, 1-16.

You, X., Shan, X., & Shi, Q. (2020). Research advances in the genomics and applications for molecular breeding of aquaculture animals. Aquaculture, 526, 735357. https://doi.org/10.1016/j.aquaculture.2020.735357

Young, W. P., Wheeler, P. A., Coryell, V. H., Keim, P., & Thorgaard, G. H. (1998). A detailed linkage map of rainbow trout produced using doubled haploids. Genetics, 148, 839–850.

Yu, Y., Wang, Q., Zhang, Q., Luo, Z., Wang, Y., Zhang, X., Huang, H., Xiang, J., & Li, F. (2019). Genome scan for genomic regions and genes associated with growth trait in Pacific white shrimp Litopenaeus *vannamei*. Marine Biotechnology, 21(3), 374–383. https://doi.org/10.1007/s10126-019-09887-w

Yu, Y., Zhang, X., Yuan, J., Li, F., Chen, X., Zhao, Y., ... & Xiang, J. (2015). Genome survey and high-density genetic map construction provide genomic and genetic resources for the Pacific White Shrimp Litopenaeus vannamei. Scientific reports, 5(1), 15612.

Yu, Z. N., & Guo, X. M. (2003). Genetic linkage map of the eastern oyster Crassostrea virginica Gmelin. Biological Bulletin, 204, 327–338.

Yue, G. H. (2014). Recent advances of genome mapping and marker-assisted selection in aquaculture. Fish and Fisheries, 15, 376–396. https://doi.org/10.1111/faf.12020

Zeng Q, Fu Q, Li Y, Liu S, Waldbieser G, Yang Y, Bao L, Yuan Z, Li N, Liu Z. (2017). Development of 690 K SNP arrays for whole genome mapping and genetic studies in catfish. Sci Rep.;7:40347.

Zhang, G., Fang, X., Guo, X., Li, L. I., Luo, R., Xu, F., ... & Wang, J. (2012). The oyster genome reveals stress adaptation and complexity of shell formation. Nature, 490(7418), 49-54.

Zhao, L., Zhang, Y., Ji, P., Zhang, X., Zhao, Z., Hou, G., ... & Sun, X. (2013). A dense genetic linkage map for common carp and its integration with a BAC-based physical map. PloS one, 8(5), e63928.

Zheng, X., Kuang, Y., Zhang, X., Lu, C., Cao, D., Li, C., & Sun, X. (2011). A genetic linkage map and comparative genome analysis of common carp (Cyprinus carpio L.) using microsatellites and SNPs. Molecular Genetics and Genomics, 286, 261-277.

Zhou, Y. L., Wang, Z. W., Guo, X. F., Wu, J. J., Lu, W. J., Zhou, L., & Gui, J. F. (2021). Construction of a high-density genetic linkage map and fine mapping of QTLs for growth and sex-related traits in red-tail catfish (*Hemibagrus wyckioides*). Aquaculture, 531, 735892. https://doi.org/10.1016/j.aquaculture.2020.735892

Zhou, Z., Bao, Z., Dong, Y., Wang, S., He, C., Liu, W., ... & Zhu, F. (2006). AFLP linkage map of sea urchin constructed using an interspecific cross between Strongylocentrotus nudus (♀) and S. intermedius (♂). Aquaculture, 259(1-4), 56-65.

12

Genomic Selection as a Tool for Advancing Selective Breeding Strategies in Aquaculture

Vishal K. Solanki*, Foram D. Vala, Anjali R. Baraiya and Nilesh H. Joshi

College of Fisheries Science, Kamdhenu University, Veraval, Gujarat (362 265) India

Abstract

Aquaculture is rapidly becoming a crucial solution to satisfy the growing demand for fish worldwide. Despite its significant potential, this industry has lagged behind terrestrial farming in implementing selective breeding techniques to enhance fish production. Recent developments in genetic technologies, including gene editing with CRISPR-Cas9, satisfy the rising demand for seafood worldwide. High-throughput sequencing offers exciting opportunities for better desirable characteristics such as growth rates and resistance against illness in various aquatic species. Research shows selective breeding might double production rates for key creatures like Nile tilapia and Atlantic salmon within just a few generations. However, the aquaculture sector faces several challenges, including the high costs of implementing these technologies and ethical concerns regarding genetic modifications. This review examines various breeding strategies, including genomic and marker-assisted selection, which enhance breeding accuracy and efficiency. The preservation of genetic materials from both wild and farmed fish is essential to aquaculture's long-term viability. The industry can improve productivity and protect biodiversity by integrating effective genetic management strategies. The insights gained from this review aim to assist researchers, industry practitioners, and policymakers in navigating the complexities of genetic improvement in aquaculture. By leveraging these advancements, the aquaculture sector can significantly contribute to global food security and ensure a resilient future for seafood production.

Keywords: *Selective Breeding, Aquaculture, Genetic Improvement, Molecular Markers, Genomic Selection*

Introduction

The method of selective breeding involves selecting particular breeding prospects to accumulate beneficial alleles and increase the prevalence of desired traits in a population. Characteristics, including growth, disease resistance, reproduction, development, and coloration, are frequently addressed in aquaculture (Maluwa, 2022). Selective breeding, whereby people are picked to have offspring with desired traits, is the foundation of genetic improvement initiatives in all organisms. This is a type of directed evolution in which the breeder determines fitness instead of the organism's innate capacity for survival and procreation. Darwin's theory of natural selection was affected by selective breeding, which existed before the mechanics of heredity were discovered. Darwin agreed that deliberate human selection used excellent outcomes (Hill, 2017). By choosing better parents, a species' desirable features can be improved. Common carp (*Cyprinus carpio*) were first cultivated in ponds in China 4,000–5,000 years ago, marking the beginning of fish farming. Environmental effects, growing competition for water, and increased demand due to population development and declining wild fish sources are some of the issues aquaculture faces as it grows. Controlled mating and lineage monitoring were two early breeding techniques that resulted in severe inbreeding and decreased output. Mendel's plant experiments established the groundwork for quantitative and contemporary genetics when they were made public in 1900 (Gjedrem and Baranski, 2010). Mendel's experiments were confirmed in 1900, marking the beginning of selective breeding for plant production. Fifteen years later, Farm animals were treated using the same methods (Hagedoorn, 1950). Genetically modified stocks are used for most plant and farm animal production. Genetic advancement in aquaculture is still sluggish, though. Although the first aquaculture family-based breeding program began in 1975, by 2010, only 8.2% of aquaculture production was based on genetically modified stocks (Neira, 2010; Rye *et al.* 2010).

In the 1920s, the first aquaculture selection studies were conducted on brook trout to enhance survival rates from 2% to 69% by reducing furunculosis mortality (Embody *et al.,* 1925). Additionally, common carp were selected at this time, leading to the formation of two fruitful strains (Kirpichnikov, 1987), and notable decreases in mortality were documented in Germany (Gjedrem and Robinson, 2014). Rainbow trout were first selected by Lauren R. Donaldson in 1932 for their enhanced growth, increased egg output, and early sexual maturation (Donaldson and Olson, 1957). Donaldson and Menasveta 1961 also reported favourable selection responses in 1949 while concentrating on chinook salmon (*Oncorhynchus tshawytscha*). Rainbow trout exhibited heritability values of 0.16 at 150 days and 0.32 at 280 days, although Nenashev 1966 estimated it to be 0.21 for fingerling weight in common carp (Aulstad *et al.,* 1972). In 1972, JE discovered that the body weight in oysters (*Crassostrea gigas*) is heritable at 0.33. Teichert-Coddington & Smitherman, 1988, found no response to their research on Nile tilapia collective selection for growth. The American oyster (*Crassostrea virginica*) showed a notable growth response; on the other hand, Haley *et al.,* 1976 suggested combining individual and family selection. Selective breeding in commercially farmed bivalves has been shown to produce 10–20 per cent increases for each generation (Newkirk 1980). In crustaceans, De Donato *et al.,* 2005 saw a 42% decrease in body weight variance at harvest and an average growth rate gain of 1.3% each generation in *P. vannamei* over 11 generations. Salmonids were first bred in Norway in 1975 using family-based programs (Gjedrem, 2010), and sib selection is currently the industry norm for aquaculture genetic enhancement (Neira, 2010; Rye *et al.*, 2010).

In aquaculture productivity, selective breeding is becoming more and more critical. Many aquatic species today benefit from genetic advancements through selective breeding programs, even though most were domesticated more recently than terrestrial ones. From early domestication, these techniques have developed into genomic selection (GS), which comprises three types of selection: marker-assisted, family, and mass. According to Meuwissen *et al.* (2013), GS uses genome-wide genetic markers to determine breeding values for quantitative traits. Advancements in SNP arrays and high-throughput sequencing have made GS feasible. Nowadays, it is used in many terrestrial species, particularly pigs (Samore and Fontanesi, 2016), agricultural products (Desta and Ortiz, 2014), dairy cattle (Boichard *et al.*, 2016), increasing the precision of genetic gains and breeding value predictions.

A significant challenge to the global food supply is balancing the United Nations' sustainable development objectives, which include ending poverty, with the world's expanding population, which is expected to increase to 10 billion by the year 2050. To preserve biodiversity and mitigate the adverse impacts on the environment of animal husbandry, nutritious food must be produced using fewer natural resources. According to Anderson *et al.,* 2017, aquaculture is the sector of food production with the fastest pace of expansion, which is crucial for both economic and food security. In aquaculture, genetic selection has significantly improved muscle quality, growth, and survival while lowering feed conversion ratios (Houston *et al.,* 2020). Numerous studies demonstrate the great potential of breeding selectively (Kocour *et al.*, 2007) and cross-breeding to enhance a variety of performance characteristics of common carp.

In Central Europe, cross-breeding is standard, although systematic selective breeding is still relatively new. According to recent studies, the generation interval has dropped from 4-5 years to 2-3 years, which is good.

Genetic Basis of Selective Breeding

Genetic Variability and Heritability

Genetic Variability

The most important raw element for breeders is variation. Utilizing genetic variety among animals is central to animal breeding theory since without it, no benefit can be obtained. In almost every trait examined in both terrestrial and aquatic species, genetic variation has been found. Accordingly, defining and quantifying genetic variation is essential to maximizing its use in changing the target population's traits (Gjedrem and Baranski, 2010).

Single Gene Traits

Certain qualities have a simple Mendelian inheritance pattern and are regulated by a single locus. These features are referred to as simple Mendelian traits. For instance, children of rainbow trout families with both parents heterozygous for the normal colour allele ('Aa') and the albino allele ('aa') will have a genotype frequency of 25% 'AA' (normal colour), 50% 'Aa' (normal colour), and 25% 'aa' (albino) (Gjedrem and Baranski 2010).

Quantitative Traits

According to Gjedrem and Baranski 2010, aquaculture qualities of economic significance are often governed by numerous individual genes and heavily impacted by environmental influences, even when there are simple Mendelian traits. These characteristics, which are known as quantitative attributes, have two primary characteristics:

- Several genes influence the phenotypic.
- The environment has an impact on the phenotypic to some extent.

Variation Between Species

There is minimal variation in protein-coding genes between species, even though they must be maintained for proper function. One significant variation in the genomic organization of different animals is the number of chromosomes. Survival rates have been comparatively high in crosses between several salmonid species, such as Arctic charr, brown trout, sea trout, and Atlantic salmon. But crosses with rainbow trout, a Pacific species, produced no surviving offspring (Refstie and Gjedrem, 1975; Gjedrem and Baranski, 2010).

Variation Within Species

In Atlantic salmon, most genetic diversity occurs within strains rather than between them. According to research by Gunnes and Gjedrem 1978, over 90% of the diversity was found within strains, with body weight differences across strains accounting for only 7–8.6% of the overall variation. This implies that individual strains contain the majority of the genetic diversity affecting characteristics like body weight.

Genetic Heritability

Heritability is a measure of how much genetic variation across individuals contributes to phenotypic trait variation within a group. The ratio of phenotypic to genetic variance is known as narrow-sense heritability, while the additive ratio of phenotypic to genetic variance is known as broad-sense heritability. Narrow-sense heritability shows how a trait should react to selection. According to, estimates might differ considerably depending on environmental factors, life stage, and genetic diversity. Heritability is essential to quantitative genetics and breeding programs because it makes it easier to estimate breeding values and forecast selection responses. Gjedrem and Baranski (2010) define broad-sense heritability (h^2T) as the ratio of genetic variation (σ^2G) to phenotypic variance (σ^2P). Estimates of heritability are only important for the populations they are generated from. However, more accurate predictions with wider applicability are produced by larger datasets with numerous full- and half-sib groups. The life stage at which a characteristic is measured can have a significant impact on its heritability. Most commercially important traits in mollusks have moderate to high narrow-sense heritability, indicating that additive genetic variance is typically adequate for selective breeding. Moreover, genetic markers are increasingly being used to enhance heritability estimates by utilizing family variance components since molecular parentage assignment reduces environmental variation by enabling families to grow up in a common environment (Nguyen *et al.*, 2014; Kong *et al.*, 2015).

Quantitative Trait Loci (QTL) and Marker-Assisted Selection (MAS)

Complex climate-resilient traits with suitable breeding goals and inbreeding control can now be selected thanks to the quick development of genetic techniques like genome editing, genomic selection, and marker-assisted selection (MAS). In contrast to conventional techniques, MAS and GS make use of both within- and between-family genetic variation by utilizing genome-wide genetic markers connected to quantitative trait loci (QTL) (Goddard 2009). These contemporary methods increase the climate change resilience of farmed animals by providing fresh approaches and chances for efficient genetic development. Aquaculture's economically important traits are controlled by a broad class of small-effect genes known as quantitative trait loci (QTL). QTL mapping locates genetic marker alleles in the genome that segregate with QTL (Lynch and Walsh, 1998). Marker alleles associated with desirable characteristics can be used to identify individuals and then choose which ones to breed. For characteristics like disease resistance or meat quality that are difficult to quantify, the The marker-assisted selection (MAS) approach is especially beneficial (Sonesson, 2007).

Genomics in Selective Breeding

Genomic-based selection, which integrates genomic information into selective breeding programs, is crucial for improving the genetics of both plants and animals (Meuwissen *et al.*, 2013; Muranty *et al.*, 2014). Modern aquaculture operations typically focus on species such as salmonids, prawns, tilapia, carp, sea bream, sea bass, turbot, flounders, sturgeons, oysters, scallops, clams, catfish, and moronids. Many aquaculture programs began with straightforward mass selection for growth and appearance; however, an increasing number are now utilizing family information to enhance genetic gains and select for complex traits like disease resistance and flesh quality. While sib selection assesses the average breeding values for families, it only utilizes half of the available additive genetic variation. This limitation can restrict genetic gains and potentially increase inbreeding, as not all families are selected for the next generation. In contrast, genomic selection has the potential to significantly boost genetic gains and can reduce inbreeding by as much as 81% (Sonesson and Meuwissen, 2009). According to Zenger *et al.* (2017), the effective population size on aquaculture farms is quite small due to the limited number of unique broodstock available to produce offspring for the entire production system. Therefore, it is possible to apply genomic selection on a family or farm-wide basis using fewer genome-wide loci than would be necessary in outbred populations.

Genomic selection systems can significantly improve genetics in aquaculture, especially when paired with controlled nucleus breeding programs. However, the effectiveness of genomic selection can vary among species due to differences in fertility, effective population size, life cycle, and breeding objectives (Zenger *et al.*, 2017). When features are linked to large-effect quantitative trait loci (QTL), marker-assisted selection (MAS) is advantageous for breeding candidates with hard-to-measure attributes (Sonesson, 2007; Robledo *et al.*, 2016). A substantial QTL, for example, was found in Atlantic salmon populations resistant to the IPN virus in Norway and Scotland (Houston *et al.*, 2008; Moen *et al.*, 2009). The use of MAS in commercial breeding programs has reduced the incidence of IPN (Moen *et al.*, 2015; Robledo *et al.*, 2016). Genomic selection (GS) has recently extended MAS to a genome-wide scale.

Global Progress in Aquaculture Selective Breeding

Early aquaculture breeding efforts primarily concentrated on individual or mass selection, with a strong emphasis on growth rates due to their economic significance and straightforward measurement. In the 1920s, selective breeding was initiated with brook trout to combat high mortality from furunculosis (Embody and Hyford, 1925) and with common carp, resulting in two robust strains (Kirpichnikov, 1987). Later advancements in Germany further improved furunculosis resistance in common carp. Only in more recent years, however, have comprehensive estimations of the genetic and phenotypic factors essential for improving the efficacy of breeding programs been generated.

Selective Breeding in Salmon Aquaculture

The farming of Atlantic salmon (Salmo salar) began in Norway in the 1970s and is now a thriving industry in key producing regions such as Norway, Chile, the UK, Canada, and Australia. After starting their life cycle in freshwater hatcheries or net pens, salmon smoltify and move to seawater, where they reach a harvestable size of about 4 kg. Selective breeding has been essential to salmon farming progress since its inception.

The Mowi strain was established in the late 1960s when the first Atlantic salmon breeding program was started using fish from Mowi AS (Glover *et al.*, 2009). By crossing Mowi fish with salmon from the Swedish and Norwegian rivers, AKVAFORSK created the AquaGen strain in the 1970s (Gjedrem 2012; Gjedrem *et al.*, 1991). Other significant strains, such as Jakta in the 1980s and Bolaks in 1974–1975, have further influenced contemporary salmon stocks.

A breeding effort with Chinook salmon began in 1949 to improve characteristics such as rapid growth, resistance to disease, early maturation, and survival (Donaldson, 1971). A complicated interaction between environmental and genetic factors determines salmon growth and sexual maturation (Gutierrez *et al.*, 2015). Combining individual and family selection has proven effective in breeding efforts (Gjedrem, 1983). Long-term selection in programs like those for Coho salmon has led to substantial genetic gains, though with some loss of genetic diversity (Hershberger, 1990).

Selected strains, particularly those derived from the wild Namsen stock, have experienced over a twofold increase in growth rate, alongside a 40% rise in feed intake and a 20% enhancement in feed conversion ratio (FCR). This significant improvement has led to a considerable reduction in production costs, as feed accounts for more than 60% of the expenses in salmon farming.

Genomic selection has outperformed traditional pedigree methods in predictive accuracy, positively impacting various traits, including growth, fatty acid profiles, fillet pigmentation, and resistance to diseases such as sea lice, amoebic gill disease, and salmon rickettsial syndrome (Tsai *et al.,* 2015; Horn *et al.,* 2020; Odegard *et al.,* 2014; Kjetsa *et al.,* 2020; Aslam *et al.,* 2020; Bangera *et al.,* 2017). Its effectiveness is particularly notable in sib-testing schemes, where the genetic similarity between reference and validation populations enhances accuracy. Future progress may involve leveraging functional variants and advanced genomic tools, including genome editing technologies (Houston *et al.,* 2020).

Selective Breeding in *Oncorhynchus mykiss* Rainbow Trout

The rainbow trout (*Oncorhynchus mykiss*), which originated in the chilly Pacific waters of Asia and North America, has become increasingly popular worldwide for aquaculture and recreational fishing (Halverson, 2010). The commercial farming of this species saw significant growth starting in the 1950s, with selective breeding programs initiated in the 1970s aimed at enhancing traits such as body weight and maturation rates, although these efforts initially lacked extensive scientific backing (Guyomard 1981; Gjedrem 1985; Morkramer *et al.*, 1985; Paisley *et al.*, 2010). Today, rainbow trout farming is widely established in Europe, the Americas, and Asia, benefiting from innovations such as pelleted feeds (Janssen *et al.,* 2017).

Selective breeding has yielded notable achievements: males reach early maturation within the first year (680 g), while females mature later, often reaching 4.5 kg by the second year (Donaldson, 1971). In Finland, a breeding program improved the rate of development and early maturation across two generations, achieving an average 7% growth gain per generation (Kause *et al.,* 2005). Culling early-maturing males helped prevent early maturation from rising in one population, while the rate increased for females in another. California experiments showed heritability for early spawning dates, with a seven-day advance per generation (Siitonen and Gall 1989).

Selection for growth rates has proven highly successful, with genetic gains per generation estimated at 13% (Gjerde, 2002). In Japan, efforts to selectively breed for resistance to the IPN virus yielded significant results, lowering mortality rates in a resistant strain to just 4.3%, in stark contrast to the 96.1% mortality observed in susceptible strains (Okamoto *et al.,* 1993).

Genetic maps (Guyomard *et al.*, 2012), SNP chips (Palti *et al.*, 2015), and access to the entire rainbow trout genome (Berthelot *et al.*, 2014) are examples of genomic advancements that have made it possible to use increasingly complex breeding techniques and enable ongoing improvements in growth, disease resistance, and other commercially valuable traits (Reis Neto *et al.*, 2019).

Selective Breeding of Nile Tilapia (*Oreochromis niloticus*)

Tilapia (*Oreochromis niloticus*), originally from Africa, has now been cultivated in over 140 countries, making it one of the fastest-growing species in aquaculture. Early selective breeding efforts faced challenges, as attempts to individually select for growth yielded minimal results, likely due to environmental variability and limited breeding populations (Teichert-Coddington, 1983; Huang and Liao, 1990). To address these issues, a significant breeding program for Tilapia (GIFT) was initiated in 1988 in the Philippines by ICLARM in collaboration with AKVAFORSK and local research organizations.

In the GIFT experiment, wild African tilapia strains—aside from the Ghana strain—performed better than locally farmed strains in the Philippines, exhibiting higher growth rates, particularly in the Kenyan and Egyptian strains (Eknath *et al.*, 1991). According to tests carried out in eleven distinct habitats, the strain-by-environment interaction was insignificant and only accounted for 1% of the variation in body weight, demonstrating broad adaptability (Eknath *et al.*, 1991). Starting with eight strains, the GIFT breeding program produced an artificial base population with a heritability of 0.15 for harvest body weight: four wild strains from Africa and four farmed strains from the Philippines (Eknath *et al.*, 2007). The complete heterosis level for body weight and strains was determined by crossing them (8 × 8 diallel cross).

Ongoing studies confirmed notable growth improvements through selective breeding in tilapia. For example, Maluwa and Gjerde 2007 recorded a 6.6% increase per generation for harvest weight in *O. shiranus* in Malawi. Selective breeding for maturation timing in the GIFT strain achieved responses of 57% and 34% for high and low frequencies of early maturation, respectively (Longalong *et al.,* 1999).

At least 20 family-based breeding programs for Nile tilapia are presently underway worldwide, with 10 of them deriving from the GIFT strain and primarily using fifth-generation breeders (Neira, 2010; Rye *et al.* 2010). In addition to helping tilapia reach its objective of becoming the most common aquaculture species, the GIFT project has had a major worldwide impact (Fitzsimmons *et al.*, 2011).

Selective Breeding Programs for Carp in Aquaculture

The major Indian carps (catla, rohu, and mrigal) along with the common carp are essential to Indian aquaculture. In the 1970s, selective breeding experiments in

Israel aimed at enhancing growth in common carp showed limited success for increased growth rates, though growth was later confirmed as highly heritable (Moav and Wohlfarth, 1973; Kocour *et al.,* 2007). Improved methods and conditions may unlock further genetic gains in carp aquaculture.

Selective Breeding Program for *Labeo rohita*

Labeo rohita is a widely recognized tropical freshwater carp and is a key species in aquaculture selective breeding efforts. It was the primary fish selected for India's first rohu selective breeding initiative, which commenced in 1992. This program was conducted under the guidance of AKVAFORSK and the Central Research Institute of Freshwater Aquaculture (CIFA) to enhance growth rates and produce the genetically modified "Jayanthi rohu." The program effectively created a diverse base population by combining five wild river strains with a local farmed stock (Reddy *et al.*, 2002).

Four generations of rohu were selected; two of these were raised in monoculture, while the other two were raised in polyculture with mrigal and catla. Growth rates showed a substantial genetic response, with a 29.6% increase per generation, indicating significant genetic variability. Monoculture rearing yielded a slightly higher growth rate (30.6%) than polyculture (27.0%) (Mahapatra *et al.,* 2004), suggesting selective breeding could double growth rates within four generations. However, crossbreeding did not enhance growth or survival (Gjerde *et al.,* 2002).

Selective Breeding Program for *Catla Catla*

The ICAR-CIFA launched the first selective breeding program for *Catla catla* in India, aiming to improve growth traits through genetic selection. As part of this initiative, nine strains were collected from various sources to assess genetic diversity using microsatellite and phenotypic markers. The study revealed a molecular variance of 58.63% between the strains, which helped establish a base population for selective breeding (Mahapatra *et al.*, 2018).

Program for Selective Common Carp (*Cyprinus carpio*) Breeding

Common carp (*Cyprinus carpio*) has the longest history of domestication and can be found across Asia and around the world. In the 19th century, several distinct strains began to emerge in Germany and the Czech Republic (Komen, 1990). In Israel, selective breeding was introduced in 1995, primarily focusing on the crossbreeding of inbred strains, with traditional selective breeding playing a modest role (Janssen *et al.*, 2015).

This species displays a variety of traits that facilitate effective crossbreeding; however, the phenomenon of heterosis, or hybrid vigour for growth, is not consistently observed across all strains. In India, the first selective breeding program for common carp was initiated in Karnataka, which involved collecting six strains from local sources as well as from other countries such as Hungary, Indonesia, and Vietnam (Basavaraju *et al.*, 2003). These selection efforts have significantly enhanced the growth rates of common carp and other species, including crayfish, tilapia, and magur, with a particular focus on increasing harvested body weight (Dong *et al.*, 2015; Lind, 2012).

Selective Breeding Program for Gilthead Sea Bream (*Sparus aurata*)

The gilthead sea bream (*Sparus aurata*) is a commercially valuable species found in warm coastal euryhaline waters, known for its migratory and demersal behaviour, as well as protandrous hermaphroditism. Mass production began in the late 1980s, supported by a consistent supply of hatchery fry and juveniles, with broodstocks sourced from various geographic origins.

Sea bream commercial operations began in the early 2000s, following selective breeding efforts in the mid-1990s (Knibb *et al.*, 1998) (Thorland *et al.*, 2006). The species was originally selected exclusively for males due to the sex shift from males to females. It was predicted that nearly half of the potential advantages of selecting both sexes would be covered by the 12.1% genetic gain for growth rate. Thorrland *et al.* (2006) estimated that the selection differential for the first generation was 22%, which was in good accord with the results.

Another crucial trait is resistance to illness, particularly in cases of pasteurellosis (photo bacteriosis). According to Antonello *et al.* (2009), the heredity of this bacterial infection was determined to be moderate (0.18–0.45). Using sequencing methods for genotyping (Robledo *et al.*

Selective Breeding Programme for European Seabass (*Dicentrarchus labrax*)

In the 1970s, aquaculture of European seabass (*Dicentrarchus labrax)* started in southern Europe, first using natural larvae and juveniles. The industry grew rapidly in the 1990s due to advancements in reproductive techniques (Divanach and Kentouri, 2000). Breeding programs were established across several countries, with growth rate being the primary trait of interest, as seen in other fish breeding programs (Vandeputte *et al.,* 2019). Recent efforts to improve genetics have focused on processing yields and feed efficiency (Besson *et al.*, 2019).

First, RAD sequencing was used in genome-wide genotyping experiments as part of the FISHBOOST European Union FP7 project (Palaiokostas *et al.*, 2018). But it's anticipated that SNP arrays will eventually take the lead in commercial genomic selection. Following the development of a 3 K Illumina

SNP Chip in 2017 (Faggion *et al.*, 2019), a French consortium produced a 57 K Thermo Fisher SNP Chip in 2018 (Griot *et al.*, 2021).

Selective Breeding Programme for *Litopenaeus vannamei*

Shrimp production has rapidly expanded globally, with *Litopenaeus vannamei* now surpassing *Penaeus monodon*, particularly in Asia. Significant inaccuracies result from the range of 0 to 1 heritability estimates for *P. vannamei* growth (Lester 1988). The base population used in well-established artificial reproduction techniques consists of four stocks, two of which have a limited genetic base. According to Godin *et al.* (1996), families are colored elastomer polymer-tagged after being grown in buckets until they weigh around 1 g. According to Fjalestad *et al.* (1997), one generation of concurrent selection for growth and resistance to the Taura syndrome virus (TSV) produced genetic gains of 12.4% for TSV survival and 4.4% for harvest body weight. *L. vannamei* currently has four breeding programs (Neira 2010), while P. monodon currently has three (Rye *et al.*, 2010).

Selective Breeding Programme for Pacific Oyster (*Crassostrea gigas*)

Because of its quick development and versatility, the Pacific oyster is the most extensively cultivated mollusc in the world (FAO, 2005). On the West Coast of the United States, a breeding program combines growth rate and survival to increase live weight production. Since the Ostreid herpesvirus 1 (OsHV-1) µVar epidemic, improving disease resistance has become a top priority (Pernet *et al.*, 2016).

Family yields from chosen broodstock improved by 0.4% to 25.6% over wild broodstock, indicating a wide range of selection responses. According to Langdon *et al.* (2003), the average genetic increase over the seven trials was 9.5% each generation. Four breeding lines in Australia produced weight gains of 18% or 9% per generation, outperforming two non-selected groups (Nell *et al.*, 1999). Notably, Barber *et al.* (1983) and Newkirk and Haley

Selective Breeding Programme of Atlantic Cod

The Atlantic cod, a commercially valuable marine species, is distributed from the U.S. East Coast to Greenland, Iceland, Norway, and Europe's west coast (Boudry 2021). Norway's national program, run by NOFIMA, seeks to enhance cod aquaculture profitability through selective breeding, modelled after Atlantic salmon. NOFIMA produces around 400,000 juveniles annually, while Havlandet Marin Yngel, a private breeding program, generates about 3 million (Boudry 2021). Resources including its fully sequenced genome, SNP chips, and linkage maps are currently available, even if the genomic selection has not yet been applied in Atlantic cod farming (Torresen *et al.*, 2017; Pocwierz-Kotus *et al.*, 2015).

Selective Breeding Programme of Channel Catfish

A bidirectional selection experiment in the 1970s targeted growth rate improvement in Georgia, yielding around 80% growth increase over six generations approximately 13% per generation. According to Dunham (2007), the best interspecific hybrid was the blue × channel F1 hybrid. Due to its increased growth, improved resilience to disease, and higher fillet yields, F1 hybrids (female channel × male blue) have become the focus of commercial production during the last ten years (Geng *et al.*, 2016).

Techniques and Approaches in Selective Breeding

Next-Generation Sequencing (NGS) in Aquaculture

Aquaculture genetics has been transformed by next-generation sequencing (NGS), which makes it possible to accurately and economically identify markers in a variety of non-model fish species (Satam *et al.*, 2023). Many microsatellite markers for freshwater and marine species have been developed thanks to NGS (Barnes *et al.*, 2014; Santos *et al.*, 2015; Villanova *et al.*, 2015).

Breeding using Marker-Assisted Selection (MAS)

Using molecular markers associated with desired qualities, MAS improves selective breeding with increased accuracy and efficiency (Amiteye, 2021). This approach involves identifying specific DNA sequences associated with target phenotypic traits, aiding in selecting breeding candidates (Amiteye 2021). Genetic mapping, which places markers like SNPs across the genome, is integral to MAS strategies (He *et al.,* 2014).

Gene Editing in Aquaculture

Gene editing and gene transfer are two forms of genetic alteration used in aquaculture. Through gene transfer via microinjection, the first genetically altered fish was produced: a goldfish (*Carassius auratus*) (Zhu *et al.*, 1985). Subsequent methods included electroporation (Inoue 1992), virus-mediated (Yee *et al.*, 1994), sperm-mediated (Khoo *et al.*, 1992), and transposon-mediated transfers (Fadool *et al.*, 1998). However, these traditional approaches depended on random vector integration, which occasionally resulted in genomic rearrangements and insertional mutagenesis.

Gene editing, however, allows targeted genome modifications, including insertions, deletions, or replacements. Technologies like ZFN, TALEN, and CRISPR offer precise edits within any genomic region, particularly in coding sequences (Robb, 2019), and are widely used for creating genetically edited organisms.

Zinc-Finger Nucleases (ZFN)

ZFNs are modified endonucleases that have FokI cleavage enzymes and DNA-binding domains. The cleavage is carried out by FokI, whilst the zinc finger motifs target particular DNA regions (Urnov *et al.*, 2010). ZFN was initially used for gene editing in zebrafish, allowing for targeted alterations in genes such as EGFP in medaka and golden (Slc24a5) and no tail/Brachyury (Ntl) (Ansai *et al.*, 2012).

Transcription Activator-Like Effector Nucleases (TALEN)

Since its introduction in 2009, TALEN has gained popularity as a gene-editing technique due to its great precision and adaptability (Joung and Sander 2013). According to Kim and Kim (2014), TALEN, which is made up of the endonuclease FokI and a transcriptional activator-like effector for DNA recognition, causes double-strand breaks that are fixed by homology-directed repair (HDR) or non-homologous end joining (NHEJ). TALEN has a wider range of editing features than ZFN.

CRISPR-Cas9 and Genome Editing in Aquaculture

According to Yang *et al.* (2024), CRISPR-Cas9 is a precise genome-editing technique that has promising uses in aquaculture to increase species' growth, resistance to illness, and tolerance to environmental conditions. For instance, Atlantic salmon and tilapia have shown faster development and greater muscle growth as a result of altering the myostatin (MSTN) gene (Edvardsen *et al.*, 2014; Luo *et al.*, 2022). Additionally, CRISPR has enhanced disease resistance, creating ESC-resistant catfish and WSSV-resistant shrimp through gene modifications critical to infection processes (Wikumpriya *et al.*, 2023).

Transcriptomics and Proteomics in Breeding

Traditional breeding relies on phenotypic selection, which can be tedious and inaccurate (Ahmar *et al.*, 2020). Transcriptomics studies all RNA transcripts produced by the genome under specific conditions, revealing gene expression patterns and responses to various stimuli (Singh *et al.*, 2018). RNA sequencing (RNA-Seq) allows for high-throughput assessment of RNA presence and expression levels by sequencing cDNA and discovering new transcripts. To validate RNA-Seq and microarray findings, quantitative PCR (qPCR) is used for accurate quantification of selected genes (Gratz *et al.*, 2022).

Mapping QTL Regions for Disease Resistance

Quantitative trait loci (QTL) represent the variability at specific loci that influence quantitative traits (Geldermann 1975). According to Davey *et al.* (2011), a QTL is a phenotypic trait that is controlled by the statistical correlation between phenotypic measures and neighbouring genetic markers. Identifying QTL enhances understanding of the genes involved in trait expression. Once mapped, fish with beneficial QTL are selected for breeding, while those with negative QTL are excluded.

Single nucleotide polymorphism (SNP) markers are vital for constructing high-density maps, helping to fine-map QTLs and identify causative genes Salmonids can be mapped for disease-resistant genes using SNP markers (Palti *et al.*, 2015). Because QTL mapping is based on molecular markers, it can be used in a variety of breeding programs. Salmon breeding already makes use of technologies like Genotyping by Sequencing (GBS), which when paired with genome-wide association studies, should improve the production of disease-resistant salmon strains for commercial aquaculture.

Applications of Selective Breeding in Aquaculture

In aquaculture, selective breeding is a deliberate approach that aims to improve the genetic quality of aquatic species by selecting particular individuals with desired qualities as parents for subsequent generations. This method substitutes human influence for natural selection, which, depending on the features chosen, can have both beneficial and detrimental effects. Successful aquaculture initiatives benefit from the integration of genetic insights and well-structured breeding programs, emphasizing careful planning to achieve desired traits effectively. While academic research has contributed significantly to genetic knowledge, aligning breeding efforts with industry objectives can enhance efficiency and relevance. Additionally, selective breeding helps the industry adapt to evolving market demands by developing improved fish stocks tailored to consumer preferences. Despite the continuous evolution of aquaculture, the sector still faces challenges, particularly a lack of sufficient genetic data for many species, which hampers the establishment of effective breeding programs. While there is sufficient evidence for certain species, such as rainbow trout and Atlantic salmon, more thorough studies are required for other important species. As competition intensifies, robust breeding programs that create superior stocks are essential. Moreover, advancements in genetic engineering, such as chromosome manipulation and novel gene introduction, complement selective breeding efforts, offering exciting possibilities for further genetic improvement in aquaculture.

Challenges and Future Perspectives

The ongoing discussions regarding the impact of genetically modified organisms (GMOs) on natural ecosystems and wild fish populations highlight important ethical and regulatory issues. A significant obstacle in aquaculture is precisely forecasting and controlling characteristics like growth rates and disease resistance, which are impacted by a variety of genetic and environmental factors. Current genome sequencing techniques also encounter significant obstacles, particularly when dealing with large and genetically complex species. Furthermore, the lack of standardized methods for collecting and sharing genetic data complicates collaboration among researchers and limits the integration of findings across various species. Addressing these challenges requires the establishment of clear protocols and the promotion of cooperative efforts within the research community. By improving data sharing and collaboration, the aquaculture industry can better navigate these complexities, ultimately leading to more effective breeding programs and sustainable practices.

Conclusion

In conclusion, aquaculture is making great strides thanks to selective breeding and genetic developments, which are enhancing growth rates, disease resistance, and sustainability for all fish and shellfish species. The potential for genetic improvement in aquaculture is impressive: species like tilapia, salmon, and carp have seen their growth rates double in as few as three to seven generations, which far exceeds similar gains in livestock. This rapid progress is mainly due to high reproductive rates in aquatic species and the larger genetic variation available in fish compared to land animals. Recent tools like genomic sequencing, marker-assisted selection, and gene editing provide new ways to enhance breeding programs. However, challenges remain, such as the high costs of these technologies, regulatory concerns, and the need for ethical guidelines. To fully benefit from these advancements, aquaculture must invest in better infrastructure, develop international partnerships, and educate more scientists and breeders in genetic techniques. Overall, selective breeding not only benefits farmers by reducing production costs but also has environmental advantages. Faster-growing or disease-resistant fish can result in more economical use of feed, water, and space, making aquaculture a more sustainable means of supplying the world's expanding food needs. With continued research and development, selective breeding and genetics offer promising solutions for a sustainable future in global aquaculture.

References

Ahmar, S., Gill, R. A., Jung, K. H., Faheem, A., Qasim, M. U., Mubeen, M., & Zhou, W. (2020). Conventional and molecular techniques from simple breeding to speed breeding in crop plants: recent advances and future outlook. International journal of molecular sciences, 21(7), 2590. https://doi.org/10.3390/ijms21072590.

Amiteye, S. (2021). Basic concepts and methodologies of DNA marker systems in plant molecular breeding. Heliyon, 7(10). https://doi.org/10.1016/j.heliyon.2021.e08093.

Anderson, J. L., Asche, F., Garlock, T., & Chu, J. (2017). Aquaculture: Its role in the future of food. In World agricultural resources and food security: International food security (pp. 159-173). Emerald Publishing Limited.https://doi.org/10.1108/S1574-871520170000017011.

Ansai, S., Ochiai, H., Kanie, Y., Kamei, Y., Gou, Y., Kitano, T., ... & Kinoshita, M. (2012). Targeted disruption of exogenous EGFP gene in medaka using zinc-finger nucleases. Development, growth & differentiation, 54(5), 546-556. doi:10.1111/j.1440-169X. 2012.01357.x.

Aslam, M. L., Boison, S. A., Lillehammer, M., Norris, A., & Gjerde, B. (2020). Genome-wide association mapping and accuracy of predictions for amoebic gill disease in Atlantic salmon (Salmo salar). Scientific Reports, 10(1), 6435.https://doi.org/10.1038/s41598-020-63423-8.

Aulstad, D., Gjedrem, T., & Skjervold, H. (1972). Genetic and environmental sources of variation in length and weight of rainbow trout (Salmo gairdneri). Journal of the Fisheries Board of Canada, 29(3), 237-241. https://doi.org/10.1139/f72-042.

Bangera, R., Correa, K., Lhorente, J. P., Figueroa, R., & Yáñez, J. M. (2017). Genomic predictions can accelerate selection for resistance against Piscirickettsia salmonis in Atlantic salmon (Salmo salar). BMC genomics, 18, 1-12. https://doi.org/10.1186/s12864- 017-3487-y.

Barber, B., Davis, C., & Hawes, R. (1998). Genetic improvement of oysters for the marine aquaculture industry. In Abstract of the first Annual Northwest Aquaculture Conference nd Exposition (Vol. 35).

Barnes, T. C., Izzo, C., Bertozzi, T., Saint, K. M., Donnellan, S., Hammer, M. P., & Gillanders, B. M. (2014). Development of 15 microsatellite loci from mulloway, Argyrosomus japonicus (Pisces: Sciaenidae) using next generation sequencing and an assessment of their cross-amplification in other sciaenids. Conservation Genetics Resources, 6, 345-348.

Basavaraju, Y., Penman, D. J., & Mair, G. C. (2003). Stock evaluation and development of a breeding program for common carp (*Cyprinus carpio*) in Karnataka, India: progress of a research project.

Berthelot, C., Brunet, F., Chalopin, D., Juanchich, A., Bernard, M., Noël, B., ... & Guiguen, Y. (2014). The rainbow trout genome provides novel insights into evolution after whole-genome duplication in vertebrates. Nature Communications, 5(1), 1-10. https://doi.org/10.1038/ncomms4657.

Boichard, D., Ducrocq, V., Croiseau, P., & Fritz, S. (2016). Genomic selection in domestic animals: principles, applications and perspectives. Comptes rendus biologies, Desta 339(7-8), 274-277. https://doi.org/10.1016/j.crvi.2016.04.007.

Boudry, P., Allal, F., Aslam, M. L., Bargelloni, L., Bean, T. P., Brard-Fudulea, S., & Houston, R. D. (2021). Current status and potential of genomic selection to improve selective breeding in the main aquaculture species of International Council for the Exploration of the Sea (ICES) member countries. Aquaculture Reports, 20, 100700.

Davey, J. W., Hohenlohe, P. A., Etter, P. D., Boone, J. Q., Catchen, J. M., & Blaxter, M. L. (2011). Genome-wide genetic marker discovery and genotyping using next-generation sequencing. Nature Reviews Genetics, 12(7), 499-510.DOI: 10.1038/nrg3012.

De Donato, M., Manrique, R., Ramirez, R., Mayer, L., & Howell, C. (2005). Mass selection and inbreeding effects on a cultivated strain of *Penaeus (Litopenaeus) vannamei* in Venezuela. Aquaculture, 247(1-4), 159-167. https://doi.org/10.1016/j.aquaculture.2005.02.005.

Desta, Z. A., & Ortiz, R. (2014). Genomic selection: genome-wide prediction in plant improvement. Trends in plant science, 19(9), 592-601.https://doi.org/10.1016/j.tplants.2014.05.006.

Divanach, P., & Kentouri, M. (2000). Hatchery techniques for specific diversification in Mediterranean finfish larviculture. Cahiers options méditerranéennes, 47, 75-87.

Donaldson, L. R. (1971, March). Selective breeding of salmonid fishes. In Proceedings of the annual workshop World Mariculture Society (Vol. 2, No. 1-4, pp. 75-83). Oxford, UK: Blackwell Publishing Ltd.

Donaldson, L. R., & Menasveta, D. (1961). Selective breeding of chinook salmon. Transactions of the American Fisheries Society, 90(2), 160-164. https://doi.org/10.1577/1548-8659(1961)90[160:SBOCS]2.0.CO;2.

Donaldson, L. R., & Olson, P. R. (1957). Development of rainbow trout brood stock by selective breeding. Transactions of the American Fisheries Society, 85(1), 93-101. https://doi.org/10.1577/1548-8659(1955)85[93:DORTBS]2.0.CO;2.

Dong, Z., Nguyen, N. H., & Zhu, W. (2015). Genetic evaluation of a selective breeding program for common carp Cyprinus carpio conducted from 2004 to 2014. BMC genetics, 16, 1-9.

Dunham, R. A. (2007). Comparison of six generations of selection, interspecific hybridization, intraspecific crossbreeding and gene transfer for growth improvement in ictalurid catfish. Aquaculture, 272, S252-S253.

Edvardsen, R. B., Leininger, S., Kleppe, L., Skaftnesmo, K. O., & Wargelius, A. (2014). Targeted mutagenesis in Atlantic salmon (Salmo salar L.) using the CRISPR/Cas9 system induces complete knockout individuals in the F0 generation. PloS one, 9(9), e108622. https://doi.org/10.1371/journal.pone.0108622.

Eknath, A. E., Bentsen, H. B., Gjerde, B. J. A. R. N. E., Tayamen, M. M., Abella, T. A., Gjedrem, T. R. Y. G. V. E., & Pullin, R. S. (1991). Approaches to national fish breeding programs: pointers from a tilapia pilot study.

Eknath, A. E., Bentsen, H. B., Ponzoni, R. W., Rye, M., Nguyen, N. H., Thodesen, J., & Gjerde, B. (2007). Genetic improvement of farmed tilapias: composition and genetic parameters of a synthetic base population of *Oreochromis niloticus* for selective breeding. Aquaculture, 273(1), 1-14.

Embody, G. C., & Hayford, C. O. (1925). The advantage of rearing brook trout fingerlings from selected breeders. Transactions of the American Fisheries Society, 55(1), 135-148.. https://doi.org/10.1577/1548-8659(1925)55[135:TAORBT]2.0.CO;2.

Fadool, J. M., Hartl, D. L., & Dowling, J. E. (1998). Transposition of the mariner element from *Drosophila mauritiana* in zebrafish. Proceedings of the National Academy of Sciences, 95(9), 5182-5186. doi:10.1073/pnas.95.9.5182.

Faggion, S., Vandeputte, M., Chatain, B., Gagnaire, P. A., & Allal, F. (2019). Population-specific variations of the genetic architecture of sex determination in wild European sea bass *Dicentrarchus labrax* L. Heredity, 122(5), 612-621. https://doi.org/10.1038/s41437-018-0157-z.

FAO (2005). Cultured aquatic species information programme. Crassostrea gigas. Cultured aquatic species information programme. Text by the helm, M.M. FAO Fisheries Division [online]. Rome. Updated 13 April 2005. http://www.fao.org/home/search/en/?q=crassostrea%20gigas.

Fitzsimmons, K., Martinez-Garcia, R., & Gonzalez-Alanis, P. (2011). Why tilapia is becoming the most important food fish on the planet.

Fjalestad, K. T., Gjedrem, T., Carr, W. H., & Sweeney, J. N. (1997). Final report: the shrimp breeding program, selective breeding of *Penaeus vannamei*. The Oceanic Institute, Waimanalo, HI, USA.

Geldermann, H. (1975). Investigations on inheritance of quantitative characters in animals by gene markers I. Methods. Theoretical and Applied Genetics, 46(7), 319-330.

Geng, X., Liu, S., Yao, J., Bao, L., Zhang, J., Li, C., & Liu, Z. (2016). A genome-wide association study identifies multiple regions associated with head size in catfish. G3: genes, genomes, genetics, 6(10), 3389-3398.https://doi. Org/10.1534/g3.116.032201.

Gjedrem, T. (1983). Genetic variation in quantitative traits and selective breeding in fish and shellfish. Aquaculture, 33(1-4), 51-72.

Gjedrem, T. (1985). Improvement of productivity through breeding schemes. GeoJournal, 10, 233-241.

Gjedrem, T. (2010). The first family-based breeding program in aquaculture. Reviews in Aquaculture, 2(1), 2-15. https://doi.org/10.1111/j.1753-5131.2010.01011.x.

Gjedrem, T. (2012). Genetic improvement for the development of efficient global aquaculture: a personal opinion review. Aquaculture, 344, 12-22.

Gjedrem, T., & Baranski, M. (2010). Selective breeding in aquaculture: an introduction (Vol. 10). Springer Science & Business Media.

Gjedrem, T., & Robinson, N. (2014). Advances by selective breeding for aquatic species: a review. Agricultural Sciences, 5(12), 1152.

Gjedrem, T., Gjøen, H. M., & Gjerde, B. (1991). Genetic origin of Norwegian farmed Atlantic salmon. Aquaculture, 98(1-3), 41-50.

Gjerde, B., Reddy, P. V., Mahapatra, K. D., Saha, J. N., Jana, R. K., Meher, P. K., ... & Rye, M. (2002). Growth and survival in two complete diallele crosses with five stocks of Rohu carp (Labeo rohita). Aquaculture, 209(1-4), 103-115.

Glover, K. A., Otterå, H., Olsen, R. E., Slinde, E., Taranger, G. L., & Skaala, Ø. (2009). A comparison of farmed, wild and hybrid Atlantic salmon (*Salmo salar* L.) reared under farming conditions. Aquaculture, 286(3-4), 203-210.

Godin, D. M., Carr, W. H., Hagino, G., Segura, F., Sweeney, J. N., & Blankenship, L. (1996). Evaluation of a fluorescent elastomer internal tag in juvenile and adult shrimp *Penaeus vannamei*. Aquaculture, 139(3-4), 243-248.

Grätz, C., Bui, M. L., Thaqi, G., Kirchner, B., Loewe, R. P., & Pfaffl, M. W. (2022). Obtaining reliable RT-qPCR results in molecular diagnostics—MIQE goals and pitfalls for transcriptional biomarker discovery. Life, 12(3), 386. https://doi.org/10.3390/life12030386.

Griot, R., Allal, F., Phocas, F., Brard-Fudulea, S., Morvezen, R., Bestin, A., ... & Vandeputte, M. (2021). Genome-wide association studies for resistance to viral nervous necrosis in three populations of European sea bass (*Dicentrarchus labrax*) using a novel 57k SNP array DlabChip. Aquaculture, 530, 735930. https://doi.org/10.1016/j.aquaculture.2020.735930.

Gunnes, K., & Gjedrem, T. (1978). Selection experiments with salmon: IV. Growth of Atlantic salmon during two years in the sea. Aquaculture, 15(1), 19-33. https://doi.org/10.1016/0044-8486(78)90069-8.

Gutierrez, A. P., Yáñez, J. M., Fukui, S., Swift, B., & Davidson, W. S. (2015). Genome-wide association study (GWAS) for growth rate and age at sexual maturation in Atlantic salmon (*Salmo salar*). PloS one, 10(3), e0119730.

Guyomard, R. (1981). Electrophoretic variation in four French populations of domesticated rainbow trout (*Salmo gairdneri*). Canadian Journal of Genetics and Cytology, 23(1), 33-47.

Guyomard, R., Boussaha, M., Krieg, F., Hervet, C., & Quillet, E. (2012). A synthetic rainbow trout linkage map provides new insights into the salmonid whole genome duplication and the conservation of synteny among teleosts. BMC genetics, 13, 1-12. https:// doi.org/10.1186/1471-2156-13-15.

Hagedoorn, A.L. (1950) Animal Breeding. Crosby Lockwood & Son Ltd., London.

Haley, L. E., GF, N., DW, W., & RW, D. (1976). A report on the quantitative genetics of growth and survivalship of the American oyster, Crassostrea virginica, under laboratory conditions. Vol. 1: pp. 221–228; 17–23.

Halverson, A. (2010). An entirely synthetic fish: how rainbow trout beguiled America and overran the world. Yale University Press.

He, J., Zhao, X., Laroche, A., Lu, Z. X., Liu, H., & Li, Z. (2014). Genotyping-by-sequencing (GBS), an ultimate marker-assisted selection (MAS) tool to accelerate plant breeding. Frontiers in plant science, 5, 484.https://doi.org/10.3389/fpls.2014.00484.

Hershberger, W. K. (1990). Selective breeding in aquaculture. Food Reviews International, 6(3), 359-372.

Hill, W. G. (2017). Selective breeding. https://doi.org/10.1016/B978-0-12-809633-8.07124-7.

Horn, S. S., Meuwissen, T. H., Moghadam, H., Hillestad, B., & Sonesson, A. K. (2020). Accuracy of selection for omega-3 fatty acid content in Atlantic salmon fillets. Aquaculture, 519, 734767. https://doi.org/10.1016/j.aquaculture.2019.734767.

Houston, R. D., Bean, T. P., Macqueen, D. J., Gundappa, M. K., Jin, Y. H., Jenkins, T. L., ... & Robledo, D. (2020). Harnessing genomics to fast-track genetic improvement in aquaculture. Nature Reviews Genetics, 21(7), 389-409. https://doi.org/ 10.1038/s41576-020-0227-y.

Houston, R. D., Haley, C. S., Hamilton, A., Guy, D. R., Tinch, A. E., Taggart, J. B., ... & Bishop, S. C. (2008). Major quantitative trait loci affect resistance to infectious pancreatic necrosis in Atlantic salmon (*Salmo salar*). Genetics, 178(2), 1109-1115. https://doi.org/10.1534/genetics.107.082974.

Huang, C. M., & Liao, I. C. (1990). Response to mass selection for growth rate in *Oreochromis niloticus*. Aquaculture, 85(1-4), 199-205. https://doi.org/10.1016/0044-8486(90)90019-J.

Inoue K. (1992). Expression of reporter genes introduced by microinjection and electroporation in fish embryos and fry. Mol Mar Biol Biotechnol.;1(4–5):266-270.

Janssen, K., Chavanne, H., Berentsen, P., & Komen, H. (2017). Impact of selective breeding on European aquaculture. Aquaculture, 472, 8-16. https://doi.org/10.1016/j.aquaculture.2016.03.012.

Janssen, K., Prchal, M., Kocour, M., Berentsen, P. B. M., & Komen, H. (2015). Common Carp—Current Status of Selective Breeding in Europe.

JE, L. (1972). Estimating heritability and predicting response to selection for the Pacific oyster, *Crassostrea gigas*. In Proc Natl Shellfish Assoc (Vol. 62, pp. 62-66).

Joung, J. K., & Sander, J. D. (2013). TALENs: a widely applicable technology for targeted genome editing. Nature reviews Molecular cell biology, 14(1), 49-55. doi:10.1038/nrm3486.

Kause, A., Ritola, O., Paananen, T., Wahlroos, H., & Mäntysaari, E. A. (2005). Genetic trends in growth, sexual maturity and skeletal deformations, and rate of inbreeding in a breeding programme for rainbow trout (*Oncorhynchus mykiss*). Aquaculture, 247(1-4), 177-187.

Khoo, H. W., Ang, L. H., Lim, H. B., & Wong, K. Y. (1992). Sperm cells as vectors for introducing foreign DNA into zebrafish. Aquaculture, 107(1), 1-19. doi:10.1016/0044-8486(92)90046-N.

Kim, H., & Kim, J. S. (2014). A guide to genome engineering with programmable nucleases. Nature Reviews Genetics, 15(5), 321-334. doi:10.1038/nrg3686.

Kirpichnikov, V. S. (1987). Selection and new breeds of pond fishes in the USSR.

Kjetsa, M. H., Odegard, J., & Meuwissen, T. H. E. (2020). Accuracy of genomic prediction of host resistance to salmon lice in Atlantic salmon (*Salmo salar*) using imputed high-density genotypes. Aquaculture, 526, 735415. https://doi.org/10.1016/j.aquaculture.2020.735415.0.

Knibb, W., Gorshkova, G., & Gorshkov, S. (1998). Selection and crossbreeding in Mediterranean cultured marine fish.

Kocour, M., Mauger, S., Rodina, M., Gela, D., Linhart, O., & Vandeputte, M. (2007). Heritability estimates for processing and quality traits in common carp (*Cyprinus carpio* L.) using a molecular pedigree. Aquaculture, 270(1-4), 43-50.

Komen, J. (1990). Clones of common carp, *Cyprinus carpio*: new perspectives in fish research. Wageningen University and Research.

Kong, N., Li, Q., Yu, H., & Kong, L. F. (2015). Heritability estimates for growth-related traits in the Pacific oyster (*Crassostrea gigas*) using a molecular pedigree. Aquaculture Research, 46(2), 499-508. https://doi.org/10.1111/are.12205.

Lande, R., & Thompson, R. (1990). Efficiency of marker-assisted selection in the improvement of quantitative traits. Genetics, 124(3), 743-756. https://doi.org/10.1093/genetics/124.3.743.

Langdon, C., Evans, F., Jacobson, D., & Blouin, M. (2003). Yields of cultured Pacific oysters *Crassostrea gigas* Thunberg improved after one generation of selection. Aquaculture, 220(1-4), 227-244.

Lester, L. J. (1988). Differences in larval growth among families of Penaeus stylirostris Stimpson and P. vannamei Boone. Aquaculture Research, 19(3), 243-251.

Lind, C. E., Ponzoni, R. W., Nguyen, N. H., & Khaw, H. L. (2012). Selective breeding in fish and conservation of genetic resources for aquaculture. Reproduction in domestic animals, 47, 255-263.

Longalong, F. M., Eknath, A. E., & Bentsen, H. B. (1999). Response to bi-directional selection for frequency of early maturing females in Nile tilapia (Oreochromis niloticus). Aquaculture, 178(1-2), 13-25.

Luo, M., Wang, J., Dong, Z., Wang, C., & Lu, G. (2022). CRISPR-Cas9 sgRNA design and outcome assessment: Bioinformatics tools and aquaculture applications. Aquaculture and Fisheries, 7(2), 121-130.

Mahapatra, K. D., Gjerde, B., Saha, J. N., Reddy, P. V. G. K., Jana, R. K., Sahoo, M., & Rye, M. (2004). Realized genetic gain for growth in rohu (Labeo rohita). Aquaculture, in manuscript.

Mahapatra, K., Sahoo, L., Saha, J. N., Murmu, K., Rasal, A., Nandanpawar, P., ... & Patnaik, M. (2018). Establishment of base population for selective breeding of catla (Catla catla) depending on phenotypic and microsatellite marker information. Journal of genetics, 97, 1327-1337.

Maluwa, A. (2022). Genomics, genetics, and breeding in aquaculture. In Sustainable Fish Production and Processing (pp. 3-43). Academic Press. https://doi.org/10.1016/B978-0-12-824296-4.00003-7

Maluwa, A. O., & Gjerde, B. (2007). Response to selection for harvest body weight of Oreochromis shiranus. Aquaculture, 273(1), 33-41.

Meuwissen, T., Hayes, B., & Goddard, M. (2013). Accelerating improvement of livestock with genomic selection. Annu. Rev. Anim. Biosci., 1(1), 221-237. https://doi.org/10.1146/annurev-animal-031412-103705.

Moav, R., & Wohlfarth, G. W. (1973). Carp breeding in Israel. Agricultural Genetics, 295-318.

Moen, T., Baranski, M., Sonesson, A. K., & Kjøglum, S. (2009). Confirmation and fine-mapping of a major QTL for resistance to infectious pancreatic necrosis in Atlantic salmon (Salmo salar): population-level associations between markers and trait. BMC genomics, 10, 1-14.doi:10.1186/1471-2164-10-368.

Morkramer, S., Hörstgen-Schwark, G., & Langholz, H. J. (1985). Comparison of different European rainbow trout populations under intensive production conditions. Aquaculture, 44(4), 303-320.

Muranty, H., Jorge, V., Bastien, C., Lepoittevin, C., Bouffier, L., & Sanchez, L. (2014). Potential for marker-assisted selection for forest tree breeding: lessons from 20 years of MAS in crops. Tree Genetics & Genomes, 10(6), 1491-1510.. doi: 10.1007/s11295-014-0790-5.

Neira, R. (2010, August). Breeding in aquaculture species: genetic improvement programs in developing countries. In Proceedings of the 9th World Congress on Genetics Applied to Livestock Production (Vol. 8).

Nell, J. A., & Hand, R. E. (2003). Evaluation of the progeny of second-generation Sydney rock oyster Saccostrea glomerata (Gould, 1850) breeding lines for resistance to QX disease Marteilia sydneyi. Aquaculture, 228(1-4), 27-35.

Nell, J. A., Smith, I. R., & Sheridan, A. K. (1999). Third generation evaluation of Sydney rock oyster Saccostrea commercialis (Iredale and Roughley) breeding lines. Aquaculture, 170(3-4), 195-203.

Nenashev, G. A. (1966). Heritability of several morphological (diagnostic) traits in the Ropsha common carp. Izvestija Gosud. nauchno-issled. Inst Ozern Rechn Rybn Khos (GosNIORKh), 61, 125-135.

Newkirk, G. F. (1980). Review of the genetics and the potential for selective breeding of commercially important bivalves. Aquaculture, 19(3), 209-228. https://doi.org/10.1016/0044-8486(80)90045-9.

Newkirk, G. F., & Haley, L. E. (1983). Selection for growth rate in the European oyster, Ostrea edulis: response of second-generation groups. Aquaculture, 33(1-4), 149-155.

Nguyen, T. T., Hayes, B. J., & Ingram, B. A. (2014). Genetic parameters and response to selection in blue mussel (Mytilus galloprovincialis) using a SNP-based pedigree. Aquaculture, 420, 295-301. https://doi.org/10.1016/j.aquaculture.2013.11.021.

Odegard, J., Moen, T., Santi, N., Korsvoll, S. A., Kjøglum, S., & Meuwissen, T. H. (2014). Genomic prediction in an admixed population of Atlantic salmon (Salmo salar). Frontiers in Genetics, 5, 402.

Okamoto, N., Tayama, T., Kawanobe, M., Fujiki, N., Yasuda, Y., & Sano, T. (1993). Resistance of a rainbow trout strain to infectious pancreatic necrosis. Aquaculture, 117(1-2), 71-76.

Paisley, L. G., Ariel, E., Lyngstad, T., Jónsson, G., Vennerström, P., Hellström, A., & Ostergaard, P. (2010). An overview of aquaculture in the Nordic countries. Journal of the World Aquaculture Society, 41(1), 1-17.

Palaiokostas, C., Cariou, S., Bestin, A., Bruant, J. S., Haffray, P., Morin, T., ... & Houston, R. D. (2018). Genome-wide association and genomic prediction of resistance to viral nervous necrosis in European sea bass (Dicentrarchus labrax) using RAD sequencing. Genetics Selection Evolution, 50, 1-11.

Palaiokostas, C., Ferraresso, S., Franch, R., Houston, R. D., & Bargelloni, L. (2016). Genomic prediction of resistance to pasteurellosis in gilthead sea bream (Sparus aurata) using 2b-RAD sequencing. G3: Genes, Genomes, Genetics, 6(11), 3693-3700.

Palti, Y., Gao, G., Liu, S., Kent, M. P., Lien, S., Miller, M. R., ... & Moen, T. (2015). The development and characterization of a 57 K single nucleotide polymorphism array for rainbow trout. Molecular ecology resources, 15(3), 662-672. DOI: 10.1111/1755-0998.12337.

Pernet, F., Lupo, C., Bacher, C., & Whittington, R. J. (2016). Infectious diseases in oyster aquacultire a new integrated approach. Philosophical Transactions of the Royal Society B: Biological Sciences, 371(1689), 20150213.

Poćwierz-Kotus, A., Kijewska, A., Petereit, C., Bernaś, R., Więcaszek, B., Arnyasi, M., ... & Wenne, R. (2015). Genetic differentiation of brackish water populations of cod Gadus morhua in the southern Baltic, inferred from genotyping using SNP-arrays. Marine genomics, 19, 17-22.

Pullin, R. S., Eknath, A. E., Gjedrem, T., Tayamen, M. M., Macaranas, J. M., & Abella, T. A. (1991). The genetic improvement of farmed tilapias (GIFT) project: the story so far.

Reddy, P. V., Gjerde, B., Tripathi, S. D., Jana, R. K., Mahapatra, K. D., Gupta, S. D., ... & Gjedrem, T. (2002). Growth and survival of six stocks of rohu (Labeo rohita, Hamilton) in mono and polyculture production systems. Aquaculture, 203(3-4), 239-250.

Refstie, T., & Gjedrem, T. (1975). Hybrids between Salmonidae species. Hatchability and growth rate in the freshwater period. Aquaculture, 6(4), 333-342. https://doi.org/10.1016/0044-8486(75)90112-X.

Reis Neto, R. V., Yoshida, G. M., Lhorente, J. P., & Yáñez, J. M. (2019). Genome-wide association analysis for body weight identifies candidate genes related to development and metabolism in rainbow trout (Oncorhynchus mykiss). Molecular Genetics and Genomics, 294, 563-571. https://doi.org/10.1007/s00438-018-1518-2.

Robb, G. B. (2019). Genome editing with CRISPR□Cas: an overview. Current Protocols Essential Laboratory Techniques, 19(1), e36. doi:10.1002/cpet.36.

Robledo Sánchez, D., Taggart, J. B., Ireland, J. H., McAndrew, B. J., Starkey, W. G., Haley, C. S., ... & Houston, R. D. (2016). Gene expression comparison of resistant and susceptible Atlantic salmon fry challenged with the Infectious Pancreatic Necrosis virus reveals a marked contrast in the immune response. BMC.

Robledo, D., Palaiokostas, C., Bargelloni, L., Martínez, P., & Houston, R. (2018). Applications of genotyping by sequencing in aquaculture breeding and genetics. Reviews in aquaculture, 10(3), 670-682. https://doi.org/10.1111/raq.12193.

Rye, M., Gjerde, B., & Gjedrem, T. (2010, August). Genetic improvement programs for aquaculture species in developed countries. In Proceedings of the 9th world congress on genetics applied to livestock production (Vol. 8).

Samorè, A. B., & Fontanesi, L. (2016). Genomic selection in pigs: state of the art and perspectives. Italian Journal of Animal Science, 15(2), 211-232. https://doi.org/10.1080/1828051X.2016.1172034.

Santos, B. S., Romana-Eguia, M. R. R., Basiao, Z. U., & Ikeda, M. (2015). Development and characterization of nine novel microsatellite markers for the milkfish Chanos chanos. Conservation Genetics Resources, 7, 451-453.

Satam, H., Joshi, K., Mangrolia, U., Waghoo, S., Zaidi, G., Rawool, S., ... & Malonia, S. K. (2023). Next-generation sequencing technology: current trends and advancements. Biology, 12(7), 997. https://doi.org/10.3390/biology12070997.

Selective Breeding for a Better Catch - CAT-Center for Aquaculture Technologies. (n.d.). Retrieved October 2, 2024, from https://aquatechcenter.com/breeding/selective-breeding-for-a-better-catch/.

Siitonen, L., & Gall, G. A. E. (1989). Response to selection for early spawn date in rainbow trout, Salmo gairdneri. Aquaculture, 78(2), 153-161.

Singh, K. P., Miaskowski, C., Dhruva, A. A., Flowers, E., & Kober, K. M. (2018). Mechanisms and measurement of changes in gene expression. Biological research for nursing, 20(4), 369-382. https://doi.org/10.1177/1099800418772161.

Sonesson, A. K. (2007). Within-family marker-assisted selection for aquaculture species. Genetics Selection Evolution, 39(3), 301. DOI: 10.1051/gse:2007005.

Sonesson, A. K., & Meuwissen, T. H. (2009). Testing strategies for genomic selection in aquaculture breeding programs. Genetics Selection Evolution, 41, 1-9. doi:10.1186/1297-9686-41-37.

Teichert-Coddington, D. (1983). Divergent selection for prematuration body weight in tilapia niloticus (Doctoral dissertation, MS Thesis, Auburn University, Auburn, AL).

Teichert-Coddington, D. R., & Smitherman, R. O. (1988). Lack of response by Tilapia nilotica to mass selection for rapid early growth. Transactions of the American Fisheries Society, 117(3), 297-300. https://doi.org/10.1577/1548-8659(1988)117%3C0297:LORBNT%3E2.3.CO;2.

Thorland, I., Papaioannou, N., Kottaras, L., Refstie, T., Papasolomontos, S., & Rye, M. (2006, June). The Kego breeding programs for sea bream (Sparus aurata) and sea bass (Dicentrarchus labrax) in Greece. In Proceedings of the 8th Hellenic Symposium on Oceanogra.

Tørresen, O. K., Star, B., Jentoft, S., Reinar, W. B., Grove, H., Miller, J. R., ... & Nederbragt, A. J. (2017). An improved genome assembly uncovers prolific tandem repeats in Atlantic cod. BMC genomics, 18, 1-23.https://doi.org/10.1186/s12864-016-3448-x.

Tsai, H. Y., Hamilton, A., Tinch, A. E., Guy, D. R., Gharbi, K., Stear, M. J., ... & Houston, R. D. (2015). Genome-wide association and genomic prediction for growth traits in juvenile farmed Atlantic salmon using a high-density SNP array. BMC genomics, 16, 1-9.https://dx.doi.org/10.1186%2Fs12864-015-2117-9.

Urnov, F. D., Rebar, E. J., Holmes, M. C., Zhang, H. S., & Gregory, P. D. (2010). Genome editing with engineered zinc finger nucleases. Nature Reviews Genetics, 11(9), 636-646. doi:10.1038/nrg2842.

Vandeputte, M., Gagnaire, P. A., & Allal, F. (2019). The European Sea bass: a key marine fish model in the wild and aquaculture. Animal genetics, 50(3), 195-206.https://doi.org/10.1111/age.12779.

Villanova, G. V., Vera, M., Díaz, J., Martinez, P., Calcaterra, N. B., & Arranz, S. E. (2015). Isolation and characterization of 20 polymorphic microsatellite loci in the migratory freshwater fish Leporinus obtusidens (Characiformes: Anostomidae) using 454 shotgun pyrosequencing. Journal of fish biology, 86(3), 1209-1217.

Wikumpriya, G. C., Prabhatha, M. W. S., Lee, J., & Kim, C. H. (2023). Epigenetic modulations for prevention of infectious diseases in shrimp aquaculture. Genes, 14(9), 1682. https://doi.org/10.3390/genes14091682.

Yang, Z., Fu, G., Lee, M., Yeo, S., & Yue, G. H. (2024). Genes for editing to improve economic traits in aquaculture fish species. Aquaculture and Fisheries.

Yee, J. K., Friedmann, T., & Burns, J. C. (1994). Generation of high-titer pseudotyped retroviral vectors with a very broad host range. Methods in cell biology, 43, 99-112.

Zenger, K. R., Khatkar, M. S., Jerry, D. R., & Raadsma, H. W. (2017). The next wave in selective breeding: implementing genomic selection in aquaculture.

Zhu, Z., He, L., & Chen, S. (1985). Novel gene transfer into the fertilized eggs of goldfish (Carassius auratus L. 1758). Journal of Applied Ichthyology, 1(1), 31-34. doi:10.1111/j.1439-0426. 1985.tb00408. x.

13

Synergy of Biosecurity and Biotechnology in Modern Aquaculture Challenges and Opportunities

Samanthula Surya Teja[1*], Lakkoju Nischal[2], Ravi Chovatia[3] and Samad Sheikh[1]

[1]*ICAR – Central Institute of Fisheries Education, Mumbai, Maharashtra (9400061) India*

[2]*College of Fishery Science, Andhra Pradesh Fisheries University, Muthukur, Andhra Pradesh (524 344), India*

[3]*Kerala University of Fisheries and Ocean Studies, Panangad, Kerala (682 506), India*

Abstract

Aquaculture is an emerging food sector which plays a crucial role in global food security and economic growth, but disease outbreaks pose significant threats to its sustainability. Biosecurity plays a vital role in preventing and controlling these outbreaks, focusing on disease prevention, early detection, and the implementation of physical, biological, and operational measures to safeguard fish health and maintain production integrity. While biosecurity enhances fish health, it faces challenges such as high operational costs and potential environmental consequences. Biotechnology offers promising solutions to improve aquaculture productivity, including genetic modification, hybridization, hormonal treatments, chromosome manipulation, and molecular techniques like RNA interference and vaccines. These innovations have the potential to optimize selective breeding, improve growth rates, enhance disease resistance, and conserve genetic resources. However, the application of biotechnology in aquaculture raises concerns about the impact of genetically modified organisms (GMOs) on existing biodiversity and ecosystem health. Effective regulation and integration of biotechnology and biosecurity measures are essential to ensure environmentally sound practices, safeguard human health, and promote sustainable development in aquaculture. This chapter explores the synergy between biosecurity and biotechnology in aquaculture, emphasizing their combined potential for enhancing productivity and sustainability while mitigating environmental risks.

Introduction

In the twenty-first century, the importance of the fisheries and aquaculture sectors to global food security and nutrition has become more and more evident. The global population is anticipated that there will be roughly 10 billion people (by 2050) on the planet, which will lead to an increase in food demand. Over the past few decades, the aquaculture sector acted as the most promising food-producing sector (FAO, 2022). According to reports, the aquaculture sector in India has suffered significant harm from the recurrent occurrence of infectious diseases with bacterial, viral and parasitic pathogens, which has hindered sustainable development. Reports have surfaced regarding the emergence and reappearance of diseases in aquaculture (Bondad-Reantaso *et al.*, 2005). Over the past twenty years, there has been a major increase in the occurrence of pathogens and disease emergence and reemergence. In addition, it has become clear that diseases transcend the boundaries of land and water when they first appear or reappear. This means that the transfer of germs and the dissemination of diseases are closely associated with fish movement, particularly with live fish intended for aquaculture or ornamental seafood sector. Therefore, biosecurity measures are essential for safeguarding our food supply, the environment, and both human and animal populations from the introduction and spread of various diseases (Austin *et al.*, 2022; Austin, 2023).

One technique with great potential for solving these aquaculture issues is biotechnology. Global aquaculture productivity and environmental quality can be greatly enhanced by the relatively new biotechnology tools available. These days, biotechnology has produced innovative new techniques for identifying the gene responsible for traits, such as resistance to disease or nutrient composition and transferring that gene into another aquatic organism (Fletcher *et al.*, 2001). Biotechnology is widely used in fish health management, and in the last ten years, major advancements have been made not only in the development of diagnostic probes and tests (Adams, 1999, Cunningham, 2004) but also in the creation of fish vaccines (Gudding *et al.*, 1999, Thompson and Adams, A, 2004). Opportunities to boost productivity and improve the both fresh and processed cultivated species quality have also been made possible by modern biotechnology. Furthermore, disease-resistant agricultural species are now being created, which could assist rural areas by lowering losses and enabling higher production on the same area. Furthermore, new technological developments, like the use of digital tools for disease diagnosis and health monitoring, have the potential to increase the efficacy of biosecurity protocols in aquaculture. Conventional studies in parasitology, virology, bacteriology and mycology are suitable for identifying the most common, easily grown pathogens; yet, these approaches can be costly for any pathogens, time-consuming, and could not provide a conclusive diagnosis to be made, even when strengthened by histological proof (Adams *et al.*, 1995). As a result, a comprehensive diagnosis of a disease frequently necessitates the use of many techniques. To identify numerous common diseases, immunodiagnostic and molecular techniques have been created.

Understanding that biosecurity measures must be put in place to lower risks to a manageable level is essential to comprehend the need to maintain biosecurity in aquaculture operations. Adopting suitable biosecurity procedures promotes trade, product quality, environmental sustainability, farm productivity, animal welfare, and, eventually, profitability. Since the weaker and more stressed members of the farmed population are more prone to viruses and parasites that can transfer to healthy people, dense populations reared nearby are more likely to experience disease outbreaks. When farm animals come into intentional or unintentional contact with animals from various habitats, the danger of disease transmission rises dramatically. Water, animals, people, equipment, and feed are among the many layers of biosecurity threats that might arise at the farm's gate, inside the farm, and at the farm exit.

Biosecurity

Biosecurity is the word used to describe techniques that minimize the risk of disease entry and subsequent transmission. The potential carriers of contamination are the first factor to be taken into account in biosecurity. Infected hosts, non-host biological carriers, and inanimate items tainted with infections are among them.

There has been an increase in the development and re-emergence of diseases and infections during the last 20 years. Regardless of land or ocean boundaries, when a disease first appears or re-emerges, it can spread throughout the area and beyond. This suggests that the transportation of aquatic animals, particularly live animals intended for aquaculture or the ornamental aquatic organism sector, is closely linked to the spread of viruses and infections. Additional pathways for the spread of pathogens include water currents, biofouling, wild reservoirs, ballast water, and microplastics. Modern aquaculture disease prevention and health management techniques need the implementation of a comprehensive strategy, incorporating complete biosecurity, given the prohibitive socioeconomic constraints that led to outbreaks of diseases and the intricate nature of their dissemination. "*A paradigm shift is needed in dealing with aquaculture biosecurity issues*" is what the FAO recently stated in this regard. In the transfer of live aquatic animals with unknown health states, usually for aquaculture development, the disease may have spread around the world (even to wild populations) by the period it has been detected and its host series established.

"*Aquaculture biosecurity*" refers to methods that reduce the possibility of disease establishment, transmission, and transfer. Among them are methods for lessening fish stress, which lowers their vulnerability to infections and diseases. *Aquaculture biosecurity* is defined as an efficient and cost-effective strategy to mitigate hazards associated with pathogens in aquaculture. It involves coordinated efforts at enterprise, nationwide and international levels, emphasizing communal responsibilities between public and private sectors, as highlighted in the FAO's Progressive Management Pathway (PMP) (Subasinghe *et al.*, 2023). Key components include (a) animal management, which focuses on maintaining healthy cultured animals and enhancing their immunity through proper farming practices; (b) pathogen control, aimed at preventing, minimizing, or eradicating disease-causing agents; (c) people management, involving education and supervision of relevant stakeholders; (d) focused research initiatives; and most critically, (e) supportive policies. These elements collectively form a robust and forward-looking biosecurity management framework.

The last three decades have demonstrated how major diseases and mass mortality events of aquatic animals, both in the wild and in agriculture, have challenged government officials, industry, and all stakeholders. For example, fish and shrimp aquaculture has been impacted by the abrupt emergence of several new illnesses and viruses in recent years. Acute hepatopancreatic necrosis disease (AHPND), snakehead rhabdovirus (SHRV), white faeces syndrome (WFS), koi herpesvirus (KHV), tilapia lake virus (TiLV), tilapia parvovirus (TiPV), lymphocystis disease virus (LCDV), red seabream iridovirus (RSIV), infectious spleen kidney necrosis virus (ISKNV), decapod iridescent virus 1 (DIV1) and newly, *Enterocytozoon hepatopenaei* (EHP), which affects prawns, crayfish and shrimp. This is caused by the use of wild-caught broodstock, broodstock exposed to wild animals, or broodstock fed tainted fresh or live feeds where the illness finds a new susceptible host. Known diseases that have migrated to new regions include infectious myonecrosis virus (IMNV) in India and Malaysia, epizootic ulcerative syndrome (EUS) in Malawi, and white spot disease (WSD) in Saudi Arabia and Australia. Depending on the industry impacted and the pathogen of concern, responses to mortalities and aquatic disease outbreaks differed. Despite the absence of regular economic evaluations, losses were typically economically considerable.

Using strategies including quarantine, screening, and surveillance, biosecurity seeks to stop the spread of new diseases among indigenous populations.

Individuals or groups who have come into contact with a disease but are not yet showing symptoms are kept in quarantine as a precautionary measure. People are identified and tested through screening to see if they are contaminated. To identify new instances and stop the spread of infection, surveillance keeps an eye on epidemics. Preventing the spread of diseases within a community through immunization, good hygiene, and personal safety gear is another essential component of biosecurity. In addition to hygienic measures like handwashing, surface cleaning, and appropriate waste disposal, vaccinations are an effective way to stop the propagation of many diseases.

Biosecurity History

The history of biosecurity is complicated and dates back to ancient times when the necessity of preventing the transmission of contagious diseases between humans and animals was first realized. The idea of biosecurity has changed over time as a result of scientific and technological breakthroughs, as well as a greater awareness of the dangers posed by disease outbreaks (Koblentz, 2010). John Snow's research during the London cholera epidemic in the 19th century helped to establish epidemiology as a legitimate scientific discipline (Tulchinsky, 2018). Developments like the discovery of antibiotics and vaccines in the 19th and 20th centuries were crucial in containing and eliminating serious infectious diseases (Echaubard *et al.*, 2018). However, infectious diseases have returned in the 20th and 21st centuries, with 75% of newly discovered human infections coming from zoonotic sources and 61% among them being zoonotic (Birhan *et al.*, 2015). Global health continues to face problems due to the growth and persistence of infectious diseases, in particular those of zoonotic origins, despite the dramatic advances in disease prevention and control brought about by biomedical sciences. In the upcoming years, the significance of biosecurity will only increase as the sector is shaped by the requirement for sustainable production methods and the worldwide danger of newly developing infectious diseases (Aly & Fathi, 2024).

Types of Aquaculture Biosecurity

Aquaculture biosecurity may be generally divided into three major categories: physical, biological, and operational. To avoid the introduction and spread of infections both within and across aquaculture practices. Sustaining the health and well-being of aquatic organisms and stopping the transmission of infections inside aquaculture farms depend heavily on these biosecurity measures (Assefa & Abunna, 2018; Aly and Fathi, 2024).

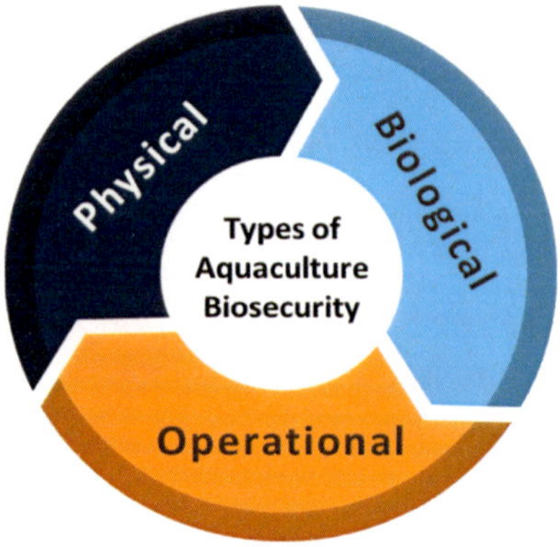

Physical Aquaculture Biosecurity Measures

Implementing physical biosecurity measures is crucial to preventing diseases, reducing the risk of disease outbreaks, and ensuring the overall health of aquatic animals in aquaculture systems (Dong & Zhou, 2023). In aquaculture, physical barriers such as fences and nets are used as part of physical biosecurity measures to keep possible disease vectors out of the system. Birds, mammals, and reptiles are vectors, thus fencing and netting often help prevent such animals from getting into aquaculture areas. Physical barriers will depend on such factors as the type and strength of the barrier, the opening size, and the location of the facility. The proximity of an aquaculture facility to wild fish populations, along with the design of farm units, significantly affects disease transmission risks through interactions between farmed and wild fish. (Bondad-Reantaso *et al.*, 2012). Physical barriers reduce the entry of pathogens into aquaculture systems and also lower the probability of farmed fish escaping, hence reducing the transfer of disease from farmed to wild fish inhabitants (Dong & Zhou, 2023). Escapees infected with pathogens, parasites, or other infections pose a serious ecological and economic threat to wild fish (Naylor *et al.*, 2005). Other physical biosecurity strategies involve closed systems: RAS (recirculating aquaculture systems) and flow-through systems. These are capable of preventing all kinds of disease-carrying pathogens, invasive species, or others from entering an aquaculture facility (Rurangwa & Verdegem, 2015). RAS, in its many forms, is a cost-effective and eco-friendly form of aquaculture that supports heavy fish densities by recycling water and treating it repeatedly. This closed system is an efficient biosecurity strategy as it reduces the possibility of disease transmission from outside sources (Schreier *et al.*, 2010).

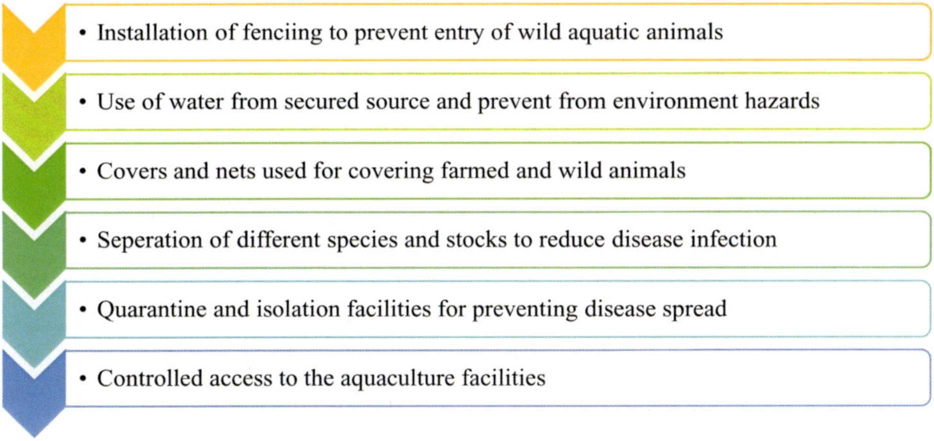

Biological Aquaculture Biosecurity Measures

One of the most important methods for controlling and stopping the spread of diseases in aquaculture is biological biosecurity. In aquaculture, one of the most crucial biological biosecurity strategies is the adoption of disease-resistant species. Due to their genetic characteristics, specific pathogen-free (SPF) and specific pathogen-resistant (SPR) animals are less susceptible to disease. Using species that are resistant to infection is a good method to lower the chance of disease outbreaks and enhance aquatic animals' general health. For instance, it has been demonstrated that using disease-resistant tilapia lowers the frequency and intensity of bacterial infections during culture (Elaswad & Dunham, 2018). Probiotics are the other key biological biosecurity measure. Austin *et al.* (2022) describe probiotics as live microorganisms that lower the disease risk and help to improve the immune response of aquatic animals. There is already proof that the survival and growth of fish, shrimp, and other cultured species in aquaculture can be raised by several folds (Hai, 2015).

In aquaculture, bacterial infections can also be cured with proper utilization of probiotics (El-Saadony *et al.*, 2021). Furthermore, another crucial biological biosecurity tool in aquaculture is the application of immunostimulants. According to Austin *et al.* (2022), immunostimulants are compounds that can strengthen aquatic animals' immune systems and make them more disease-resistant. It has been demonstrated that using immunostimulants increases aquatic animals' growth and survival rates in a variety of aquaculture settings (Wang *et al.*, 2017). In general, the promotion of sustainable aquaculture requires biological biosecurity measures.

Strengthening aquatic animals' immunity, preventing disease epidemics and enhancing health as well as the aquaculture systems productivity can be achieved through the use of disease-resistant animal species, probiotics, immunostimulants, vaccines, bacteriophages and other microbial-based therapies.

Operational Aquaculture Biosecurity Measures

According to Oidtmann *et al.* (2011), operational biosecurity measures are aimed at preserving the health and welfare of aquatic organisms as well as stopping the transmission of diseases inside aquaculture systems. These precautions include appropriate feed management, frequent water quality monitoring and the use of stringent hygiene and sanitation guidelines. Operational biosecurity precautions must include proper feed management (Bera *et al.*, 2018). Aquatic animals' health and immunity can be preserved by using premium meals devoid of pollutants and infections (Palić *et al.*, 2015). Feed usage that is specially prepared for the species reared, frequent inspections for contamination or spoiling, and appropriate feed storage are all important aspects of feed management techniques (White *et al.* 2018).

Water quality plays a major key role in maintaining the health of aquatic. Water quality parameters like dissolved oxygen, temperature, pH and ammonia levels can be regularly supervised to help spot possible issues beforehand they convert to serious outbreaks. Deprived water quality can stress aquatic creatures, making them more vulnerable to infection (Ina-Salwany *et al.*, 2019). Water quality may be preserved, and the likelihood of disease outbreaks will decrease by using RAS and other treatment methods for influent and effluent water (Ahmed & Turchini, 2021).

Another crucial component of operational biosecurity measures is the application of stringent hygiene and sanitation procedures. Pathogens can be kept from spreading within and between aquaculture facilities by properly washing and disinfecting trucks, holding units, and equipment. Additionally, staff members must get training on good hygiene habits, such as washing their hands and wearing personal protection equipment (PPE). Workers and aquatic animals can avoid spreading diseases by using personal protective equipment (PPE), for example, gloves and boots (Muthu *et al.*, 2020).

Along with these tactics, the effectiveness and efficacy of operational biosecurity measures may be increased by utilizing cutting-edge technology like drones, and artificial intelligence (AI) and sensors. Drones may be used to assess aquaculture amenities and find any hazards (Ubina & Cheng, 2022), while sensors can be used to track water quality indicators and identify real-time fluctuations (Mustapha *et al.* 2021). To create prediction models that can assist in spotting possible disease outbreaks before they happen, artificial intelligence (AI) can be utilised to evaluate data from various sources, such as sensors and drones (Daoliang & Chang, 2020; Mustapha *et al.*, 2021).

Progressive Management Pathway (PMP) / Aquaculture Biosecurity (AB) Approach

It is essential to have a thorough grasp of the four variables and routes that lead to endemic and exotic illnesses, which are as follows:

- ***Live animals and their products trade:*** Aquatic organisms (aquatic plants and animals) are sold internationally as food supplies; if a country's biosecurity is inadequate, diseases may spread along with host migration.
- ***Knowledge regarding pathogens and their hosts:*** Lack of information about new pathogens (including their pathology and modes of spread), vulnerable hosts (such as life stages, affected species, immunity and genetics), and diagnostic techniques (which must be quick, sensitive, and specific) limit the industry's rapid growth and the wide range of species grown under various farming systems. During times when there are no outbreaks, a sluggish understanding of emerging hazards demonstrates complacency.
- ***Aquatic animal health management and disease control:*** This includes coordinating among the different organizations involved in aquaculture and aquatic health management, enforcing international standards and biosecurity guidelines, and responding effectively to emergencies. All this is being hindered by a lack of institutional and technical capacities that can effectively implement health supervision and biosecurity measures.
- ***Ecosystem fluctuations:*** Human activity both directly and indirectly affects dynamic aquatic ecosystems. According to Subasinghe and Shinn (2023), aquaculture is constrained in such dynamic settings by the physiological condition of the aquatic animals, the appearance of opportunistic diseases and the shifting geographic distributions of bacteria, parasites and wild animals.

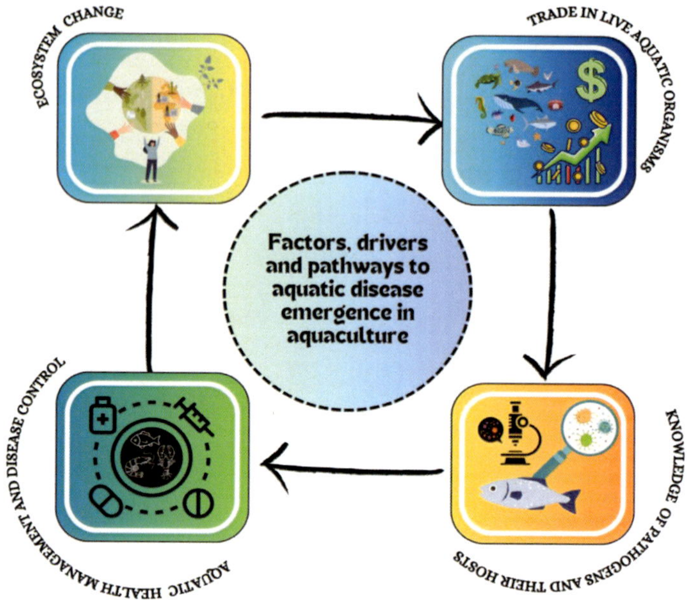

Fig. 1: Progressive Management Pathway for biosecurity in aquaculture practices (PMP-AB)

There are Four Steps of PMP/AB for Enhancing Aquatic Biosecurity

The PMP/AB is a risk-based, cooperative and progressive program. Strong input from public and commercial stakeholders is incorporated into the four different stages of PMP/AB to encourage the sector-level implementation of risk management as part of a national strategy. Countries choose where to start, how far to go, and how quickly to go on to the next phase. Since each nation has a wide variety of farmed species, the aquaculture industry may develop independently, at various rates, or with different objectives; yet, they all require close collaboration between the public and private sectors, such as through public-private sector partnerships (PPPs). Clarity about roles and duties, the identification of critical gaps that call for enhanced infrastructure and capacity, and a greater understanding of the costs and benefits of biosecurity across the complete value chain are all made possible by this approach (Subasinghe *et al.*, 2023).

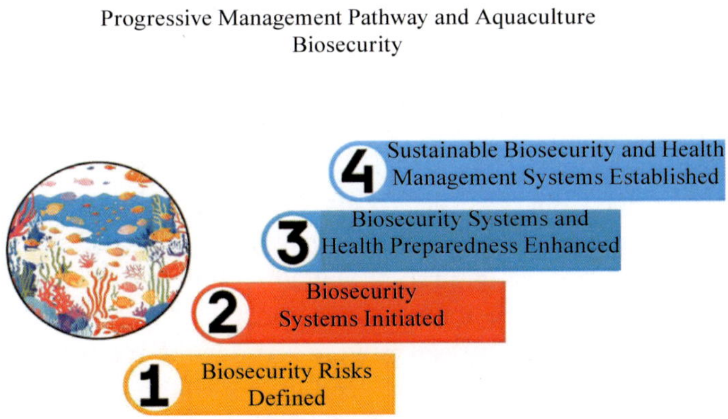

Fig. 2: Four stages of progressive management pathway for aquaculture biosecurity

One strategy that may be investigated is the PMP/AB as it provides the chance to:

1. Lessen the burden of illness
2. Enhance health on farms and at the national level
3. Prevent the transmission of diseases worldwide
4. Optimizing aquaculture's socioeconomic advantages,
5. Attracting investment possibilities, and
6. Achieving One Health goals.

A sustainable future for aquaculture is essential to global food security, making advancements in aquatic health management crucial to its ongoing and increasing role in food production worldwide.

Aquacultural Biotechnology

Biotechnology shows promise for aquaculture and may provide an incomplete answer to the difficulty of feeding the growing population, as significant production intensifications are required to satisfy this need. Aquaculture cannot be environmentally sustainable if it depends on extra resources such as more water, fish for meal and oil, and excessive chemical usage. However, advances in contemporary biotechnology provide new potential to upsurge productivity and improve the quality of both fresh and processed cultivated species (Muhammet *et al.*, 2013). In this chapter, different biotechnological methods are discussed which are being used in aquaculture.

Genetically Modified Species (Transgenics)

Transgenics, a primary usage of contemporary biotechnology, are new species variations created by crossing two strains to advance the hereditary features of the species in question by transferring desired qualities from each strain into the new variety in the aquaculture industry. Selectively transferring one or more genes for desirable features from one variety to another is known as transgenics. These characteristics could include increased size, better development rates, more effective feed conversion to muscle, and regulation of sexual maturation (Elzaeems, 2004). Several fish species, including salmon, trout, and tilapia, have successfully received growth hormone genetic factors from human or animal sources, leading to growth that is several times quicker than that of the fish species' natural equivalents. This approach to developing new types is quicker and more precise. Gene transfer between species is another option if conventional breeding is unable to provide the required features. For instance, enhancing resistance to harsh conditions and improving tolerance to environmental stressors are crucial uses to create distinctive traits and generate a useful biological product, such as the transfer of the antifreeze protein gene (AFP) in fish to help them adapt to a freezing environment (Hew *et al.*, 1992).

Aquaculture's intensification and long-term viability depend on preventing disease; hence biotechnology is crucial for increasing pathogen resistance and enhancing the health of farmed species by selecting for resistance to particular diseases (El-Zaeem and Aseem, 2004). Additionally, disease transmission between wild and cultivated populations can be decreased by employing this strategy. Fish genetic material can be transferred using two primary methods. These include electroporation, which uses an electrical current to transfer DNA into embryos or directly into tissues, and micro-injection, which involves injecting genetic material into freshly fertilized fish eggs. Transgenic fish has offered new species for aquaculture expansion but any escapes into the environment can risk wild populations and biodiversity.

Hybridization

Hybridization is a widely used genomic technique in aquaculture, involving the breeding of males from one species with females from another. This method helps improve growth rates, combine desirable traits, and enhance resistance to environmental conditions. It can also produce mono-sex populations, benefiting from sexual dimorphism, as seen in tilapia hybrids. Additionally, hybridization is useful in stocking programs to prevent unwanted reproduction by creating sterile or mono-sex fish populations, while also improving environmental tolerance. Examples include hybrid striped bass, hybrid catfish in the USA and Thailand respectively, and various salmonid hybrids worldwide. Because of environmental problems, especially the possibility of uncontrolled reproduction, sterile hybrid fish are thought to be advantageous in marine fish farming. The genetic composition of the parent species still affects hybridization, despite its widespread adoption. These difficulties include unexpected and undesirable results in hybrid progeny, like decreased survival rates, colour pattern changes, and the inability to attain sterility. Genetic changes result from the transfer of genes from one species to another, a process known as hybridization. These hybrids have the potential to seriously jeopardize the aquaculture sector by affecting biodiversity and upsetting genetic resources if they break free into the wild.

Hormonal Applications

Fish breeding and the early growth of aquatic species can be induced by the use of biotechnological technologies. According to Alok *et al.* (2000), gonadotropin-releasing hormone (GnRH) is currently widely utilized hormone in fish-induced breeding and is marketed commercially worldwide. Hormonal stimulation makes it possible for economically valuable species to produce gametes and fry throughout the year. Aquatic species must be domesticated for aquaculture, and the number of farm aquatic species is continuously growing quickly. During domestication, hormone medication is used to enhance and regulate reproductive cycles. This approach can also offer strategies to enhance the reproductive success and increased survival rates of endangered species, contributing to the conservation of biodiversity in natural habitats. In aquaculture, promoting faster growth rates remains a primary objective for farmers. To achieve this, various growth hormones are administered to fish and other aquatic organisms. Additionally, hormone treatments are utilized to induce sterility and triploidy, further supporting aquaculture management goals.

Chromosome Set Manipulation

Chromosome manipulation techniques have been widely employed in fish breeding to achieve gonadal sterilization, sex control, and clonal line production. These methods play a vital role in aquaculture by enabling the induction of polyploidy and uniparental chromosome inheritance. Uniparental inheritance occurs in two forms: gynogenesis, which involves maternal inheritance, and androgenesis, characterized by paternal inheritance. Polyploidy is commonly used to produce triploid fish, which are sterile. Additionally, tetraploidy induction facilitates the generation of sterile triploids through interploidy crosses between tetraploid and diploid individuals. This approach is also valuable for producing homozygous fish lines in species such as tilapia, cyprinids, and salmonids (Muhammet *et al.*, 2013).

Cryopreservation

Cryopreservation preserves biological materials, such as fish sperm, at extremely low temperatures (−196°C in liquid nitrogen), halting physiological and biochemical activities to maintain long-term viability. Cryoprotectants prevent cellular damage during freezing and thawing. This technology enables flexible breeding schedules, overcomes limited sperm availability, supports year-round gamete production, and facilitates genetic resource conservation through gene banking, benefiting both aquaculture and biodiversity preservation (Muhammet *et al.*, 2013).

Use of dsRNA to Activate the Host RNAi Mechanism

In the 2000s, Paul Gross' team and Skol Panyim's group pioneered the usage of dsRNA to protect shrimp against viral diseases like WSD and YHD, respectively in the U.S. and Thailand. This approach has shown success against various shrimp viruses and has also advanced the study of shrimp protein functions, including immune-related ones. While using dsRNA as a feed additive in shrimp farming shows promise, cost-effective large-scale production and efficient oral delivery methods remain challenges (Flegel 2019). However, it is already practical for clearing viruses, such as IHHNV, from breeding stocks, potentially enhanced by combining with ivermectin. If proven effective, this technique could support selective breeding programs by ensuring virus-free shrimp stocks.

Impact of Biotechnology on Fish Health Management

Vaccines

Intensive fish farming often leads to stress and disease, resulting in the usage of antibiotics and other chemicals for treatment. However, prioritizing disease prevention through optimal husbandry practices, vaccination, and immunostimulants is essential. This approach addresses growing concerns about environmental pollution from chemical treatments and the rising challenge of antibiotic-resistant pathogens, making infection control more effective and sustainable (Chinabut and Puttinaowarat, 2005).

For a vaccine to be effective in aquaculture, it must be cost-efficient, safe, and capable of providing long-term protection under intensive farming conditions. Key factors include targeting all serotype variants, timing of administration, delivery method, and vaccine type (e.g., whole-killed, attenuated, sub-unit or recombinant). Most marketable vaccines use inactivated pathogens, but live attenuated vaccines are sometimes necessary, despite concerns about potential reversion to virulence (Thompson *et al.*, 2004). Sub-unit vaccines, which isolate specific disease-causing components, and recombinant vaccines, where antigen-encoding genes are cloned and expressed in bacterial DNA, are emerging alternatives for safer and more effective disease control (Benmansour and De Kinkelin, 1997).

Most of the commercial fish vaccines are multivalent, meaning they target the major farmed species, such as salmon and trout. Nevertheless, there is growing potential for applications on Mediterranean marine species, such as European seabass and sea bream; and in the Asia-Pacific region, temperate and warm-water species such as grouper, Asian seabass, Japanese yellowtail, amberjack and tilapia (Sommerset *et al.*, 2005). Most vaccines currently available are against bacterial and some viral diseases. To date, not one vaccine has been developed and approved against fish parasites. Gradually, a recombinant DNA technology is applied in vaccine development. For instance, a marketable recombinant vaccine against infectious pancreatic necrosis virus (IPNV) is already being used in Norway (Heppell *et al.*, 1998). DNA vaccination is perhaps one of the most significant achievements in fish vaccinology and has been demonstrated to successfully induce cellular immune responses. The first commercial DNA fish vaccine against IPNV is currently on the market in Canada (Lee *et al.*, 2001). Other new technologies include proteomics and epitope mapping, which provide new opportunities for discovering vaccine antigens and peptide-based vaccines. Besides that, other adjuvants like interleukin-1 (Staata *et al.*, 1999) and delivery systems like PLGA micro-particles (Anderson and Shive, 1997) and antigen protection vehicles for oral vaccines are also under development for enhancing vaccine efficacy and delivery.

Immunological Methods for the Pathogen's Detection

Immunological techniques, including fluorescent antibody technique (FAT), indirect fluorescent antibody technique (IFAT), immunohistochemistry (IHC), ELISA and western blot, allow rapid and precise pathogen recognition without requiring prior pathogen isolation (Adams *et al.*, 1995). Monoclonal antibodies, widely available commercially, serve as standardized reagents for these assays. FAT and IFAT are more sensitive, quick approaches requiring specialized microscopes and skilled operators but are prone to non-specific background staining. They are particularly effective for detecting viruses like ISAV. IHC, an extension of histology, requires longer processing (2–3 days) but utilizes standard light microscopes and offers visualization of both pathogens and associated tissue pathology (Adams and Mateo, 1994). While less sensitive than IFAT, IHC provides valuable pathological context. ELISA, commonly used for pathogen detection and serology, is highly quantitative, allows high-throughput analysis, and can be automated (Adams and Thompson, 2005). However, it requires standards for quantification and is less effective in detecting sub-clinical infections.

Molecular Methods

Despite the sensitivity of immunological methods, pathogen detection can sometimes fail if their concentration in water or fish tissue is below detection thresholds or if antigen expression changes during different life cycle stages, as seen in parasites. PCR technology has revolutionized pathogen detection by enabling the amplification of small DNA or RNA sequences, making them detectable even at low concentrations. Unlike antigen-based methods, PCR is unaffected by antigen variation; however, false positives or negatives can occur due to contamination or inhibitors (Morris *et al.*, 2002).

PCR has also advanced pathogen detection in water, soil, and food samples, contributing significantly to epidemiological studies by identifying and differentiating closely related strains. Bacterial pathogens are often detected using conserved and variable regions of rRNA genes (5S, 16S, and 23S) (Puttinaowarat *et al.*, 2000). Standard PCR is useful for detecting pathogens; however, advanced techniques such as nested PCR, RAPD, RT-PCR, and RT-PCR enzyme hybridization assays are necessary for species-level identification and distinguishing between live and dead pathogens (Wilson and Carson, 2003). These variations improve sensitivity, specificity, and throughput. Additionally, in-situ hybridization and in-situ hybridization-PCR allow the localization of exact RNA and DNA sequences within cells, aiding in mapping infections, gene expression, and pathogen entry points, primarily for research purposes (Morris *et al.*, 2000).

Future Directions

The word 'Omics' has gained popularity in the last ten years, including disciplines like proteomics and genomics. In clinical and veterinary medicine, proteomics a combination of mass spectrometry and two-dimensional gel electrophoresis has made significant strides (Harry *et al.*, 2000). This technique is extremely relevant to fish health research since it has great potential for both illness diagnosis and vaccine development. The lack of aquaculture-specific data in existing databases, however, is a significant drawback that calls for the extensive gathering of datasets from sick fish and the infections that affect them. Furthermore, even though 2D electrophoresis is still the gold standard for analyzing protein expression, it is expensive, time-consuming, and frequently unreproducible. Its primary detection of numerous proteins, even when combined with mass spectrometry, emphasizes the need for better technology. Protein arrays, especially antibody microarrays, may offer more efficient and reliable solutions in the future (MacBeath, 2022).

Future Prospects

Certain genetic features of certain pathogen-resistant (SPR) and pathogen-free (SPF) species reduce their vulnerability to disease. Utilizing species resistant to illness is a good strategy to lower the possibility of disease outbreaks and enhance aquatic animals' general health. For instance, it has been demonstrated that using disease-resistant tilapia can lower the frequency and intensity of bacterial infections in aquaculture (Elaswad & Dunham, 2018). The proficiency and efficacy of operational biosecurity measures can be enhanced by the application of cutting-edge technology like sensors and drones and further with the use of artificial intelligence (AI). Water quality metrics may be tracked using sensors, and changes can be identified instantly (Mustapha et al. 2021; Kumar & Aravindh, 2020). Rapid advancements in molecular technology have led to the development of techniques with low risk of cross-contamination and high sample quantity, such as real-time PCR and nucleic acid-based sequence amplification (NASBA) (Overturf *et al.*, 2001, Starkey *et al.*, 2004). While vaccinations have made it possible to effectively control many bacterial infections, viral diseases continue to pose serious transmittable concerns for salmonid and marine finfish, and there are currently very few commercially available effective vaccines for these diseases. Similar to this, around 20 distinct virus DNA vaccines targeting viruses including rhabdoviridae, orthomyxoviridae, togaviridae, and nodaviridae have been produced experimentally for use in fish as a preventative measure. While most rhabdoviridae DNA vaccines have moderate to low efficacies, some, including VHSV and IHNV, have demonstrated significant levels of efficacy (Dalmo, 2017).

References

Adams, A. (1999). Application of antibody probes in the diagnosis and control of fish diseases. In Aquaculture and Biotechnology (Karunasagar, I., Karunasagar, I. and Reilly, A., eds), pp. 1–12, Oxford & IHB Publishing Co. PVT. Ltd.

Adams, A. and Mateo, M.D., (1994). Immunohistochemical detection of fish pathogens.

Adams, A. et al. (1995). Development and use of monoclonal antibody probes for immunohistochemistry, ELISA and IFAT to detect bacterial and parasitic fish pathogens. Fish Shellfish Immunol. 5, 537–547.

Adams, A., Thompson, K.D., Morris, D., Farias, C. and Chen, S.C., (1995). Development and use of monoclonal antibody probes for immunohistochemistry, ELISA and IFAT to detect bacterial and parasitic fish pathogens. Fish & Shellfish Immunology, 5(8), pp.537-547.

Ahmed, N., & Turchini, G. M. (2021). Recirculating aquaculture systems (RAS): Environmental solution and climate change adaptation. Journal of Cleaner Production, 297, 126604.

Alok, D., Talwar, G.P. and Garg, L.C., (2000). In vivo activity of salmon gonadotropin-releasing hormone (GnRH), its agonists with structural modifications at positions 6 and 9, mammalian GnRH agonists and native cGnRH-II on the spawning of an Indian catfish. Aquaculture International, 7, pp.383-392.

Aly, S. M., & Fathi, M. (2024). Advancing aquaculture biosecurity: a scientometric analysis and future outlook for disease prevention and environmental sustainability. Aquaculture International, 1-27.

Anderson, J.M. and Shive, M.S., (1997). Biodegradation and biocompatibility of PLA and PLGA microspheres. Advanced drug delivery reviews, 28(1), pp.5-24.

Assefa, A., & Abunna, F. (2018). Maintenance of fish health in aquaculture: review of epidemiological approaches for prevention and control of infectious disease of fish. Veterinary Medicine International, 2018(1), 5432497.

Assem, S.S. and El-Zaeem, S.Y., (2005). Application of biotechnology in fish breeding. II: production of highly immune genetically modified redbelly tilapia, Tilapia zillii. African journal of biotechnology, 4(5).

Austin B (2023). The impact of disease on the sustainability of aquaculture. Sustainable Aquatic Research 2(1):74–91. https://doi.org/10.5281/zenodo.7882040

Austin B, Lawrence A, Can E, Carboni C, Crockett J, Demirtas N, Schleder D, Adolfo J, Kayis S, Karacalar U, Kizak V, Kop A, Thompson K, Ruiz CAM, Serdar O, Seyhaneyildiz Can S, Watts S, Yucel Gier G (2022). Selected topics in sustainable aquaculture research: current and future focus. Sustainable Aquatic Research 1(2):74–125. https://doi.org/10.5281/zenodo.7032804

Austin, B. (2023). The impact of disease on the sustainability of aquaculture. Sustainable Aquatic Research, 2(1), 74-91.

Austin, B., Lawrence, A., Can, E., Carboni, C., Crockett, J., Demirtas, N., ... & Gier, G. Y. (2022). Selected topics in sustainable aquaculture research: Current and future focus. Sustainable Aquatic Research, 1(2), 74-125.

Benmansour, A. and De Kinkelin, P., (1997). Live fish vaccines: history and perspectives. Developments in biological standardization, 90, pp.279-289.

Bera, K. K., Karmakar, S., Jana, P., Das, S., Purkait, S., Pal, S., & Haque, R. (2018). Biosecurity in aquaculture: an overview. Aqua International, 18(12), 42-46.

Bianchi, M., Hallström, E., Parker, R. W., Mifflin, K., Tyedmers, P., & Ziegler, F. (2022). Assessing seafood nutritional diversity together with climate impacts informs more comprehensive dietary advice. Communications Earth & Environment, 3(1), 188.

Birhan, G., Alebie, A., Admassu, B., Shite, A., Mohamed, S., & Dagnaw, B. (2015). A Review on Emerging & Re-Emerging Viral Zoonotic Diseases. Int. J. Basic Appl. Virol, 4, 53-59.

Bondad-Reantaso, M. G., Arthur, J. R., & Subasinghe, R. P. (2012). Improving biosecurity through prudent and responsible use of veterinary medicines in aquatic food production. FAO Fisheries and Aquaculture Technical Paper, (547), I.

Bondad-Reantaso, M. G., Subasinghe, R. P., Arthur, J. R., Ogawa, K., Chinabut, S., Adlard, R., ... & Shariff, M. (2005). Disease and health management in Asian aquaculture. Veterinary parasitology, 132(3-4), 249-272.

Chinabut, S. and Puttinaowarat, S., (2005). The choice of disease control strategies to secure international market access for aquaculture products. Developments in biologicals, 121, pp.255-261.

Cunningham, C.O. (2004). Use of molecular diagnostic tests in disease control: making the leap from laboratory to field application. In Molecular Aspects of Fish and Marine Biology: Current Trends in the Study of Bacterial and Viral Fish and Shrimp Diseases (Vol.3) (Leung, K-Y., ed.), pp. 292–312, World Scientific Publishing Co.

Dalmo, R.A., (2017). DNA vaccines for fish: review and perspectives on correlates of protection. J. Fish Dis. 41 (1), 1 9.

Daoliang, L. I., & Chang, L. I. U. (2020). Recent advances and future outlook for artificial intelligence in aquaculture. Smart agriculture, 2(3), 1.

14

Advancements in Molecular Cytogenetics for Sustainable Fisheries and Aquaculture

***Bhargawi Mishra*[1] *and Gautam Geeta Jiwatram*[2*]**

[1]*UMS, Biology, Government of Madhya Pradesh, Singrauli, Madhya Pradesh (486 886), India*

[2]*Zoology section, Mahila Mahavidyalaya, Banaras Hindu University, Varanasi Uttar Pradesh (221 005), India*

Abstract

Fisheries and aquaculture are vital for global food security, supplying a substantial share of the world's protein needs. As natural fish stocks decline and the global population grows, sustainable aquaculture becomes increasingly essential. Molecular cytogenetics has emerged as a critical tool in advancing aquaculture by improving the understanding of genetic traits such as growth, reproduction, and environmental resilience. Traditional cytogenetic techniques, including chromosome manipulation, gynogenesis, and hybridization, have long been used to enhance species like sea bream, sea bass, and sturgeon. Recent innovations, such as Fluorescence In-situ Hybridization (FISH), have allowed for precisely mapping DNA sequences, offering more profound insights into chromosome structure, gene families, and evolutionary processes in fish. These advancements have supported species improvement, disease management, and biodiversity conservation, fostering sustainable aquaculture practices. Key chromosomal manipulations like induced triploidy and tetraploidy are achieved through physical and chemical treatments such as heat shock, hydrostatic pressure, and cytochalasin B. Triploidy, in particular, has become critical for producing sterile, fast-growing fish populations. In contrast, though promising, tetraploidy remains challenging due to higher mortality rates in early development.

Keywords: *Molecular cytogenetics, Genome, Chromosome, Fish, Aquaculture*

Introduction

Fisheries and aquaculture are crucial for global food security, providing a significant portion of the world's animal protein supply. With the rapid growth of the human population and the overfishing of natural fish stocks, it is more important than ever to increase aquaculture production sustainably. Advances in molecular cytogenetics are crucial for promoting sustainable growth and innovation in fisheries and aquaculture. These advancements offer deeper insights into the genetic factors that govern key traits like growth, reproduction, and environmental resilience, enabling aquaculture to address global food demands while safeguarding biodiversity and maintaining genetic integrity. The integration of these techniques in fisheries and aquaculture has led to significant advancements in the genetic improvement of cultured species, disease management, and biodiversity conservation in recent years. Moreover, these scientific advancements offer innovative methods for effective disease control and the preservation of aquatic biodiversity, all of which contribute to aquaculture's long-term viability and growth. Over the past two decades, fish karyotype studies have been combined with genome sequencing, revealing more remarkable genome plasticity and variability in fish compared to other vertebrates. Comparative fish genome analysis evaluates divergence rates and explores sex-determination genes and gene family evolution in teleosts. Currently, over 80 fish species have their genomes annotated in the Ensembl database (https: //www.ensembl.org/index.html, accessed on 6 April 2021), and around 150 are assembled to the chromosome level in the NCBI database (https://www.ncbi.nlm.nih.gov/genome/). However, the quality of many assembled genomes has been criticized for not meeting reference genome standards. Fish genome sequencing is expected to grow significantly by 2028 due to initiatives such as the Earth BioGenome Project. This chapter will explore the chronological development of cytogenetic techniques used in aquaculture and fisheries, emphasizing their role in improving quality, ensuring disease resistance, and promoting environmentally sustainable practices.

Traditional Cytogenetics Techniques in Fisheries and Aquaculture

Cytogenetic techniques in fisheries focus on taxonomy, chromosomal polymorphism, hybridization, embryogenesis, chromosome manipulation, and sex determination. These methods have been applied to study taxonomy, evolution, and chromosomal polymorphism in economically significant Pacific salmon (genus *Oncorhynchus*) (Gorshkova, 2006). Chromosome set manipulation is a technique in fish genetics and aquaculture that alters the number or structure of chromosomes to enhance desirable traits. It is commonly applied to boost growth rates, produce sterile fish, create all-female populations, and improve specific genetic characteristics in aquaculture species. Species like Sea Bream (*Sparus aurata*) and Sea Bass (*Dicentrarchus labrax*), in The National Centre for Mariculture (NCM, Israel) pioneered chromosome set manipulation for genetic improvement, producing inbred lines, sterile fish, all-female populations, and growth-enhanced triploids (Gorshkov, 2006). Gynogenesis is a form of asexual reproduction in which offspring develop solely from the mother's genetic material, with no genetic input from the father. In this process, the sperm initiates the activation of the egg's development, but its DNA is either inactivated or excluded from the embryo. As a result, the offspring inherit solely the mother's genome, which often leads to the creation of clones. This technique is widely used in fish genetics for producing all-female populations, creating inbred lines, and studying maternal inheritance, as well as for selective breeding programs to enhance desirable traits in aquaculture. Example: Gynogenesis in Sea Bass. Cold and heat shocks on fertilized eggs produced gynogenetic progeny, verified by mass karyotyping and cytometric analysis. (Manan *et al.,* 2022). Sturgeon hybrids is another technique in which offspring are produced by crossbreeding different species within the sturgeon family (Acipenseridae). Sturgeons are ancient, long-living fish, and hybridization among different sturgeon species has been widely practised in both the wild and aquaculture to combine desirable traits such as faster growth rates, improved resistance to environmental stress, and enhanced meat or caviar quality. Karyotyping of hybrids between Russian sturgeon (Acipenser) and Beluga sturgeon (*Huso huso*) confirmed polyploid origin and assisted in conservation management in Israel. (Gorshkova *et al.,* 1996). Cytological techniques also proved highly effective in studying early embryogenesis in cultured marine fish. A study on the white grouper (*Epinephelus aeneus*) aimed to uncover the genetic causes of high egg and larval mortality. The research found that 35.5% to 79% of embryos had chromosomal abnormalities, varying significantly between different spawning. The study suggested that these chromosomal disorders may be a key genetic factor affecting survival during the early stages of development and a potential cause of high egg and larval mortality in White Grouper (*Epinephelus aeneus*). The karyotype of *Dicentrarchus labrax* was studied, revealing 46 acrocentric and 2 small subtelocentric chromosomes. A notable finding (Faggion *et al.,* 2021) was the sex chromosome heteromorphism-males typically exhibited a heteromorphic pair of subtelocentric chromosomes, while females showed a homologous pair. This sex-associated difference was highly significant.

Advances in Molecular Techniques in Aquaculture and Fisheries

Fluorescence *In Situ* Hybridization (FISH)

Fluorescence in situ hybridization (FISH) is a method employed to pinpoint DNA sequences within metaphase and interphase cells. This technique enables the identification of chromosomes, the detection of chromosomal abnormalities, and the mapping of genes to particular chromosomal sites. Commonly utilized in human genetics, FISH is capable of visualizing both repetitive and unique sequences, allowing for the "painting" of specific chromosomal regions or entire chromosomes in hybrids between different species. (Cui *et al* ., 2016)

Centromere and Telomere-specific Sequences

Eukaryotic genomes have two main types of repetitive DNA: tandem repeats at centromeres and telomeres, and dispersed repeats throughout the genome. The telomeric sequence (TTAGGG)n is conserved across vertebrates, while centromeric and subtelomeric sequences vary by species.

Centromeric and telomeric sequences have been studied in fish species such as *Chiondraco hamatus* (Capriglione *et al.,* 1994), *Hoplias malabaricus* (Haaf *et al.,* 1993), *Sparus aurata* (Garrido-Ramos *et al.,* 1994), and *Salvelinus alpinus* (Hartley and Davidson, 1994). Notable findings include:

- A 186 bp centromeric sequence in *S. aurata*, with variations in a 31 bp region.
- A 355 bp centromeric sequence in *H. malabaricus*, with subfamily A on most chromosomes and subfamily B linked to a sex chromosome Reed, 1996).
- A 127 bp centromeric sequence in *S. alpinus* and related sequences in *S. namaycush* and other *Salvelinus* species.
- The B2-repetitive sequence in *Oncorhynchus tshawytscha* localized to subtelomeric regions, aiding in stock identification and hybrid studies (Reed and Phillips, 1995)

The conserved telomeric sequence (TTAGGG)n has been identified in telomeres of various fish species, with occasional non-telomeric localization.

Histone Genes

Histone genes encode essential proteins responsible for organizing DNA into nucleosomes. These structures play a critical role in DNA packaging, gene expression regulation, replication, and repair. The organization and distribution of histone genes are therefore critical for maintaining genomic stability and proper cellular function.

In salmonids, the primary histone gene cluster has been mapped in several species, including brown trout (*Salmo trutta*), Atlantic salmon (*Salmo salar*), rainbow trout (*Oncorhynchus mykiss*), and *Coregonus lavaretus* (Pendas *et al.*, 1994). As descendants of an ancestral tetraploid event, salmonids exhibit partial diploidization—a process where some duplicate gene loci are retained on two chromosomal pairs, while others are either silenced or scattered across different chromosomes (Allendorf and Thorgaard, 1984). Similarly, ribosomal genes have also undergone diploidization or have been redistributed across multiple chromosomes (Phillips *et al.*, 1989).

Fluorescence in situ hybridization (FISH) studies have shown that histone genes are localized at a single chromosomal site in Atlantic salmon, brown trout, and rainbow trout. In contrast, duplicate histone gene loci are still present in *Coregonus* species, indicating differences in how these species have managed their duplicate gene clusters following tetraploidization.

In general, the evolutionary dynamics of histone genes in salmonids reveal a complex interplay of retention, silencing, and chromosomal redistribution.

Chromosome Elimination Technique

In several Japanese hagfish species (*Eptatretus okinoseanus*, *E. burgeri*, and *Myxine garmani*), researchers have identified a remarkable phenomenon where certain chromosomes are selectively eliminated during the early stages of embryonic development (Nakai *et al.*, 1991)

This chromosome elimination process appears to play a role in differentiating somatic and germline cells. Specifically, *E. okinoseanus* possesses germline-specific supernumerary B chromosomes, which are additional chromosomes not present in somatic cells (Kubota *et al.*, 1992).

Analysis of DNA from germ cells revealed digestion fragments of 90 bp and 180 bp, corresponding to two repetitive DNA families measuring 95 bp and 85 bp. These repetitive sequences constitute about 19% of the DNA eliminated during embryogenesis. Further investigation using fluorescence in situ hybridization (FISH) demonstrated that these sequences are localized on small C-band positive chromosomes, which are exclusively found in germ cells of *E. okinoseanus* (Kubota *et al.*, 1993). Overall, this process likely plays a crucial role in preserving the functional differences between germline and somatic cell lineages.

Chromosome Manipulation Technique

Chromosome set manipulation refers to techniques used to alter the number and combination of homo- and hetero-specific genomes or chromosome sets (Arai, and Fujimoto, 2018). This field, first explored in amphibians in the early 20th century from a basic biological perspective, expanded in the 1950s to 1970s with pioneering research on fish species. Studies during this period examined the effects of ploidy elevation and uniparental development, conducted by scientists from the Soviet Union, the UK, the US, Norway, Hungary, Japan, and other countries. Although chromosome manipulation techniques may seem traditional given their research history. Arai, 2001 argued that they should not be dismissed as outdated in the future breeding of aquatic organisms. He emphasized that combining these methods with modern genetic technologies, such as hypervariable DNA markers, transgenics, and germ cell technology, could lead to further advancements. Recently, several aspects of this prediction have been realized.

1. Homozygous strain's genome annotation. (Lien *et al.,* 2016)
2. Creation of transgenic inbred lines via induced androgenesis (Nam, 2000)
3. Use of sterile triploids as hosts for surrogate production. (Okutsu, 2007)

Induced Triploidy in Fish and Shellfish

Producing triploid fish involves inhibiting the release of the second polar body during meiosis II, resulting in zygotes with two chromosome sets from the female and one from the male. Common induction methods include physical treatments such as cold, heat, or hydrostatic pressure shocks, following species-specific protocols. Homospecific triploids are referred to as autotriploids, while interspecific hybrids are known as allotriploids (Purdom, 1983; Yamazaki, 1983).

In shellfish, chemical treatments (e.g., cytochalasin B, caffeine) or combinations of chemical and physical methods are often used. Triploid induction is typically done after meiosis II, as meiosis I inhibition often causes aneuploidy. Verification methods include DNA content analysis by flow cytometry, chromosome counting, and karyotyping, with earlier techniques like Ag-NOR staining now less common. (Piferrer *et al.*, 2009; Guo *et al.*, 2009; Arai and Okumura, 2013)

Induced Tetraploidy in Fishes

Fertile tetraploid fish produce diploid gametes (2n eggs and sperm), which are valuable for ploidy manipulation. Tetraploids can be established by fertilizing diploid eggs with diploid sperm, while sterile triploids result from fertilizing diploid eggs with haploid sperm. Diploid gametes can also generate higher polyploids, such as pentaploids and hexaploids, through chromosome manipulation techniques. (Arai *et al.,* 2001; Chourrout *et al.,* 1986 & Chourrout *et al.*, 1987, Zou *et al.,* 2004; Nam *et al.,* 2004).

Tetraploids are induced via endomitosis by suppressing early cleavage, but producing viable, fertile tetraploid adults is challenging, with high mortality during early developmental stages. Successful examples include species like rainbow trout, bluntnose bream, and mud loach. Optimization of tetraploid induction involves adjusting factors like temperature, hydrostatic pressure, chemical treatments, and timing. However, outcomes often vary due to biological differences and lack of gamete synchronization.

Somatic Cell Nuclear Transfer (SCNT) is another interesting cloning technique where the nucleus of a somatic cell from a donor organism is transferred into an egg cell that has been enucleated, creating a genetically identical organism or embryo after cell division. Although early developmental success was low, SCNT remains a promising cloning technique (Fatira *et al.* 2018). The position of the donor cell within the egg is crucial for SCNT success, and using multiple somatic cells increases the chances of development. This approach was successfully tested in Russian sturgeon using sterlet eggs (Fatira *et al.,* 2019). The nucleus of Primordial germ cells (PGCs) are used for this purpose. These cells are essential for gametogenesis, gonadal formation, and sex determination, form during gastrulation in many fish species (Ewen-Campen *et al.,* 2010; Slanchev *et al.,* 2005; Goto *et al.,* 2012). Earlier studies required killing the fish or using time-consuming histological techniques to study PGCs, which is unsuitable for endangered species like sturgeon. But thanks to Saito and Psenicka 2015 who used fluorescein isothiocyanate (FITC)-dextran, fluorescent marker to label the vegetal hemisphere of sterlet embryos, enabling a real time in vivo visualization of PGCs.

Sex Determination Based on Chromosome Manipulation

Fish exhibit an extraordinary diversity of sex determination systems, including the commonly observed XX female–XY male and ZW female–ZZ male systems, as well as more complex arrangements like XX female–XO male and X1X2 female–X1X2Y male systems. While key sex-determination genes have been identified in a few species (e.g., *dmy* in medaka, *amhr2* in puffer fish, *sdY* in rainbow trout), most fish lack well-characterized sex chromosomes or markers. Despite the diversity of these systems, the genetic basis of sex determination remains poorly understood in most fish. Only a few sex-determining genes have been identified so far. For instance, the *dmy* gene has been discovered in medaka (*Oryzias latipes*) (Matsuda *et al.*, 2002), the *amhr2* gene in pufferfish (*Takifugu rubripes*) (Kamiya *et al.*, 2012), and the *sdY* gene in rainbow trout (*Oncorhynchus mykiss*) (Yano *et al.*, 2012). These genes play crucial roles in initiating male-sex differentiation.

Environmental factors particularly temperature also influence sex determination, with female-to-male sex reversal observed in species such as Japanese flounder (Yamamoto, 1999) and red sea bream (Kato *et al.,* 2002), with similar findings in tilapia through chromosome manipulation (Devlin and Nagahama, 2002). Genetic factors, including recessive sex-reversal genes, have been identified in tilapia and common carp (Komen *et al.,* 1992).

In the majority of fish species, neither clear sex-determining genes nor reliable sex-linked genetic markers have been identified. Among teleosts, various sex chromosome systems have been identified, including the common XX female-XY male and ZW female-ZZ male systems, as well as more complex systems like XX female-XO male, X1X2 female-X1X2Y male, and others (Devlin and Nagahama, 2002; Pandian, 2011). While some sex determination genes have been identified in medaka (dmy) (Matsuda *et al.,* 2002) puffer fish *Takifugu rubripes* (amhr2) (Kamiya, *et al.*, 2012), and rainbow trout (sdY) (Yano *et al.*, 2012). Most fish species lack known sex-determination genes or sex-linked markers.

In previous literature, gynogenesis and androgenesis have been used to investigate sex determination in different species. (Stanley 1976; Pandian *et al.*, 1998; Arai, 2001). In species with a ZW-ZZ system, meiotic gynogenetic diploids can result in ZZ males, WW superfemales, and ZW females due to recombination between Z and W chromosomes. Depending on the recombination frequency, progeny may consist of 67% females or be entirely female. Gynogenetic (double haploid) DHs are expected to produce ZZ males and WW superfemales, which can generate all-female offspring when crossed with ZZ males. By analyzing the sex ratios of progeny from gynogenesis or androgenesis, researchers can infer the sex determination system of a species, even without identified sex chromosomes or genes. For instance, an XX-XY system has been suggested based on all-female gynogenetic diploids in grass carp. (Shelton 1986; Rothbard *et al.,* 2000) Atlantic halibut (Tvedt *et al.,* 2006 spotted halibut (Ji *et al.,* 2010), Eurasian perch (Rougeot *et al.,* 2005) and Atlantic cod (Whitehead *et al.,* 2012). Similarly, a ZW-ZZ system has been inferred from meiotic gynogenetic diploids and gynogenetic DHs in several sturgeon species, paddlefish (Shelton and Mims, 2012), turbot (Meng *et al.,* 2016), and half-smooth tongue sole (Ji *et al.,* 2010). Zebrafish studies also suggest a ZW-ZZ system based on androgenetic DH experiments (Hou *et al.,* 2015). Sex ratios in polyploid fish are more complex. In species with male heterogamety, triploid progeny can result in females (XXX) and males (XXY), while tetraploids may produce XXXX females and XXYY males. In rainbow trout, male-biased sex ratios were observed in the second generation of tetraploids (Chourrout *et al.,* 1986). In female heterogametic species, triploid progeny can include ZZZ males and ZWW females, while tetraploids may produce ZZZZ males and ZZWW females, though these genotypes have not been experimentally confirmed. The mechanisms governing sex ratios in polyploid species remain poorly understood.

Genome Transplantation in Sturgeon

Surrogacy in fish involves transplanting germline stem cells from donor gonadal tissues into host larvae, enabling the host to produce donor-derived gametes. This method is effective for aquaculture and species conservation, particularly for species with long reproductive cycles, like sturgeons. Germline stem cells are injected into the host's intraperitoneal cavity at the larval stage, where they migrate to the genital ridges to initiate gametogenesis. Applications include cryopreservation, gene banking, and surrogate reproduction. (Goto & Saito, 2019; Yoshizaki & Yazawa 2019; Takeuchi *et al.,* 2022).

Successful examples include germline chimaera formation in endangered Siberian sturgeon and Chinese sturgeon, achieved by transplanting germ cells into closely related species. Somatic cell nuclear transfer (SCNT) is another conservation tool, where somatic cell nuclei are introduced into eggs to produce clones, as demonstrated in Russian sturgeon using sterlet eggs. Though SCNT has low success rates, it remains promising for cloning critically endangered species. (Takeuchi *et al.,* 2003; Yoshizaki *et al.,* 2003; Okutsu *et al.,* 2006; Psenicka *et al.,* 2015; Kazama-Wakabayashi *et al.,* 1999; Yoshizaki *et al.,* 2003; Ye *et al.,* 2017; Fatira *et al.,* 2018; Fatira *et al.* 2018)

Primordial germ cells (PGCs), vital for gametogenesis and sex determination, are now studied in vivo using FITC-dextran labelling in sterlet embryos, providing a non-lethal alternative for studying endangered species. (Ewen-Campen *et al.,* 2010; Slanchev *et al.,* 2005; Goto *et al.,* 2012)

Genome Editing

The advent of genome-editing tools like CRISPR/Cas9 has opened new avenues for precise genetic modifications in aquaculture species. Genome editing allows for the targeted introduction or removal of specific genes to advance traits such as growth and disease resistance. For instance, CRISPR/Cas9 has been used to knockout the myostatin gene in tilapia, resulting in increased muscle development and improved fillet yield. (Zhu *et al.,* 2024).

Use of molecular Techniques in Disease Diagnostics and Management of Fisheries and Aquaculture

Disease outbreaks pose a significant threat to aquaculture production, causing substantial economic losses. Molecular and cytogenetic tools are invaluable in diagnosing and managing diseases in fisheries and aquaculture, helping detect pathogens, assess genetic health, and improve disease resistance through selective breeding. These tools provide precise, rapid, and reliable methods to identify genetic abnormalities, pathogens, and chromosomal aberrations that can affect fish health, productivity, and survival. By quantifying pathogen load, farmers can adjust treatment protocols to reduce mortality (Loch, 2015). DNA barcoding involves sequencing short, standardized regions of DNA (such as the mitochondrial cytochrome c oxidase I gene) to identify species and detect pathogens. It's useful for detecting parasitic infections and ensuring the genetic integrity of fish stocks.

DNA barcoding is used to detect parasitic copepods in farmed fish, which can cause significant health issues and economic losses (Hebert *et al.,* 2003). Barcoding animal life: cytochrome c oxidase subunit 1 divergences among closely related species. NGS allows for high-throughput genome sequencing of entire transcriptomes, providing detailed information about the presence of pathogens and the genetic character of aquaculture species. NGS has been used to identify novel viruses in shrimp aquaculture, such as the *Decapod iridescent virus 1 (DIV1)*, which can cause mass mortality in shrimp farms. (Liao *et al.,* 2019).

Table 1. describing techniques for disease detection

Technique	Disease	Caused by	Species
PCR	*Infectious Hematopoietic Necrosis Virus* (IHNV)	Virus	Salmonids
	Viral Hemorrhagic Septicemia Virus (VHSV)		
qPCR	*Flavobacterium psychrophilum*	Bacterial	Rainbow trout

Marker-Assisted Selection (MAS)

Marker-Assisted Selection (MAS) is an advanced breeding technique that uses molecular markers linked to specific genetic traits to guide selection decisions. Unlike traditional breeding, which depends primarily on observable traits, MAS allows breeders to identify and select individuals with desirable genetic characteristics early in their development. This method is especially valuable for improving complex traits influenced by multiple genes, such as disease resistance, growth performance, and environmental adaptability.

MAS represents a groundbreaking approach in aquaculture breeding, providing accuracy, efficiency, and sustainability in developing resilient and disease-resistant strains. Its implementation not only boosts productivity but also strengthens the long-term stability of aquaculture systems against disease-related challenges.

In aquaculture, diseases remain a major challenge, often causing significant economic losses and threatening sustainable production. MAS has proven to be an effective tool for addressing these issues by enabling the development of disease-resistant strains. For instance, in Atlantic salmon (Salmo salar), MAS has been successfully applied to improve resistance to infectious pancreatic necrosis (IPN), a viral disease responsible for high mortality rates in farmed populations. By identifying genetic markers associated with IPN resistance, breeders can selectively cultivate fish with stronger immunity, reducing dependence on antibiotics and vaccines while enhancing overall stock health.

Molecular Diagnostics

Polymerase chain reaction (PCR)-based techniques and next-generation sequencing (NGS) have revolutionized pathogen detection in aquaculture. These methods enable the rapid and accurate identification of pathogens, including viruses, bacteria, and parasites. Early detection of pathogens allows for timely intervention, reducing the impact of disease outbreaks. For instance, real-time PCR assays have been developed to detect viral haemorrhagic septicaemia virus (VHSV) in fish, facilitating effective disease management.

Future Perspectives

Molecular and cytogenetic techniques have become essential tools in fisheries and aquaculture, offering valuable insights into fish biology, genetics, and reproduction while enabling more efficient breeding strategies. These advancements contribute to increased productivity, conservation of endangered species, and sustainable resource management. Notably, a novel cold shock method for inducing androgenesis eliminates the need for egg irradiation, making the process more accessible for species with large eggs, such as salmonids and sturgeons. Furthermore, distant hybridization has proven effective in producing unreduced diploid gametes and has been successfully utilized in China to develop innovative aquaculture strains. Collectively, these chromosomal manipulation techniques deepen our understanding of fish sex determination systems and hold significant potential for advancing aquaculture practices globally.

Conclusion

The integration of molecular cytogenetics into fisheries and aquaculture has significantly advanced sustainable practices by enhancing our understanding of key genetic traits that influence growth, reproduction, and resilience. Traditional cytogenetic techniques such as chromosome manipulation, gynogenesis, and hybridization have laid the groundwork for improving commercially important species, while innovative methods like Fluorescence In Situ Hybridization (FISH) have provided more detailed insights into genetic structures. Key chromosomal manipulations, including induced triploidy and tetraploidy, continue to play a vital role in producing high-performance, sterile fish populations, though challenges like early mortality in tetraploidy remain. The emergence of genome editing technologies, particularly CRISPR/Cas9, has opened new frontiers in genetic enhancements, allowing precise modifications for better growth, disease resistance, and reproductive control. These molecular tools also offer powerful solutions for species conservation, disease diagnostics, and biodiversity management. As research and technological advancements continue to evolve, molecular cytogenetics will remain a cornerstone in fostering the sustainability, productivity, and conservation of aquaculture, ensuring its vital role in global food security.

References

Arai, K. (2001). Genetic improvement of aquaculture finfish species by chromosome manipulation techniques in Japan. Aquaculture 197, 205–228. https://doi.org/10.1016/S0044-8486(01)00588-9

Arai, K. and Fujimoto, T., (2018). Chromosome manipulation techniques and applications to aquaculture. Sex control in Aquaculture, 137-162. https://doi org/10. 1002/9781119127291.ch6

Arai, K. and Okumura, S. (2013). Aquaculture-oriented genetic researches in abalone: Current status and future perspective. African Journal of Biotechnology 12(26), 4044–4052.

Chourrout, D., Chevassus, B., Krieg, F., Happe, A., Burger, G., & Renard, P. (1986). Production of second generation triploid and tetraploid rainbow trout by mating tetraploid males and diploid females—potential of tetraploid fish. Theoretical and Applied Genetics, 72, 193-206. https://doi.org/10.1007/BF00266992

Chourrout, D., Nakayama, I. (1987). Chromosome studies of progenies of tetraploid females rainbow trout. Theoretical and Applied Genetics 74, 687–692. https://doi.org/10.1007/BF00247542

Cui, C., Shu, W. and Li, P., (2016). Fluorescence in situ hybridization: cell-based genetic diagnostic and research applications. Frontiers in cell and developmental biology, 4, 89. https://doi.org/10.3389/fcell.2016.00089

Devlin, R. H. and Nagahama, Y. (2002). Sex determination and sex differentiation in fish: an overview of genetic, physiological, and environmental influences. Aquaculture 208, 191–364. https://doi.org/10.1016/S0044-8486(02)00057-1

Faggion, S., Vandeputte, M., Vergnet, A., Clota, F., Blanc, M.O., Sanchez, P., Ruelle, F. and Allal, F., (2021). Sex dimorphism in European sea bass (*Dicentrarchus labrax* L.): New insights into sex-related growth patterns during very early life stages. PloS one, 16(4). https://doi.org/10.1371/journal.pone.0239791

Fatira, E., Havelka, M., Labbé, C., Depincé, A., Pšenička, M., & Saito, T. (2019). A newly developed cloning technique in sturgeons; is an important step towards recovering endangered species. Scientific Reports, 9(1), 10453. https://doi.org/10.1038/s41598-019-46892-4

Fatira, M., & Nasution, A. W. (2019). Boosting the welfare of the business community: implementing the model of Islamic Micro Bank of Waqf in Pesantren. AL-FALAH: Journal of Islamic Economics, 4(1). https://doi.org/10.29240/alfalah.v4i1.771

François. A & Nguyen, H. N. (2022) Genomic Selection in Aquaculture Species. 469–491. https://doi.org/10.1007/978-1-0716-2205-6_17

Gomelsky, B. (2003). Chromosome set manipulation and sex control in common carp: a review. Aquatic Living Resources 14, 327–334. https://doi.org/10.1016/S0990-7440(03)00085-8

Gorshkova, G. (2006). Practical Use of Cytogenetics in Fish Biology and Aquaculture. National Centre for Mariculture, Israel Oceanographic & Limnology The Israeli Journal of Aquaculture - Bamidgeh 58(4). https://doi.org/10.46989/001c.20461

Gorshkova, G.V., Protas, Y., Ben-Atia, S. and Gorshkov, S., (2001). Cytogenetic examination of early embryonic development in the white grouper. Journal of Applied Ichthyology, 17, (1-6). https://doi.org/10.1046/j.1439-0426.2002.00300.x

Goto, R., Saito, T., Takeda, T., Fujimoto, T., Takagi, M., Arai, K., & Yamaha, E. (2012). Germ cells are not the primary factor for sexual fate determination in goldfish. Developmental Biology, 370(1), 98-109. https://doi.org/10.1016/j.ydbio.2012.07.010

Guo, X., Wang, Y., Xu, Z., & Yang, H. (2009). Chromosome set manipulation in shellfish. In New technologies in aquaculture (pp. 165-194). Woodhead Publishing. https://doi.org/10.1533/9781845696474.1.165

Hebert, P. D., Ratnasingham, S., & De Waard, J. R. (2003). Barcoding animal life: cytochrome c oxidase subunit 1 divergences among closely related species. Proceedings of the Royal Society of London. Series B: Biological Sciences, 270(suppl_1), S96-S99. https://doi.org/10.1098/rsbl.2003.0025

Hershberger, W.K., Hostuttler, M. A. (2005). Variation in time to first cleavage in rainbow trout *Oncorhynchus mykiss* embryos: a major factor in induction of tetraploids. Journal of the World

Hou, J., Fujimoto, T., Saito, T., Yamaha, E., & Arai, K. (2015). Generation of clonal zebrafish line by androgenesis without egg irradiation. Scientific Reports, 5(1), 13346. https://doi.org/10.1038/srep13346

Ji, X.S., Chen, S.L., Yang, J.F., Ma, H.Y., & Jiang, Y.L. (2010). Artificial gynogenesis & assessment of homozygosity in meiotic gynogens of spotted halibut (Verasper variegatus). Aquaculture international, 18, 1151-1161. https://doi.org/10.1007/s10499-010-9330-y

Ji, X. S., Tian, Y. S., Yang, J. F., Wu, P. F., Jiang, Y. L., & Chen, S. L. (2010). Artificial gynogenesis in Cynoglossus semilaevis with homologous sperm and its verification using microsatellite markers. Aquaculture Research, 41(6), 913-920. https://doi.org/10.1111/j.1365-2109.2009.02373.x

Kamiya, T., Kai, W., Tasumi, S., et al. (2012). A trans-species missense SNP in Amhr2 is associated with sex determination in the tiger pufferfish, Takifugu rubripes (fugu). PLoS Genetics 8, e1002798. https://doi.org/10.1371/journal.pgen.1002798

Karayücel, İ., Ezaz, T., Karayücel, S., McAndrew, B. J., & Penman, D. J. (2004). Evidence for two unlinked "sex reversal" loci in the Nile tilapia, Oreochromis niloticus, and for linkage of one of these to the red body colour gene. Aquaculture, 234(1-4), 51-63. https://doi.org/10.1016/j.aquaculture.2003.12.016

Kato, K., Hayashi, R., Yuasa, D., Yamamoto, S., Miyashita, S., Murata, O., & Kumai, H. (2002). Production of cloned red sea bream, Pagrus major, by chromosome manipulation. Aquaculture, 207(1-2), 19-27. https://doi.org/10.1016/S0044-8486(01)00769-4

Kazama-Wakabayashi, M., Yamaha, E., & Yamazaki, F. (1999). The elimination and duplication of the lower part of blastoderm effects on the number of primordial germ cells in goldfish. Fisheries science, 65(4), 577-582. https://doi.org/10.2331/fishsci.65.577

Knibb, W. R., Gorshkova G. & S. Gorshkov, G. (1998). Genetic improvement of cultured marine fish: case studies. 111 137. In: S. De Silva (ed.). Tropical Mariculture. Academic Press Ltd. 487. https://doi.org/10.1016/B978-012210845-7/50004-0

Komen, H., Yamashita, M., Nagahama, Y. (1992). Testicular development induced by a recessive mutation during gonadal differentiation of female common carp (*Cyprinus carpio* L.). Development, Growth and Differentiation 34, 535–544. https://doi.org/10.1111/j.1440-169X.1992.00535.x

Kubota, S., Kuro-O, M., Mizuno, S. and Kohno, S., (1993). Germ line-restricted, highly repeated DNA sequences and their chromosome localization in a Japanese hagfish (*Eptatretus okinoseanus*). Chromo soma, 102:163-173. https://doi.org/10.1007/BF00387731

Kubota, S., Nakai, Y., Kuro-O, M., and Kohno, S., (1992). Germline restricted supernumerary (B) chromosomes in *Eptatretus okinoseanus*. Cytogenet. Cell Genet., 60:224-228. https://doi.org/10.1159/000133345

Lien, S., Koop, B. F., Sandve, S. R., Miller, J. R., Kent, M. P., Nome, T., ... & Davidson, W. S. (2016). The Atlantic salmon genome provides insights into rediploidization. Nature, 533(7602), 200-205.

Matsuda, M., Nagahama, Y., Shinomiya, A., Sato, T., Matsuda, C., Kobayashi, T., & Sakaizumi, M. (2002). DMY is a Y-specific DM-domain gene required for male development in the medaka fish. Nature, 417(6888), 559-563. https://doi.org/10.1038/nature751

Meng, Z., Liu, X., Liu, B., Hu, P., Jia, Y., Yang, Z., ... & Lei, J. (2016). Induction of mitotic gynogenesis in turbot Scophthalmus maximus. Aquaculture, 451, 429-435. https://doi.org/10.1016/j.aquaculture.2015.09.010

Nakai, Y., Kubota, S. and Kohno, S., (1991). Chromatin diminution and chromosome elimination in four Japanese hagfish species. Cytogenet. Cell Genet., 56:196-198. https://doi.org/10.1159/000133087

Nam, Y. K., Cho, Y. S., Kim, D. S. (2000). Isogenic transgenic homozygous fish induced by artificial parthenogenesis. Transgenic Research 9, 463–469. https://doi.org/10.1023/A:1026596422225

Nam, Y. K., Choi, G. C., Kim, D. S. (2004). An efficient method for blocking the 1st mitotic cleavage of fish zygote using combined thermal treatment, exemplified by mud loach (*Misgurnus mizolepis*). Theriogenology 61, 933–945. https://doi.org/10.1016/S0093-691X(03)00258-9

Nam, Y.K. and Kim, D.S. (2004). Ploidy status of progeny from the crosses between tetraploid males and diploid females in mud loach (*Misgurnus mizolepis*). Aquaculture 236, 575–582 https://doi.org/10.1016/j.aquaculture.2003.12.026

Nam, Y.K., Choi, G.C., Park, D.J., et al. (2001). Survival and growth of induced tetraploid mud loach. Aquaculture International 9, 61–71. https://doi.org/10.1023/A:1012540024333

Okutsu, T., Shikina, S., Kanno, M., Takeuchi, Y., & Yoshizaki, G. (2007). Production of trout offspring from triploid salmon parents. Science, 317(5844), 1517-1517. https://doi.org/10.1126/science.1145626

Pandian, T. J. (2011). Sex determination in fish. Science Publishers, CRC Press, New York. https://doi.org/10.1201/b11400

Pandian, T. J. and Koteeswaran, R. (1998). Ploidy induction and sex control in fish. Hydrobiologia 384, 167–243. https://doi.org/10.1023/A:1003332526659

Piferrer, F., Beaumont, A., Falguière, J. C., Flajšhans, M., Haffray, P., & Colombo, L. (2009). Polyploid fish and shellfish: production, biology and applications to aquaculture for performance improvement and genetic containment. Aquaculture, 293(3-4), 125-156. https://doi.org/10.1016/j.aquaculture.2009.04.036

Pšenička, M., Saito, T., Linhartová, Z., & Gazo, I. (2015). Isolation and transplantation of sturgeon early-stage germ cells. Theriogenology, 83(6), 1085-1092. https://doi.org/10.1016/j.theriogenology.2014.12.010

Purdom, C. E. (1983). Genetic engineering by the manipulation of chromosomes. Aquaculture 33, 287–300. https://doi.org/10.1016/0044-8486(83)90409-X

Reed, K.M. and Phillips, R.B., (1995). Molecular characterization and cytogenetic analysis of highly repeated DNAs of lake trout, *Salvelinus namaycush*. Chromosoma, 104(4), 242-251.

Rothbard, S., Shelton, W. L., Rubinshtein, I., Yaniv Hinits, Y. H., & David, L. (2000). Induction of all-female triploids in grass carp (Ctenopharyngodon idella) by integration of hormonal sex inversion and ploidy manipulation.

Rougeot, C., Ngingo, J. V., Gillet, L., Vanderplasschen, A., & Mélard, C. (2005). Gynogenesis induction and sex determination in the Eurasian perch, Perca fluviatilis. Aquaculture, 243(1-4), 411-415.

Slanchev, K., Stebler, J., de la Cueva-Méndez, G., & Raz, E. (2005). Development without germ cells: the role of the germ line in zebrafish sex differentiation. Proceedings of the National Academy of Sciences, 102(11), 4074-4079. https://doi.org/10.1073/pnas.0407475102

Shelton, W. L. (1986). Broodstock development for monosex production of grass carp. Aquaculture 57, 311–319. https://doi.org/10.1016/0044-8486(86)90209-7

Shelton, W. L. and Mims, S. D. (2012). Evidence for female heterogametic sex determination in paddlefish *Polyodon spathula* based on gynogenesis. Aquaculture 356–357, 116–118. https://doi.org/10.1016/j.aquaculture.2012.05.029

Stanley, J. G. (1976). Female homogamety in grass carp (*Ctenophayryngodon idella*) determined by gynogenesis. Journal of the Fisheries Research Board of Canada 33, 1373–1374. https://doi.org/10.1139/f76-175

Takeuchi, Y., Yazawa, R., & Yoshizaki, G. (2020). Chapter 17 intraperitoneal germ cell transplantation technique in marine teleosts. Reproduction in aquatic animals: From basic biology to aquaculture technology, 357-379. https://doi.org/10.1007/978-981-15-2290-1_18

Tvedt, H. B., Benfey, T. J., Martin-Robichaud, D. J., McGowan, C., & Reith, M. (2006). Gynogenesis and sex determination in Atlantic halibut (Hippoglossus hippoglossus). Aquaculture, 252(2-4), 573-583. https://doi.org/10.1016/j.aquaculture.2005.06.042

Yamamoto, E. (1999). Studies on sex manipulation and production of cloned populations in hirame, *Paralichthys olivaceus* (Temminck et Schlegel). Aquaculture 173, 235–246. https://doi.org/10.1016/S0044-8486(98)00448-7

Yamazaki, F. (1983). Sex control and manipulation in fish. Aquaculture 33, 329–354. https://doi.org/10.1016/0044-8486(83)90413-1

Yano, A., Guyomard, R., Nicol, B., Jouanno, E., Quillet, E., Klopp, C., ... & Guiguen, Y. (2012). An immune-related gene evolved into the master sex-determining gene in rainbow trout, Oncorhynchus mykiss. Current Biology, 22(15), 1423-1428. https://doi.org/10.1016/j.cub.2012.05.045

Yoshizaki, G., & Yazawa, R. (2019). Application of surrogate broodstock technology in aquaculture. Fisheries science, 85, 429-437. https://doi.org/10.1007/s12562-019-01299-y

Zhu, M., Sumana, S.L., Abdullateef, M.M., Falayi, O.C., Shui, Y., Zhang, C., Zhu, J. and Su, S., (2024). CRISPR/Cas9 Technology for Enhancing Desirable Traits of Fish Species in Aquaculture. International Journal of Molecular Sciences, 25(17). https://doi.org/10.3390/ijms25179299

Zou, S., Li, S., Cai, W., Zhao, J., & Yang, H. (2004). Establishment of fertile tetraploid population of blunt snout bream (Megalobrama amblycephala). Aquaculture, 238(1-4), 155-164. https://doi.org/10.1016/j.aquaculture.2004.05.020

15

Commercial Applications of Genetics and Biotechnology in Modern Fisheries and Aquaculture

Zaheer Abass**, *Tasaduq Hussain Shah and Rinkesh Nemichand Wanjari

Faculty of Fisheries, Sher-e-Kashmir University of Agricultural Sciences and Technology, Rangil, Ganderbal, Jammu and Kashmir (190006), India

Abstract

Biotechnology offers transformative potential for enhancing sustainability, disease management, and production efficiency in the fisheries and aquaculture industries. Key applications include genetic engineering, marker-assisted selection, and selective breeding to improve desirable traits in fish and shellfish, such as feed efficiency, disease resistance, and growth rates, ultimately enhancing fillet quality and conversion efficiency. Advances in biotechnology also significantly bolster disease prevention and control; DNA/RNA-based diagnostic techniques allow for rapid and precise pathogen identification, reducing the reliance on antibiotics and chemical treatments. Additionally, the development of biotechnology-based vaccines provides effective means to combat a wide range of viral and bacterial pathogens, further advancing the efficiency and sustainability of aquaculture practices.

Keywords: *Biotechnology, Genetic engineering, Sustainability, Fisheries, Aquaculture*

Introduction

By 2050, experts anticipate the world's population will surpass 9 billion, resulting in a 70 percent increase in the need for food, feed, and fiber. This increase will coincide with significant lifestyle and consumption shifts primarily driven by urbanization. A notable decline in the consumption of grains and pulses is anticipated, while there will be a marked increase in the intake of vegetables, fruits, meat, dairy, and fish. With its long-standing contribution to human nutrition, aquaculture is key in supplying high-quality proteins (Nash 2010; Gui *et al.*, 2018). As the population grows, resource scarcity and environmental degradation become increasingly urgent. The necessity for land resources to meet societal demands has turned attention toward oceans as a critical frontier for human survival. Fish, in particular, are emerging as an important source of high-quality protein for the human food (Baiden *et al.,* 2007). Over the last few decades, aquaculture has developed into the very fastest developing in the agricultural sector. In 2020, fisheries production surpassed 60 percent of what was recorded in the 1990s, driven by an unprecedented aquaculture output of 122.6 million metric tons (MT) (FAO 2022). Given this context, aquaculture and fisheries are expected to be pivotal in ensuring global nutritional security (Subasinghe, 2017). Fish provide around 20% of the world's animal protein, serving as a healthy option for wealthier populations due to their high polyunsaturated fatty acid (PUFA) content. Simultaneously, small indigenous fish species are vital for the nutrition of lower socioeconomic groups, offering essential proteins, oils, vitamins, and minerals (Mohanty *et al.,* 2010). Worldwide, the fishing and aquaculture industries play a crucial role in enhancing nutrition and sustaining economies. These sectors provide vital nutrients to approximately 3 billion individuals and constitute more than half of the animal protein consumed by 400 million people in economically disadvantaged areas. Fisheries and aquaculture play a crucial role in the livelihoods of over 500 million individuals in developing nations. The growth of aquaculture has been particularly noteworthy, with an annual growth rate of 7%, and fish products account for over 37 percent of the global food trade by volume. As wild fish stocks continue to decline, the cultivation of aquatic species, such as fish, crustaceans, mollusks, and aquatic plants, has become a cornerstone of the global seafood supply. As concerns rise over future food shortages and increasing prices, Fish farming and other forms of aquaculture serve as dependable sources of vital nutrients and protein, playing a significant role in enhancing global food availability and supporting efforts to combat malnutrition worldwide. Furthermore, the sector fosters promoting rural development by creating employment opportunities and enhancing local economic initiatives .In response to the rising demand for aquatic products, significant increases in aquaculture production will be necessary. Biotechnology offers a range of tools to facilitate the sustainable growth of aquaculture, fisheries, and the broader food sector. With the growing public demand for seafood and the degradation of marine habitats, researchers are increasingly focused on biotechnological solutions to enhance marine food production. This positions aquaculture as a rapidly evolving field of research (Melamed *et al.,* 2002). Through biotechnology, researchers can identify and merge genetic characteristics in aquatic organisms to enhance their productivity and quality. Current research emphasizes identifying genes that enhance natural growth factors and bolster defense mechanisms against microbial infections in fish. Despite its advancements, biotechnology presents new challenges and must be viewed as a complement to, rather than a replacement for, traditional methods. Its application should be guided by specific needs rather than technological availability. Genetic modification and biotechnology hold significant promise for improving the quantity and also the quality of fish in aquaculture (Mayekar *et al.,* 2013). As demand for aquaculture products increases, biotechnology will play a crucial role in meeting this demand. Biotechnological innovations in aquaculture will also undergo stringent regulatory scrutiny to ensure safety and efficacy before entering the market. Moreover, biotechnology offers environmental advantages, especially when integrated with other agricultural technologies. This synergy is critical to meeting the needs of a rapidly growing and urbanizing global population. However, the successful application of biotechnology depends on a robust interdisciplinary research foundation, including genetics, biology, physiology, and pathology. Without sustained investment in basic research, the full potential of biotechnological tools in aquaculture cannot be realized. Biotechnology must be embedded within a strong research framework to ensure its success and should not be pursued in isolation. Biotechnology holds substantial potential to enhance aquaculture productivity, improve food security, and promote environmental sustainability. Several advanced biotechnological tools—such as protein expression, microsatellites, gene mapping, genomics, Deoxyribonucleic acid (DNA) vaccines, proteomics, and transgenic technology present valuable opportunities. These technologies enable genetic manipulation, embryo manipulation, and the development of genetically engineered diagnostics and immunoprophylactic agents (Muhammet *et al.,* 2013). These techniques can greatly improve various aspects of aquaculture, including selective breeding, hybridization, productivity, health, growth, nutrition, cryopreservation, and genetic resource conservation. Ultimately, these advancements contribute to strengthening global food security.

Induced Breeding Techniques: A Biotechnological Revolution

In fish reproduction, gonadotropin-releasing hormone (GnRH) has emerged as a crucial biotechnological instrument for stimulating breeding. This hormone functions as the primary controller and trigger of reproductive processes across all vertebrate species (Bhattacharya *et al.*, 2002). Discovered in the hypothalami of pigs and sheep, this decapeptide stimulates the release of luteinizing hormone (LH) and follicle-stimulating hormone (FSH) from the pituitary gland (Schally *et al.*, 1973). GnRH regulates LH and FSH release in placental mammals, including humans. In contrast, twelve structural

GnRH variants exist in non-mammalian species, with seven to eight forms identified in fish (Halder *et al.*, 1991; Sherwood *et al.*, 1993; Jimenez-Linan *et al.*, 1997). Carolsfeld *et al.* (2000) purified and characterized the most recent variants. Numerous synthetic analogs of GnRH, developed based on structural differences and biological activity, have been created. One prominent example is the salmon GnRH analog, widely used in fish breeding under the commercial Ovaprim. This GnRH technology has significantly advanced induced breeding in fish, playing a vital role in improving aquaculture practices and fisheries management.

Surrogate Brood Stock Technology

This innovative approach holds immense promise for aquaculture and the conservation of species that are challenging to breed in captivity (Yoshizaki & Yazawa 2019). Researchers have introduced the concept of producing large fish, such as Tuna, from smaller fish like Mackerel or Trout from Salmon (Yoshizaki & Yazawa 2019), generating significant interest in hatchery seed production for various important species in conservation and aquaculture. Surrogate breeding, a cutting-edge aquaculture technology, enables the production of sperm and eggs of a target species within a closely related and easily farmable species. This technique involves using one species to produce gametes—spermatozoa and eggs—of another species through surrogate methods. Understanding the mechanisms of gonad development is crucial for successfully applying this technology. Gonads originate from Primordial Germ Cells (PGCs), which can differentiate into spermatogonia or oogonia, leading to the development of testes or ovaries. In this process, PGCs from a donor fish are implanted into the embryo of a recipient fish. Once the recipient matures, it produces gametes that are genetically identical to those of the donor species. This approach provides a practical solution for breeding larger, hard-to-culture species like Tuna using smaller, more manageable species like Mackerel. It offers considerable advantages, including reduced space, lower feeding costs, and shorter maturation times, making it a highly efficient method for both aquaculture and conservation efforts.

Engineering Fish Chromosomes: Methods for Sterility, Sex Control, and Beyond

Chromosome sex manipulation techniques, such as induced polyploidy and uniparental chromosome inheritance, have been widely employed in breeding cultured fish species (Lakra and Das, 1998). These techniques, which provide quick fixes for gonadal sterilization, sex control, improved hybrid viability, and clonal propagation, are crucial for progressing fish breeding. While polyploid creatures have extra sets of chromosomes (triploids have three, tetraploids have four, and so on), most vertebrates are diploid, having two complete sets of chromosomes in their somatic cells. In particular, induced triploidy is considered the best method for creating sterile fish, a valuable characteristic in aquaculture and fisheries management. It is widely known how triploidy and other chromosomal changes in fish are induced, and aquaculture uses these techniques (Chourrout, 1987; Thorgaard, 1983)—aquaculture, providing cutting-edge resources for conservation, genetic advancement, and selective breeding. For example, simple interploidy crossings between tetraploids and diploids can produce vast numbers of sterile triploid fish, making tetraploid breeding lines very promising (Chourrout *et al.*, 1986; Guo *et al.*, 1996). However, the survivability of these tetraploid individuals has frequently been limited, even though tetraploidy induction has been accomplished in several finfish species (Rothbard *et al.*, 1997). However, chromosomal manipulation and polyploidy effectively enhance fish breeding and develop aquaculture technologies. Sterile induction methods for teleost fish include triploidy induction (Thorgaard, 1983) and exogenous hormone therapies (Hunter and Donaldson, 1983), offering a variety of choices for aquaculture improvement. However, government laws frequently limit the use of hormone therapies, and consumers' acceptance of fish products treated with hormones is also limited. In contrast, triploidy is accomplished by subjecting fertilized eggs to chemical or physical treatments as soon as possible after fertilization to stop the second polar body from extruding (Purdom, 1983; Thorgaard, 1983). Because triploid fish are usually infertile because of the incorrect synapsis of homologous chromosomes during the first meiotic division. Triploid fertilized eggs can be created by applying hydrostatic pressure, temperature shocks (hot or cold), or chemicals such as nitrous oxide, cytochalasin-B, or colchicine. Additionally, triploids can be produced by crossing tetraploids with diploids. Tetraploid induction involves fertilizing eggs with normal sperm and then applying physical or chemical treatments to the diploid zygote to inhibit the first mitotic division. Gynogenesis is a reproductive process involving exclusive maternal inheritance, making it particularly valuable in fish breeding because it allows for significant inbreeding in a single generation. This method can assist elucidate fish sex determination mechanisms and create all-female populations in species with female homogamety. For sex inversion experiments, all-female gynogenetic progenies are favored above typical bisexual progenies. Techniques that combine hormonal sex inversion with induced gynogenesis have been used to a variety of aquaculture species (Gomelsky *et al.*, 2000). In aquaculture, androgenesis, the opposite of gynogenesis, involves paternal inheritance and has potential for economic use. This method can recover lost genotypes from cryopreserved sperm and produce homozygous fish lines. Several fish species, including salmonids, cyprinids, and cichlids, have been successfully generated as androgenetic fish (Bongers *et al.*, 1994). These chromosome manipulation methods are essential to contemporary aquaculture because they provide cutting-edge resources for conservation, genetic advancement, and selective breeding.

Biotechnological Interventions in Health Management in Aquaculture

The disease remains a major challenge and a significant concern in aquaculture, with an estimated annual cost of over USD 6 billion (Leung and Bates, 2013). Managing diseases in aquaculture is complex, but biotechnological interventions at various stages—such as diagnosis, prevention, and treatment—provide valuable support to aquaculturists. The industry increasingly demands fast-growing, disease-resistant species, rapid and precise disease diagnostic tools, cost-effective vaccines, probiotics, and reliable cell lines to improve fish health and sustainability. Prevention is always better than cure, which is true in aquaculture. Monitoring pathogens at various stages of production is crucial for maintaining farm biosecurity. Therefore, there is a strong need for simple, sensitive, reliable, and rapid methods to detect infections. Molecular techniques for pathogen detection, including those based on biochemical features, serology, and histology, offer more incredible speed and sensitivity than traditional methods. Polymerase Chain Reaction (PCR) technology has revolutionized the field of diagnostics. Various versions of PCR can detect even a single copy of a pathogen that other methods might otherwise miss. Over the past 15 years, PCR-based diagnostic methods have been extensively used to identify numerous diseases that pose challenges in aquaculture. Practically, PCR-based techniques are available for diagnosing all OIE-listed pathogens, and they play a vital role in testing Specific Pathogen-Free (SPF) seeds, ensuring healthier stock and improved biosecurity in aquaculture operations. The aquaculture sector is rapidly advancing with biotechnological innovations aimed at genetic improvements in aquatic species. One key application of modern biotechnology is the development of transgenic species. By crossing different strains, new varieties are created to transfer desirable traits from each parent into the offspring, enhancing the genetic attributes of aquaculture species.

Genetically Modified Species (Transgenics)

In order to give particular features, such as increased growth rates, size, improved feed conversion into muscle, and controlled sexual maturation, transgenic techniques entail the selective transfer of one or more genes (Maclean and Laight, 2000). For example, Salmon, trout, and tilapia have effectively incorporated growth hormone genes from human or animal sources, leading to noticeably faster growth than their native equivalents. This strategy provides a quicker and more accurate way to breed new types. Gene transfer between species can also be used when conventional breeding techniques cannot produce desired features. For example, transferring genes that enhance tolerance to environmental stresses or resistance to extreme conditions can establish unique characteristics. Transferring the antifreeze protein gene (AFP) into fish to help them adapt to freezing conditions is one noteworthy use (Hew and Fletcher, 1992). However, as environmental changes may affect the sustainability and efficacy of these biotechnological developments, it is also necessary to take into account the possible effects of climate change on aquaculture. Addressing climate change issues and improving the aquaculture industry's resilience and adaptability depend heavily on biotechnology. For instance, one method to increase aquatic creatures' capacity to flourish in shifting settings is to use transgenic techniques to create genetic characteristics that give them resistance to low oxygen levels. Effective disease prevention is critical to aquaculture's intensification and sustainable growth. Through selective breeding for disease resistance, biotechnology plays a critical role in improving the health of farmed species and boosting resistance to infections (El-Zaeem and Aseem, 2004). Furthermore, by reducing disease transmission between wild and cultivated populations, these methods can improve aquatic health in general. Additionally, biotechnology creates new avenues for the development of novel aquatic species welfare goods and markets.

Table 1. Key Transgenic Innovations in Food and Ornamental Fish, Source: (Rasal *et al.*, 2016)

S. No.	Species	Target genes
1	*Labeo rohita* (Rohu)	GFP
2	*Labeo rohita* (Rohu)	Mylz
3	*Ictalurus punctatus* (Channel catfish)	GH
4	*Danio rerio* (Zebrafish)	Color genes
5	*Labeo rohita* (Rohu)	GH
6	*Catla catla* (Catla)	GH
7	*Gymnocorymbus ternetzi* (Black tetra)	Mylz2
8	*Oryzias latipes* (Japanese rice fish)	Vitellogenin
9	*Oncorhynchus kisutch* (Coho salmon)	GH
10	*Oncorhynchus tshawytscha* (Chinook salmon)	GH
11	*Danio rerio* (Zebrafish)	Mylz2
12	*Heteropneustes fossilis* (Asian stinging catfish)	GH
13	*Misgurnus anguillicaudatus* (Pond loach)	GH
14	*Oreochromis niloticus* (Nile tilapia)	GH

15	*Oncorhyncus mykiss* (Rainbow trout)	GH
16	*Haliotus rufescens (*Red abalone)	GH
17	*Cyprinus carpio* (Common carp)	GH
18	*Megalobrama amblycephala* (Wuchang bream)	GH

Cryopreservation of Gametes (gene banking)

Cryopreservation is a technique used to preserve biological materials over long periods by storing them at extremely low temperatures, typically around -196°C, the liquid nitrogen temperature. This process halts cells' physiological and biochemical activities, keeping them viable for extended durations. The method of cryopreserving fish sperm (milt) was adapted from animal husbandry and first succeeded when Blaxter (1953) used frozen-thawed semen to fertilize herring (*Clupea harengus*) eggs. Since then, sperm cryopreservation has been achieved for nearly all cultivable fish species (Casas and Flores, 2013). Cryopreservation addresses several challenges in aquaculture, such as the asynchronous maturation of males and females, enabling selective breeding, improving stock quality, and aiding in conservation efforts (Harvey, 1996). While gene banks have been widely used to develop plant varieties, aquatic gene banks face unique hurdlesAt present, the cryopreservation of finfish reproductive cells is limited to male gametes. No effective method has been developed to successfully preserve eggs or embryos from finfish species. However, promising advances have been made, including the successful cryopreservation of shrimp embryos by Subramanian and Newton 1993 and Diwan and Kandaswami 1997. These developments highlight the importance of establishing gene banks for cultivated and cultivable aquatic species to support future breeding, genetic improvement, and conservation initiatives.

Bioaugmentation and Probiotics

In aquaculture, the use of bacteria as probiotics for the gut and water is increasingly popular for controlling pathogen proliferation and maintaining a healthy gut microbiota. Water probiotics help reduce organic load and convert harmful compounds like ammonia into less harmful forms, which is crucial for maintaining pond health. In modern shrimp aquaculture, bioaugmentation is a common practice. Another important aspect of aquaculture health management is the development of vaccines. Recombinant DNA technology has been employed in various studies to create vaccines. Currently, there is only one DNA vaccine available in Canada for Infectious Hematopoietic Necrosis Virus (IHNV). Although RNA interference (RNAi)-based technology has shown promise in providing pathogen protection, it is not yet commercially available. Despite these advancements not being widely adopted, future aquaculture must incorporate these technologies to prevent disease outbreaks and enhance overall health management.

Proteomics

In aquaculture, proteomics has become an essential tool for expanding our knowledge of aquatic species' biology, leading to high productivity and improved product quality. Proteomics offers insights into the dynamic and mechanical changes in protein activity, such as post-translational modifications (PTMs), whereas genomics provides a comprehensive picture of the genetic code. Recent studies have increasingly concentrated on commercially significant species like tilapia, cod, and salmon, even though complete genomic sequences are available for a few model species, including zebrafish, stickleback, medaka, coelacanth, fugu, and Tetraodon. Nevertheless, despite advancements, proteomic research in aquaculture still faces difficulties, especially about protein identification. To overcome these obstacles, combining transcriptomic, proteomic, and metabolomic data is a potent strategy that has worked well for reassembling the networks and pathways governing fish growth (Alsagaby, 2019). Research on human diseases, including studies on cancer, neurology in toxicology, infectious disorders, and medication development, has significantly benefited from the use of fish models. Proteomics in aquaculture is still in its infancy, but new technologies like mass spectrometry (MS) imaging have the potential to advance the area. These developments may open the door to a proteomics-based understanding of fish quality, which could lead to the inclusion of quality parameters in selective breeding initiatives. By creating biomarker-guided breeding techniques, aquaculture could improve fish health, growth, feed conversion efficiency, and overall product quality. This would open up new avenues for enhancing aquaculture results worldwide.

Nanotechnology

Nanotechnology has emerged as an up-and-coming field that could transform aquaculture and related sectors (Guttmacher & Collins, 2002). Leveraging its advanced capabilities, nanotechnology can revolutionize the fisheries and aquaculture industries through various innovative applications. One key application is in rapid disease detection. Nanotechnology enables the development of compassionate tools for identifying pathogens, which can lead to quicker and more accurate diagnoses. Additionally, nanotechnology enhances the ability of fish to absorb drugs, hormones, vaccines, and nutrients more effectively. This is facilitated by the technology's capacity for biomolecular analysis and the creation of non-viral vectors for gene therapy, including transport vehicles for DNA, proteins, or cells. DNA nano-vaccines represent a significant advancement, utilizing nanoparticle carriers such as chitosan and poly-lactide-co-glycolide acid (PLGA) to deliver vaccine antigens. These nanoparticles and mild inflammatory inducers protect against bacterial and viral diseases in fish and shellfish. Mass vaccination can be achieved using nanocapsules that are resistant to digestion and degradation. These nanocapsules, containing short strands of DNA, are absorbed by fish cells when applied to water. The ultrasound then ruptures the capsules, releasing the DNA and eliciting an immune response. Oral administration of these vaccines and site-specific release mechanisms for active agents can significantly reduce the cost and effort involved in disease management. New carrier systems for gene delivery, such as chitosan-based vectors, provide an effective alternative to synthetic vectors with fewer associated toxicities. Enhancing gene delivery through receptor-mediated endocytosis further improves efficiency. Nanotechnology also plays a role in delivering antibiotics, probiotics, and pharmaceuticals through feed or injection systems. Innovative nanotech delivery systems, which are pre-programmed and time-controlled, can precisely deliver probiotics, hormones, chemicals, and vaccines, improving overall health management and reducing intervention costs. Integrating nanotechnology into aquaculture has great potential for advancing disease management, enhancing nutrient absorption, and improving fish health and productivity (Shah and Mraz, 2020).

Conclusion

The biotechnology field has seen promising developments that have contributed significantly to the expansion of aquaculture systems. These innovations enable the production of superior, rapidly growing, and robust aquatic organisms in an eco-friendly way. Enhanced communication, teamwork, and collaboration among scientists and producers are essential to propel this vital industry further. By tackling obstacles and capitalizing on new possibilities, the sector can achieve a balance of sustainable production, economic success, environmental conservation, and social acceptance. This section summarizes recent biotechnological breakthroughs, emphasizing crucial areas such as genetic preservation, transgenic techniques, chromosomal manipulation, synthetic hormone use in fish reproduction, and the application of biotechnology in fisheries.

References

Alsagaby, S. A. (2019). Understanding the fundamentals of proteomics. Curr Top Pept Protein Res, 20(3), 25-33.

Baiden, R. Y. ; Rhule, S. W. A. ; Otsyina, H. R. ; Sottie, E. T. ; Ameleke, G., (2007). Performance of West African Dwarf sheep and goats fed varying levels of cassava pulp as a replacement for cassava peels. Livest. Res. Rural Dev., 19 (35)

Bhattacharya, S., Dasgupta, S., Datta, M., & Basu, D. (2002). Biotechnology input in fish breeding.

Bongers, A. B. J., In't Veld, E. P. C., Abo-Hashema, K., Bremmer, I. M., Eding, E. H., Komen, J., & Richter, C. J. J. (1994). Androgenesis in common carp (Cyprinus carpio L.) using UV irradiation in a synthetic ovarian fluid and heat shocks. Aquaculture, 122(2-3), 119-132.

Carolsfeld, J., Powell, J. F., Park, M., Fischer, W. H., Craig, A. G., Chang, J. P., ... & Sherwood, N. M. (2000). Primary structure and function of three gonadotropin-releasing hormones, including a novel form, from an ancient teleost, herring. Endocrinology, 141(2), 505-512.

Casas, I., & Flores, E. (2013). Gene banking: the freezing strategy. In Boar reproduction: fundamentals and new biotechnological trends (pp. 551-588). Berlin, Heidelberg: Springer Berlin Heidelberg.

Chourrout, D., & Nakayama, I. (1987). Chromosome studies of progenies of tetraploid female rainbow trout. Theoretical and Applied Genetics, 74, 687-692.

Chourrout, D., Chevassus, B., Krieg, F., Happe, A., Burger, G., & Renard, P. (1986). Production of second generation triploid and tetraploid rainbow trout by mating tetraploid males and diploid females—potential of tetraploid fish. Theoretical and Applied Genetics, 72, 193-206.

El-Zaeem, S.Y., Aseem, S.S. Application of biotechnology in fish breeding: 1 – production of highly immune genetically modified Nile, tilapia Orechromis niloticus with accelerated growth by direct injection of Shark

FAO. (2022). The state of world fisheries and aquaculture 2022 (p. 266). Towards Blue Transformation.

Gomelsky, M., Horne, I. M., Lee, H. J., Pemberton, J. M., McEwan, A. G., & Kaplan, S. (2000). Domain structure, oligomeric state, and mutational analysis of PpsR, the Rhodobacter sphaeroides repressor of photosystem gene expression. Journal of bacteriology, 182(8), 2253-2261.

Gui, J. F., Tang, Q., Li, Z., Liu, J., & De Silva, S. S. (Eds.). (2018). Aquaculture in China: success stories and modern trends. John Wiley & Sons.

Guo, X., DeBrosse, G. A., & Allen Jr, S. K. (1996). All-triploid Pacific oysters (Crassostrea gigas Thunberg) produced by mating tetraploids and diploids. Aquaculture, 142(3-4), 149-161.

Guttmacher, A. E., & Collins, F. S. (2002). Genomic medicine—a primer. New England Journal of Medicine, 347(19), 1512-1520.

Halder, S., Sen, S., Bhattacharya, S., Ray, A. K., Ghosh, A., & Jhingran, A. G. (1991). Induced spawning of Indian major carps and maturation of a perch and a catfish by murrel gonadotropin releasing hormone, pimozide and calcium. Aquaculture, 97(4), 373-382.

Hunter, G. A., Donaldson, E. M., Stoss, J., & Baker, I. (1983). Production of monosex female groups of chinook salmon (Oncorhynchus tshawytscha) by the fertilization of normal ova with sperm from sex-reversed females. Aquaculture, 33(1-4), 355-364.

Jimenez-Liñan, M., Rubin, B. S., & King, J. C. (1997). Examination of guinea pig luteinizing hormone-releasing hormone gene reveals a unique decapeptide and existence of two transcripts in the brain. Endocrinology, 138(10), 4123-4130.

Leung, T. L., & Bates, A. E. (2013). More rapid and severe disease outbreaks for aquaculture at the tropics: implications for food security. Journal of applied ecology, 215-222.

Maclean, N., & Laight, R. J. (2000). Transgenic fish: an evaluation of benefits and risks. Fish and Fisheries, 1(2), 146-172.

Mayekar, T. S., Salgaonkar, A. A., Koli, J. M., Patil, P. R., Chaudhari, A., Pawar, N., ... & Kapse, P. (2013). Biotechnology and its applications in aquaculture and fisheries. Aquafind-Aquat Fsh Database.

Melamed, P., Gong, Z., Fletcher, G., & Hew, C. L. (2002). The potential impact of modern biotechnology on fish aquaculture. Aquaculture, 204(3-4), 255-269.

Mohanty, B., Mohanty, S., Sahoo, J., & Sharma, A. (2010). Climate change: impacts on fisheries and aquaculture. Climate change and variability, 119, 978-53.

Muhammet, A., Zerife, P., Ramazan, S., Adem, T. A., & Volkan, K. (2013). Biotechnology and aquaculture in sustainable development.

Nash, C. (2010). The history of aquaculture. John Wiley & Sons.

Purdom Jr, P. W. (1983). Search rearrangement backtracking and polynomial average time. Artificial intelligence, 21(1-2), 117-133.

Lakra, W., & Das, P. (1998). Genetic engineering in aquaculture. The Indian Journal of Animal Sciences, 68(8).

Schally, A. V., Redding, T. W., & Arimura, A. K. I. R. A. (1973). Effect of sex steroids on pituitary responses to LH-and FSH-releasing hormone in vitro. Endocrinology, 93(4), 893-902.

Shah, B. R., & Mraz, J. (2020). Advances in nanotechnology for sustainable aquaculture and fisheries. Reviews in Aquaculture, 12(2), 925-942.

Sherwood, P. W., Tsang, S. V. M., & Osley, M. A. (1993). Characterization of HIR1 and HIR2, two genes required to regulate histone gene transcription in Saccharomyces cerevisiae. Molecular and cellular biology, 13(1), 28-38.

Subasinghe, R. (2017). World aquaculture (2015): a brief overview. FAO Fisheries and Aquaculture Report, (1140).

Thorgaard, G. H. (1983). 8 chromosome set manipulation and sex control in fish. In Fish physiology (Vol. 9, pp. 405-434). Academic Press.

Thorpe, A., Reid, C., van Anrooy, R., Brugere, C., & Becker, D. (2006). Asian development and poverty reduction strategies: integrating fisheries into the development discourse. Food Policy, 31(5), 385-400.

Yoshizaki, G., & Yazawa, R. (2019). Application of surrogate broodstock technology in aquaculture. Fisheries science, 85, 429-437.

Abbreviations

Abbreviated form	Full form
×	Denotes a hybrid cross (e.g., blue × channel catfish)
AFLP	Amplified Fragment Length Polymorphism
AHPND	Acute Hepatopancreatic Necrosis Disease
AI	Artificial Intelligence
AMH	Anti-Mullerian Hormone
ATCC	American Type Culture Collection
BAC	Bacterial Artificial Chromosomes
BMP	Best Management Practices
Cas9	CRISPR-associated protein 9
CCV	Channel catfish virus
cDNA	Complementary Deoxyribonucleic Acid
CIM	Composite Interval Mapping
cM	Centimorgan
CNV	Copy Number Variation
COI	Cytochrome oxidase I
CRISPER	Clustered Regularly Interspaced Short Palindromic Repeats
CRISPR-Cas9	Clustered Regularly Interspaced Short Palindromic Repeats Cas9
DAGs	Directed Acyclic Graphs
DNA	Deoxyribonucleic Acid
EBP	Earth Biogenome Project
EEZ	Exclusive Economic Zone
ENSO	El Nino – Southern Oscillation
ESC	Enteric Septicemia of Catfish
EST	Expressed Sequence Tag
EVBs	Estimated Breeding Values
FAO	Food and Agriculture Organisation
FCR	Feed Conversion Ratio
FP7	Framework Programme 7
GBS	Genotyping by Sequencing
GCGB	Grass Carp Genome Database

GH	Growth Hormone
GIFT	Genetic Improvement of Farmed Tilapias
GM	Genetically Modified
GMO	Genetically Modified Organisms
GMOs	Genetically Modified Organisms
GMT	Genetically Male Tilapia
GS	Genomic Selection
GWAS	Genome Wide Association Studies
h^2T	Broad-Sense Heritability
HDR	Homology-Directed Repair
ICAR	Indian Council of Agricultural Research
ICSASG	International Cooperation to Sequence the Atlantic Salmon Genome
ICZM	Integrated Coastal Zone Management
IHNV	Infectious hematopoietic necrosis virus
IoT	Internet of Things
IPN	Infectious Pancreatic Necrosis
IPNV	Infectious pancreatic necrosis virus
irf9	Autosomal Interferon Regulatory Factor 9
IUCN	International Union for Conservation of Nature
LOD	Logarithm of Odds
MAS	Marker-Assisted Selection
MHC	Major Histocompatibility Complex
MSTN	Myostatin
mt-DNA	Mitochondrial DNA
NCBI	National Centre for Biotechnology Information
NGS	Next-Generation Sequencing
NHEJ	Non-Homologous End Joining
NOFIMA	Norwegian Institute of Food, Fisheries and Aquaculture Research
ONT	Oxford Nanopore Technologies
OsHV-1	Ostreid Herpesvirus 1
PacBio	Pacific Biosciences
PCR	Polymerase Chain Reaction
PIC	Polymorphism Information Content
qPCR	Quantitative Polymerase Chain Reaction
QTL	Quantitative Trait Loci
RAD	Restriction-site Associated DNA
RAD-seq	Restriction Site-Associated DNA Sequencing
RAPD	Rapid Amplification of Polymorphic DNA
RFLP	Restriction Fragment Length Polymorphism

RILs	Recombinant Inbred Lines
RNA	Ribonucleic Acid
RNA-Seq	RNA Sequencing
sdy	Sex-Determining Gene
SIM	Simple Interval Mapping
SMRT	Single Molecule Real-Time
SNP	Single Nucleotide Polymorphism
SNPs	Single Nucleotide Polymorphisms
SOFIA	State of World Fisheries and Aquaculture
SOLiD	Sequencing By Oligonucleotide Ligation and Detection
SSR	Simple Sequence Repeats
SSRs	Simple sequence repeats
TALENs	Transcription Activator-Like Effector Nucleases
tRNA	Transfer Ribonucleic Acid
TSV	Taura Syndrome Virus
UN	United Nations
VNNV	Viral Nervous Necrosis Virus
WSSV	White Spot Syndrome Virus
ZFN	Zinc-Finger Nucleases
ZFNs	Zinc Finger Nucleases
σ^2G	Genetic Variance
σ^2P	Phenotypic Variance

Glossary

ABI DNA sequencing: Applied Biosystems genetic analysis systems are a trusted standard for Sanger sequencing and fragment analysis by capillary electrophoresis

Adaptation: The process by which a species fits its environment and is the result of natural selection acting upon heritable variation over several generations.

Aquaculture: The farming of aquatic organisms such as fish, crustaceans, mollusks, and aquatic plants in controlled environments for food production and other uses.

Aqua-transgenesis: Aqua-transgenesis is the process by which new genes are transferred into aquatic organisms, such as fish, crustaceans, and mollusks, to improve their genetic traits.

Biotechnology: Biotechnology is the application of biological systems, living organisms, or parts thereof to develop or create products and processes for specific uses. It combines biology with technology, often involving genetic engineering and other molecular techniques, to address various issues in fields like medicine, agriculture, and industry.

Breeding: Breeding is the process of sexual reproduction that produces offspring, typically in animals or plants.

Chiasma: A chiasma is a point of contact between homologous chromosomes, where genetic material is exchanged during crossover.

Chromosomes: Chromosomes are threadlike structures made of protein and a single molecule of DNA that serve to carry genomic information from cell to cell.

Coupling Phase: The coupling phase occurs when two genes on the same chromosome are positioned such that the recessive alleles of one gene are on the other homologue and the alleles that are dominant of both genes are on the same homologue. Another name for this arrangement is the cis configuration.

CRISPR-Cas9: A gene editing tool derived from a natural bacterial defense mechanism that enables precise modifications to an organism's genome by cutting DNA at specific locations. It is widely used owing to its simplicity, affordability, and accuracy.

Crossing over: Crossing over is a cellular process that occurs during meiosis when chromosomes of the same type are lined up.

Cytogenetics: Cytogenetics is a branch of genetics and cell biology that studies how chromosomes relate to cell behavior, particularly during mitosis and meiosis.

Deoxyribonucleic acid (DNA): Deoxyribonucleic acid (DNA) is a molecule that contains genetic information for the development, growth, and reproduction of organisms.

DNA barcoding: DNA barcoding is a method for identifying specimens through the analysis of short, standardized segments of DNA.

DNA markers: DNA, genetic, or molecular markers refer to distinct DNA, gene, or molecular sequences found at specific locations on chromosomes that can be easily identified.

Ecological engineering: This is defined as the design, restoration, or creation of ecosystems, with a strong emphasis on ecosystem self-design and self-organization.

Environmental Impact Assessment: This is a process that evaluates the environmental effects of a proposed project or development before it is approved.

Environmental Tolerance: The ability of aquaculture species to withstand environmental stressors such as temperature fluctuations, changes in salinity, or low oxygen levels. Gene editing can be used to improve the environmental tolerance of farmed species.

EST: An Expressed Sequence Tag (EST) refers to short sequences or partial cDNAs that serve as markers for genes when the gene name is known

Ethical Considerations: The moral implications of genetic modifications in aquaculture include concerns about animal welfare, environmental impact, and the potential consequences of releasing gene-edited organisms into the wild.

Feed Conversion Efficiency: The ability of farmed species to convert feed into body mass efficiently. Improved feed conversion results in faster growth and reduced feed costs, making the farming process more sustainable.

Flesh Quality: Characteristics of fish meat, such as fat content, texture, color, and nutritional value. Flesh quality is an important factor in consumer preference and marketability of aquaculture products.

FokI: A restriction enzyme used in gene editing technologies, such as TALENs and ZFNs, to create double-strand breaks in DNA at targeted sites for genetic modification.

Food Security: The availability of sufficient, safe, and nutritious food to meet the dietary needs and preferences of individuals at all times. Gene editing in aquaculture aims to improve food security by enhancing the productivity and sustainability of seafood production.

Gene Editing: A biotechnology process that allows precise alterations to an organism's DNA to enhance or suppress specific traits, often using tools like CRISPR-Cas9, TALENs, and ZFNs.

Genes: The basic building block of heredity that contains genetic information is genes.

Genetic Improvement: The process of enhancing the genetic traits of farmed species to improve productivity, disease resistance, growth rate, and adaptability to environmental stress.

Genetic mapping: Genetic mapping is defined as the process of identifying the physical location of a genetic factor on chromosomes using recombination and genetic markers.

Genetic Marker: A particular DNA sequence with a recognized position on a chromosome is called a genetic marker. This served as a reference point for mapping other genes or DNA sequences.

Genetic markers: Genetic markers, often known as DNA markers, are known DNA sequences located in the genome that are used in molecular genetics to identify alterations or DNA sequences from unknown DNA samples.

Genetically Modified Organism: It is an animal, plant, or microbe whose DNA has been altered using genetic engineering techniques.

Genetically Modified Organisms (GMOs): Organisms whose genetic material has been altered in a way that does not occur naturally through mating or natural recombination. GMOs are often created using biotechnological techniques, including gene editing, to introduce desired traits.

Genetics: The study of genetics focuses on how genes are inherited and how they contribute to variation among organisms.

Genome Sequencing: Genome Sequencing is the process of determining the DNA sequence of an organism's genome.

Genome: The genome is an organism's complete set of genetic instructions.

Genomic Selection: A modern breeding approach that uses genetic markers across an organism's genome to predict the genetic potential for desirable traits, allowing for more efficient selection in breeding programs.

Genomics: Genomics is the study of an organism's genome, or complete set of DNA, and how it functions, evolves, and interacts with the environment

Genotyping: Genotyping is the process of determining differences in an individual's genetic makeup (genotype) by examining the individual's DNA sequence using biological assays and comparing it to another individual's sequence or a reference sequence.

gRNA (gRNA): A small RNA molecule that directs the CRISPR-Cas9 system to the correct location in the genome for targeted gene editing. This helps ensure that the Cas9 enzyme cuts the DNA at the precise desired location.

Growth Hormone (GH): A hormone that regulates growth in organisms. In aquaculture, genetic modifications to growth hormone-regulating genes can accelerate the growth rate of farmed fish, leading to faster production and higher yield.

Hatchery: A facility in which fish are bred, hatched, and grown under controlled conditions is known as a fish hatchery.

Hi-C technology: Hi-C is a technology based on 3C and utilizes high-throughput sequencing to detect chromatin interactions across the entire genome

Horizontal Gene Transfer (HGT): The transfer of genetic material between organisms in a manner other than traditional reproduction. HGT is often considered in genetic engineering and gene editing when introducing modified genes into organisms.

Hybridization: Hybridization is the process of combining two atomic orbitals to create a new type of hybridized orbital.

In situ: *In situ* sequencing (ISS) is a new method in which mRNA is directly sequenced in a fixed tissue or cell sample section.

Inheritance: The reception of genetic qualities by transmission from parent to offspring all of the genetic characters or qualities transmitted from parent to offspring

Linkage Groups: A linkage group is a group of genes situated closely on a chromosome and often inherited together unless they are separated by crossover.

Linkage: Linkage as related to genetics and genomics refers to the closeness of genes or other DNA sequences to one another on the same chromosome.

LOD Score: An LOD score is a statistical measure used to assess whether two genes are connected. It calculates the likelihood that two genes are inherited together because of linkage rather than chance.

Major histocompatibility complex: The major histocompatibility complex (MHC) is a group of genes that encodes proteins located on cell surfaces, assisting the immune system in identifying foreign substances.

Marker Assisted Selection: The breeding method known as "Marker-Assisted Selection" (MAS) selects animals or plants with desired characteristics, such as resistance to disease, increased yield, or drought tolerance, by using molecular markers.

Microsatellite markers: Microsatellite markers are short repetitive DNA sequences that exhibit significant variability among individuals in a population. They are utilized in genetics for various purposes, including population studies, genetic mapping, and forensics, because they offer valuable insights into genetic diversity and relationships.

Microsatellites: Microsatellites, as related to genomics, refer to a short segment of DNA, usually one to six or more base pairs in length, repeated multiple times in succession at a particular genomic location.

Mitigation: This is a sustained action that reduces or eliminates the long-term risk to people and property from natural hazards and their effects.

Mitosis: Mitosis is a process of cell duplication, in which one cell divides into two genetically identical daughter cells.

Molecular breeding: Molecular or marker-assisted breeding (MB) is a technique that employs DNA markers instead of phenotypic selection to accelerate the breeding process and release superior cultivars.

Messenger RNA (Messenger RNA): A molecule that carries genetic instructions from DNA to the ribosome, where proteins are synthesized. In gene editing applications, mRNA can be engineered to direct specific protein production in aquaculture species.

Mutation: A mutation is a modification of the nucleotide sequence within a specific area of a genome.

NGS technologies: NGS can sequence millions of DNA fragments at once, providing detailed information about the structure of genomes, genetic variations, gene activity, and changes in gene behaviour

Off-Target Effects: Unintended genetic modifications that occur when gene-editing tools, such as CRISPR-Cas9, make changes to areas of the genome other than the target site, potentially leading to unintended consequences.

Omega-3 Fatty Acids: Essential fatty acids that are important for human health, commonly found in fish. Gene editing can be used to enhance omega-3 content in farmed fish, thereby improving their nutritional value.

Pathogen Susceptibility: The likelihood of an organism being affected by diseases caused by pathogens such as viruses, bacteria, and parasites. Reducing pathogen susceptibility in aquaculture species is crucial for improving their health and productivity.

Phenomics: Phenomics is the systematic study of phenomes (physical and biochemical traits of organisms) and their interactions with genomes and environments

Polymerase Chain Reaction (PCR): Polymerase chain reaction (PCR) is a popular molecular biology method for amplifying particular DNA sequences. It enables scientists to produce millions of copies of a DNA segment from a small sample. Since its creation by Kary Mullis in 1983, PCR has become essential in biotechnology, forensic science, medical diagnostics, and genetic research.

Polyploidy: When a cell or organism has more than two complete sets of chromosomes, it is considered to be polyploid.

Prime Editing: A next-generation gene-editing technology that offers more precise DNA modifications than traditional CRISPR-Cas9 by enabling targeted changes without causing double-strand breaks in the DNA.

Proteins: Proteins are large biomolecules and macromolecules comprising one or more long chains of amino acid residues.

Public Acceptance: The level of approval or support from the public regarding the use of biotechnologies such as gene editing in food production. Acceptance is influenced by factors such as perceived safety, ethical considerations, and transparency in the application of technology.

Pure Line: In aquaculture, a "fish pure line" is a breeding strategy in which fish are chosen and bred over a number of generations to create progeny with consistent, desired features.

QTL's: Area of the genome linked with quantitative traits in a population. Quantitative traits are measurable phenotypes, for example, height, weight, or yield, which show a range of variations and are influenced by multiple genes, as well as the environment.

REAPR: REAPR is a tool for evaluating the accuracy of a genome assembly using mapped paired-end reads without using a reference genome for comparison.

Recombination: Recombination is a genetic process wherein segments of DNA are broken and reassembled to create new combinations of alleles.

Reference Genome: The reference genome is a standardized representative sequence created from the genomes of multiple individuals. It serves as a genetic baseline for comparing DNA changes.

Regulatory Challenges: The legal and policy obstacles that arise in the approval and commercialization of genetically edited organisms vary across regions and may involve concerns regarding safety, environmental impact, and consumer acceptance.

Repulsion Phase: The genetic configuration known as "repulsion" occurs when each homologous chromosome carries one recessive and one dominant allele for two distinct genes. Another name for this arrangement is the trans configuration.

Selective Breeding: A traditional breeding method where individuals with desirable traits are selected to produce offspring with enhanced characteristics. This typically involves multiple generations of breeding.

Single Nucleotide Polymorphisms (SNPs): Variations at a single nucleotide position in the genome. SNPs are often used in genomic selection to identify traits that can be targeted to improve aquaculture breeding programs.

Sterility: A condition in which farmed fish are unable to reproduce. Inducing sterility is particularly important in preventing farmed fish from interbreeding with wild populations and reducing ecological risks.

Stratification: It is a natural phenomenon that occurs when water separates into distinct layers with different properties in a water body.

Sustainable Aquaculture: Aquaculture practices that promote social, economic, and environmental sustainability to increase capacity building and make efficient use of land for the aquaculture industry.

Therapeutics: Treatment and care of a patient for both preventing and combating disease or alleviating pain or injury.

Transcription Activator-Like Effector Nucleases (TALENs): A gene editing technology that uses custom-designed enzymes to recognize and cut specific DNA sequences, offering high precision for targeted genetic modifications.

Zinc Finger Nucleases (ZFNs): One of the first gene-editing technologies, consisting of a DNA-binding domain made up of zinc finger motifs and a nuclease domain that cuts the DNA at a targeted site, allowing for precise genetic alterations.

Index